CELLULOSOME

(IN: MOLECULAR ANATOMY AND PHYSIOLOGY OF PROTEINACEOUSE MACHINES)

CELLULOSOME

(IN: MOLECULAR ANATOMY AND PHYSIOLOGY OF PROTEINACEOUSE MACHINES)

VLADIMIR UVERSKY
AND
IRINA A. KATAEVA
EDITORS

Nova Science Publishers, Inc.
New York

For permission to use material from this book please contact us:
Telephone 631-231-7269; Fax 631-231-8175
Web Site: http://www.novapublishers.com

LIBRARY OF CONGRESS CATALOGING-IN-PUBLICATION DATA

Cellulosome.
p. cm. -- (Molecular anatomy and physiology of proteins)
Includes index.
ISBN 1-59454-950-8
1. Bacteria. 2. Proteins. 3. Medical bacteriology. I. Series.
QR74.8.C45 2006
616.9'201--dc22 2006000251

Published by Nova Science Publishers, Inc. ✢New York

CONTENTS

PREFACE

A combination of biochemical, biophysical, immunochemical and ultrastructural techniques, followed by molecular biological verification, led to the definition and proof of the cellulosome concept. The birth of the discrete, multi-enzme cellulosome complex was thus documented. Today, cellulosomes have been confirmed in numerous but not all cellulolytic bacteria. The cellulosome-producing strains exhibit surprising diversity in the composition and architecture of the component parts.

The clostridial co-culture system is potentially capable of dramatically increasing the use of ethanol as a major liquid fuel using renewable photosynthetic biomass as feedstock. Production of ethanol by a co-culture of thermophilic, anaerobic microorganisms (i.e., a cellulolytic strain such as Clostridium thermocellum and a saccharolytic strain, such as Thermoanaerobacterium thermosaccharolyticum) offers the following advantages: (i) thermophiles are thought to be robust and contain stable enzymes; (ii) anaerobes generally have a low cellular growth yield, hence more of the substrate is converted to ethanol; (iii) thermophilic fermentations are less prone to detrimental effects of contamination; (iv) growth at higher temperatures may facilitate the removal and recovery of volatile products such as ethanol. Even more important are the advantages of cellulase production in situ and the high rates of metabolism of cellulose and hemicellulose. In principle, the concept of a thermophilic ethanol fermentation is a very simple one involving a high-temperature fermentation with reduced need for power-consuming cooling and agitation/aeration of reactor vessels and with the four biologically-mediated events involved in ethanol production (cellulase and hemicellulase formation, cellulose and hemicellulose hydrolysis, hexose fermentation, and pentose fermentation) consolidated in a single process step. Crucial to the use of the co-culture system is the cellulosome, the unique, multienzyme complex dedicated to organized, concerted, synergistic and efficient catalysis. The history of cellulosomes, the interactions between their cohesins and dockerins, their catalytic makeup and their genomics are the subjects of this book.

The discovery of the cellulosome is a story of serendipity, a story of imagination, persistence, and a triumph over dogma. The authors in chapter two simply sought a "cellulose-binding factor" on the cell surface of the anaerobic thermophilic bacterium, Clostridium thermocellum, which we inferred would be responsible for the observation that the bacterium attaches strongly to the insoluble cellulose substrate prior to its degradation. Using a then unconventional experimental approach, they unexpectedly isolated a very large multi-subunit supramolecular complex, instead of a small protein. Rather than discarding the uninvited material, they were alert enough to follow up this surprising finding experimentally. A combination of biochemical, biophysical, immunochemical and ultrastructural techniques, followed by molecular biological verification, led to the definition and proof of the cellulosome concept. The birth of the discrete, multi-enzyme cellulosome complex was thus documented. Today, cellulosomes have been confirmed in numerous but not all cellulolytic bacteria. The cellulosome-producing strains exhibit surprising diversity in the composition and architecture of the component parts.

Cellulosome architecture is defined by the intermodular interaction between cohesins and dockerins, which usually display very high affinities. Within and among the different cellulosome-producing microbial species, the sequences of both cohesins and dockerins separate into distinct types, reflecting their phylogenetic relationship as dicussed in chapter three.

The crystal structure of the type-I cohesin module revealed an elongated, conical molecule, consisting of a nine-stranded β-sandwich with an extensive aromatic core. Structures of type-II cohesins were also determined recently, which showed a jelly-roll topology with 3 additional secondary structures: namely an α helix and two "β-flaps" that disrupt the normal course of a β-strand. In contrast, the dockerin module comprises a duplicated sequence of about 22 amino acids (both required for binding), connected by a linker segment. The solution structure of the ~70-residue type-I dockerin from C. thermocellum Cel48S showed a pair of calcium-binding loop-helix motifs, related by a 2-fold pseudo-symmetry axis. Calcium induces protein folding and is required for stable tertiary structure and cohesin-dockerin interaction.

The recently solved three-dimensional structure of the type-I cohesin-dockerin complex from demonstrated that the overall structure of the cohesin module is essentially unchanged. The dockerin, however, appears to form 3 α-helices in a conformation defined by a loop-helix motif followed by a helix-loop-helix motif. Helices 1 and 3 are closer together in the complex compared to the free form. The cohesin interacts predominantly with α-helix 3 of the dockerin and to a lesser extent with α-helix 1. The protein-protein contact is mediated mainly by hydrophobic interactions and relatively few intermolecular hydrogen bonds.

Comparative binding analysis between the cohesins and dockerins of C. thermocellum versus C. cellulolyticum suggested that inter-species binding is nonselective, whereas intra-species interaction is specific. The dockerin residues that appear to account for the observed specificity are at positions 10, 11, 18 and 22 in both of the duplicated sequences. In the case of the cohesin, only three residues (Ala36, Asn37 and Glu-131) appear to be involved in interspecies specificity between the two bacteria, representing but 3 of the 16 contact residues that participate in the cohesin-dockerin interaction.

The authors' current knowledge of the sequences, the structures and interactions among the different types of cohesins and dockerins is still rather sparse. Current inroads in genome sequencing and continued progress in structure determination should provide vital information regarding the diversity of cohesin-dockerin interactions.

Based on the nearly complete sequence of the Clostridium thermocellum genome (http://genome.jgi-psf.org/draft_microbes/cloth/cloth.home.html), some 70 genes of this bacterium encode polypeptides that include a dockerin domain and may thus be included within the quaternary structure pf the cellulosome. In most cases, these polypeptides include readily identified modules such as catalytic domains or carbohydrate-binding modules. Most of these modules belong to widely distributed families and may serve as paradigms to investigate the structure and function of a vast variety of related enzymes. In chapter four, the authors review the structure and mechanism of the best studied examples of catalytic domains and carbohydrate-binding modules present in components of the cellulosome. Interestingly, it appears that glycosyl hydrolases splitting different substrates often share the same basic framework and mode of action.

Cellulose hydrolysis must be considered as a microbial as well as enzymatic phenomenon in order to understand cellulose utilization in natural environments and realize the potential of biotechnological processes based on microbial conversion. Moreover, there are increasing indications that the role of microorganisms in cellulose hydrolysis extends far beyond serving merely as a source of cellulase enzymes. In chapter five, cellulose utilization is examined for the thermophilic anaerobe Clostridium thermocellum. Specific topics include the ecophysiology of this organism, the mechanism of cellulose hydrolysis, bioenergetics, control of cellulase synthesis, and mechanisms of adhesion. In addition, they suggest that it may be fruitful to apply tools and insights from the biofilm field to microbial cellulose utilization, which may be viewed as a reacting biofilm in which cellulose serves as both the solid substratum and the carbon source for heterotrophic growth. The authors close by summarizing recent progress addressing previously-posed questions, and by articulating some new questions.

C. thermocellum is a very effective cellulose degrading bacterium which hydrolyzes a wide range of polysaccharides from lignocellulosic biomass. It is thus of increasing interest for the industrial utilization of lignocellulosic biomass in biotechnology. This bacterium ferments only the sugars derived from cellulose. Its extracellular, cell-bound multienzyme complex (cellulosome) consists of 13 major and an unknown number of minor catalytic components which are attached to structural proteins by protein-protein interaction. The major components of the cellulosome not only comprise the cellulases Cel48S, Cel8A, Cel9R, Cel9K, Cbh9A, Cel9N and Cel5G, but also the xylanases Xyn10C, Xyn10D and Xyn10Z as well as xyloglucanase Xgh74A, chitinase Chi18A. C. thermocellum has a potential to degrade a number of non-cellulosic polysaccharides, among them pectin by pectate lyases and polygalacturonases. Putative pectin- and xylan-esterases are connected to hydrolytic modules and support depolymerization as do debranching enzymes like the putative β-xylosidase, α-arabinofuranosidase and glucuronosidase activities for the hydrolysis of xylan. C. thermocellum may thus play a key role in crossfeeding of other bacteria in saprophytic ecosystems with breakdown products from hemicellulose and other non-cellulosic polysaccharides.

Newly detected enzyme components, putative genes and their structure are discussed on the basis of the genomic sequence and recent experimental findings. Besides the cellulosomal (cell-bound) enzyme system C. thermocellum seems to possess a non-cellulosomal (soluble) cellulase system. The potential utilization of the energetically more feasible larger cellodextrins in the context of the enzymatic activities of the major cellulosomal components is discussed in chapter six.

The Clostridium cellulovorans cellulosome is an extracellular enzyme complex that can degrade plant cell wall material very efficiently. It is comprised of a non-enzymatic scaffolding protein called CbpA and a number of cellulosomal glycosyl hydrolase enzymes capable of degrading cellulose, hemicellulose, and pectin. The CbpA has a mass of 189,000 kDa and a cellulose binding domain (CBD) at its N-terminus, four conserved hydrophilic domains (HLDs), two near the N-terminus and two near the C-terminus, and nine homologous cohesin (Coh) domains. The CBD preferentially binds to crystalline cellulose and the cohesins bind the cellulosomal enzymes through their dockerin domains. Current studies indicate that the HLDs bind to the substrate as well as to the cell surface as indicated in chapter seven. The cellulosomal enzymes work in synergy with CbpA as well as with each other and with non-cellulosomal glycosyl hydrolase enzymes to degrade the various components of plant cell walls. The cellulosomal enzyme EngE binds not only to the CbpA, but also to the cell surface through its surface layer homology (SLH) domains and tightly binds the cellulosome to the cell surface and allows an efficient uptake of hydrolytic products. One large cluster of cellulosomal genes occurs, and also there are a number of cellulosomal genes unlinked to the cluster. The complex is useful for basic studies on plant cell wall degradation, for construction and expression of designer cellulosomes, for converting non-cellulose degraders to cellulose utilizers, and for various aspects of biotechnology involving the conversion of biomass to useful products.

Catalytic components of the Clostridium josui cellulosome are quite similar to those of the well-characterized Clostridium cellulolyticum cellulosome although the size of the scaffolding protein of the former is smaller than that of the latter, resulting in the formation of smaller cellulosome. C. josui is closely related to C. cellulolyticum judging from their 16S rDNA sequences, but the optimal growth temperature of the former (45 °C) is higher than that of the latter (33-35 °C), suggesting that properties of their cellulosomes are different from each other. Species specificity was observed in the cohesin-dockerin interaction between C. josui and Clostridium thermocellum. The authors, in chapter eight, constructed a dockerin-cohesin affinity purification system for recombinant proteins expressed in the baculovirus expression system.

SDS-PAGE and Zymogram analyses of the cellulolytic complex isolated from the culture supernatant of Ruminococcus albus F-40 revealed fifteen proteins including eight endoglucanase-active and nine xylanase-active proteins. Immunological analysis done with the antiserum raised against Cel5D-dockerin confirmed that assembly of the R. albus cellulosome was made by the cohesin-dockerin interactions.

Cellulosome chimeras designate hybrid minicellulosomes composed of a miniscaffoldin containing two (or more) cohesins of different specificities that anchors two (or more) engineered cellulolytic enzymes appended with the corresponding dockerin. Two important observations made it possible for the construction of such minicomplexes: the bacterial cohesins and dockerins are genuine modules keeping their biological properties when they are produced alone or grafted to a heterologous protein, and the high affinity cohesin / dockerin interaction is species-specific between some cellulosome-producing bacterial species. Compared to natural cellulosomes, cellulosome chimeras exhibit three major differences: they contain a limited number of enzymes, they are homogeneous, and the position of each enzyme along the miniscaffoldin is controlled. These characteristics make the cellulosome chimeras a convenient tool to dissect the functioning of natural cellulosomes, and this strategy was applied to investigate the cellulosomes produced by Clostridium cellulolyticum as shown in chapter nine. It was observed that organization of most tested C. cellulolyticum enzyme pairs into cellulosome chimeras induced enhanced activities on recalcitrant cellulose, assigned to both enzyme proximity on the resulting complex that triggers additional synergies, and to the cellulose-binding module of the scaffoldin which anchors the whole complex at the surface of the substrate. The most efficient hybrid minicellulosomes remain, however, much less active than natural cellulosomes on crystalline cellulose. For this reason, future prospects involve the construction of new hybrid scaffoldins which are able to bind onto specific locations three or more different enzymes. Besides these fundamental studies, the cellulosome chimeras may also be exploited for biotechnological applications in various fields.

Ruminococcus flavefaciens is a cellulolytic anaerobic bacterium that is one of the main agents of plant cell wall breakdown in the gut of herbivorous mammals, notably in the rumen. Plant cell wall degrading enzymes in this species display an elaborate form of cellulosome organisation involving at least four distinct dockerin:cohesin specificities. Three cohesin-containing structural proteins (ScaA, B, C) are encoded by linked genes within the sca cluster. The primary scaffoldin ScaA binds to one type of dockerin that is found in many enzymes as well as in the structural protein ScaC. ScaC apparently functions as an adaptor, since its unusual cohesin binds an entirely different group of polypeptides. Connection of the ScaA complex to the cell surface occurs via binding of ScaA to the larger ScaB, which in turn appears to bind to a component of the bacterial cell wall. Not all enzyme dockerins bind either ScaA or ScaC, however, suggesting the existence of further interactions and scaffolding proteins. ScaA lacks an identified cellulose binding module. Cellulose-binding modules are however present in a number of cellulosomal enzymes and in at least one specialised binding protein; cellulose binding may also involve the yellow affinity substance after which R. flavefaciens is named. ScaA and ScaB homologues have recently been detected in the related strain R. flavefaciens FD1, and information from the ongoing genome sequencing project on FD1 promises to expand our picture of cellulosomal complexes in this species in the near future as discussed in chapter ten.

The enzymatic system of Clostridium cellulolyticum that hydrolyzes plant cell wall polysaccharides comprises multi-enzyme complexes, named cellulosomes, and non-cellulosomal components and is discussed in chapter eleven. Each cellulosome consists of a scaffolding protein CipC complexed with a maximum of eight enzymes. CipC is composed of a cellulose binding module at the N terminus followed by one hydrophilic X2 module, seven highly homologous cohesin modules, a second X2 module, and a final cohesin module. Individual cellulosomes range in mass from about 600,000 to 700,000 Da. Cellulosomes prepared from cellulose culture revealed three major components, which were identified to be CipC, the processive endocellulase Cel48F, and the cellobiohydrolase Cel9E, in addition to numerous other components with molecular masses ranging from 100 to 30 kDa. Enzymatic components of the complexes were all found to contain one specialized docking module (called dockerin) able to specifically interact with the CipC cohesins. Clostridium cellulolyticum synthesized at least 22 different dockerin-bearing proteins. Thus the complexes may vary widely in their composition. This composition is regulated by the relative amounts of the enzymes available to be incorporated into the complexes. The composition of the cellulosomes was efficiently engineered in vivo by structural gene or regulatory gene expression. The gene encoding CipC is the first gene of a large cluster spanning 26 kb. It is immediately followed by cel48F, and then by nine other genes (cel8C-cel9G-cel9E-orfX-cel9H-cel9J-man5K-cel9M-rgl11Y-cel5N) coding for seven cellulases, one mannanase and one pectinase, also found to be components of the cellulosomes. The cellulases encoded by the genes located in this cluster were found to be essential for the building of cellulosomes efficient in crystalline cellulose degradation. Studies focused on the metabolism of C. cellulolyticum in culture with cellulose or cellobiose as substrate indicated that the strain is well adapted, and even restricted to a low carbon flow, which is characteristic of growth on its natural substrate, cellulose. The strain is not able to regulate soluble sugar entry and thus the carbon flow through glycolysis. Under these conditions, the nutrients or products of its metabolism accumulate intracellularly to toxic levels. Improvement of the wild-type strain was obtained by heterologous expression of an artificial operon to eliminate excess intracellular pyruvate.

C. thermocellum produces a cellulase system consisting of a huge protein complex called the cellulosome. The cellulosome contains a large number of subunits, most of them glycosyl hydrolases. The core of the cellulosome is a 250-kDa non-catalytic subunit, CipA, which serves as a scaffold for the catalytic subunits. CipA contains a series of nine highly homologous domains, termed cohesin domains, which serve as receptors for the catalytic subunits. Binding to the cohesin domains is mediated by a highly conserved duplicated sequence of 22 amino acid residues called the dockerin domain, which is found mostly at the C-terminus of each cellulosomal catalytic subunit. CipA further contains a cellulose-binding domain (CBD), which anchors the array of the catalytic components on the cellulose surface. In addition to the cellulosome, C. thermocellum produces many free, non-cellulosomal glycosyl hydrolases.

Most studies on the C. thermocellum cellulase system have so far focused on molecular cloning and characterization of glycosyl hydrolases as well as the structure-function relationship of the cellulosomal and noncellulosomal components. Little is known about how the biosynthesis of these proteins is regulated. The task is obviously hampered by the large number of the genes and proteins involved. The cellulosome components can be classified into three categories: 1) the scaffolding protein (CipA); 2) the dockerin-containing catalytic subunits; and 3) the scaffoldin-anchorage proteins, which anchor the cellulosome or individual enzymes to the cell surface (such as OlpA, OlpB, SdbA, and Orf2p). Analysis of the C. thermocellum draft genome sequence indicates that category 2 alone comprises at least 72 different genes. The study is further complicated by many non-cellulosomal enzyme components produced by the bacterium. The question of how the bacterium coordinates the biosynthesis of such a large number of biomass-degrading enzymes is intriguing. Elucidation of the regulatory mechanisms is important not only for understanding the cellulase biology, but also for engineering deregulated, hypercellulolytic strains. The knowledge obtained will also serve as a paradigm for the cellulosomes produced by other microorganisms since the C. thermocellum complex is the most extensively studied cellulosome so far.

The authors have submitted the C. thermocellum genomic DNA to the DOE Joint Genome Institute (JGI) for sequencing and a draft genome sequence is now available [http://genome.jgi-psf.org/draft_microbes/cloth/cloth.home.html]. This provides an unprecedented opportunity for studying the regulatory mechanism at the genomic level. In chapter twelve, they briefly review the recent progress, highlighting the results of some studies facilitated by the genome sequence.

Anaerobic fungi discovered by Orpin (J. Gen. Microbiol. 1975, 91, 270) live in close contact with bacteria and other microorganisms in the rumen and caecum of herbivorous animals, where they digest ingested plant food. There are seventeen morphologically distinct anaerobic fungal species described belonging to five genera. They have been isolated from over fifty different herbivorous animals and are discussed in chapter thirteen. Anaerobic fungi have efficient system to solubilize lignocellulose and hydrolyze plant carbohydrates. They produce all necessary enzymes for degradation of cellulose and hemicelluloses. The enzymes are exocellular and found free or bound to large cellulosomal and polycellulosomal complexes. Enzymes recognized in the cellulosomal complexes include endo- and exocellulases, β-glucosidases, xylanases, β-glucanases, lichenases, mannanases, α-methylglucuronidases, acetyl-xylan esterases, and feruloyl esterases. All enzymes associated with cellulosomes from anaerobic fungi are modular containing in addition to the catalytic modules duplicated docking modules in similarity with enzymes of the cellulosomes of anaerobic bacteria. The presence of scaffolding proteins are also recognized but a scaffoldin from an anaerobic fungus has still to be isolated. Serpins, which are inhibitors of trypsin-like proteinases, are associated with the cellulosomes. The non-catalytic docking domains (NCDD) although having the same functions as dockerins of cellulosomes of bacteria, do not have any sequence homology to them. Judging sequences of the cellulosomal enzymes it is apparent that they originate from different sources including anaerobic and aerobic bacteria, and aerobic fungi. Gene transfer between the anaerobic fungi seems prevalent as is gene duplication.

In: Cellulosome
Eds: V. Uversky and I. A. Kataeva, pp. 1-9

ISBN 1-59454-950-8

Chapter I

INTRODUCTORY CHAPTER: THE CRUCIAL IMPORTANCE OF BIOETHANOL AND THE CONTRIBUTION OF ANAEROBIC BACTERIA AND THEIR CELLULOSOMES

Arnold L. Demain[*]
Research Institue for Scientists Emeriti (R.I.S.E.), Drew University, Madison, NJ

ABSTRACT

The clostridial co-culture system is potentially capable of dramatically increasing the use of ethanol as a major liquid fuel using renewable photosynthetic biomass as feedstock. Production of ethanol by a co-culture of thermophilic, anaerobic microorganisms (i.e., a cellulolytic strain such as *Clostridium. thermocellum* and a saccharolytic strain, such as *Thermoanaerobacterium thermosaccharolyticum*) offers the following advantages: (i) thermophiles are thought to be robust and contain stable enzymes; (ii) anaerobes generally have a low cellular growth yield, hence more of the substrate is converted to ethanol; (iii) thermophilic fermentations are less prone to detrimental effects of contamination; (iv) growth at higher temperatures may facilitate the removal and recovery of volatile products such as ethanol. Even more important are the advantages of cellulase production in situ and the high rates of metabolism of cellulose and hemicellulose. In principle, the concept of a thermophilic ethanol fermentation is a very simple one involving a high-temperature fermentation with reduced need for power-consuming cooling and agitation/aeration of reactor vessels and with the four biologically-mediated events involved in ethanol production (cellulase and hemicellulase formation, cellulose and hemicellulose hydrolysis, hexose fermentation, and pentose fermentation) consolidated in a single process step. Crucial to the use of the co-culture

[*] Correspondence concerning this article should be addressed to Arnold L. Demain, Drew University, R.I.S.E., HS-330, Madison, NJ 07940, U.S.A.; ademain@drew.edu

system is the cellulosome, the unique, multienzyme complex dedicated to organized, concerted, synergistic and efficient catalysis. The history of cellulosomes, the interactions between their cohesins and dockerins, their catalytic makeup and their genomics are the subjects of this book.

Keywords: Cellulase, Hemicellulase, Clostridia, Anaerobes, Ethanol, Cellulosomes, Cohesins, Dockerins, Energy, Thermophiles.

The energy problem in the world is getting more and more precarious. The price of petroleum has risen sharply in the last few years from about 20 dollars per barrel to almost 70 dollars. It is very possible that ethanol production could help solve at least part of the problem. As a transportation fuel, ethanol is excellent and has some properties that are superior to gasoline. Neat (unblended) ethanol burns cleaner, can be burned with greater efficiency, has a higher octane rating, and produces lesser amounts of ozone precursors, thus decreasing urban air pollution. Furthermore, it puts much less net CO_2 into the atmosphere. Since the carbon dioxide produced during fermentation and fuel use of ethanol is recycled by growth of plants, ethanol does not contribute to carbon dioxide accumulation in the atmosphere and global warming.

Ethanol is considerably less toxic to humans than is gasoline. Its high heat of vaporization, high octane and low flame temperature yield good engine performance. Furthermore, ethanol offers the Western countries a more favorable trade balance, enhanced energy security and a major new crop for depressed agricultural economies. Brazil produces 3.8 billion gallons from sugar per year [25]. 90% of new cars in Brazil use neat ethanol while the remainder utilize a blend of 25% ethanol with gasoline. Ethanol accounts for 40% of all the vehicle fuel used in that country. In the U.S.A. in 1987, 340 million bushels of corn (8.6 metric tons) were used to make 850 million gallons of anhydrous ethanol for blending with some gasoline at a 10% level. Blended fuel represented 8% of total gallons of gasoline used in the U.S.A. Although small, this figure had risen from below 1% in 1981. By 2004, 3.5 billion gallons of ethanol were being made by yeast from corn and other high-starch crops, amounting to 1% of US automotive requirements [6]. This year, the figure is 4 billion gallons and this will increase to 7 billion gallons according to recent U.S. energy legislation (L.R. Lynd, personal communication). Today in the U.S.A., ethanol from corn fermentations is mixed with gasoline as a fuel and has also replaced tetraethyl lead as an octane enhancer. Furthermore, ethanol is used as an oxygenate to reduce automobile emissions. With the current and impending phase-out of methyl *tert*-butyl ether (MTBE) as oxygenate in many states in the U.S.A., ethanol is filling the void. In 1990, the US consumed 112 billion gallons of gasoline. Over 62% of all petroleum used in the U.S.A. for transportation and industry was imported in 2004 [6], costing over $100 billion per year. These figures increased to $140 billion and 71% in 2005 (L.R. Lynd, personal communication). Petroleum import represents the largest single part of the US trade deficit. A big problem is that there is not enough corn available to make a significant dent in the world's dependence on petroleum for energy.

For possible conversion into fuel, lignocellulose is inexpensive, plentiful and renewable; it is the most abundant renewable natural substrate which can be converted into fuel. Available cellulosic feedstocks from agriculture and other sources amount to about 180 million tons per year. Cellulosic biomass in the early 1980s at $20-30/dry ton was much cheaper than corn at $110/dry ton, yet cellulose hydrolysis was not competitive with that of starch to produce sugar. The problems included cellulose crystallinity and lignin content requiring expensive pretreatment. Also, corn usage provided oil and protein byproducts which were of much higher value than lignin. In 2003, cellulosic material at $44 per dry ton cost the same as petroleum at $6-7 per barrel based on equivalent mass or $12-13 per barrel based on equivalent energy content [34]. Today, the cost of cellulosic feedstock is thought to be about $40 per dry ton [21] whereas the price of petroleum has skyrocketed. It makes sense that ethanol from cellulose could become competitive with gasoline if processing costs of the former could be lowered. In addition, tremendous amounts of additional cellulose are available as municipal and industrial wastes which today contribute to our pollution problems. Thus, great interest exists in the use of cellulosic biomass as a renewable source of energy via breakdown to sugars that can then be converted to a liquid fuel

It is a promising approach towards an alternate energy source from economic and environmental viewpoints. Biomass can be provided by agricultural resources, as well as urban and industrial cellulosic wastes which are inundating the earth. The availability of landfills is becoming less and less, and landfill costs are escalating. The potential quantity of ethanol that could be produced from cellulose is over an order of magnitude larger than that produceable from corn. In contrast to the corn to ethanol conversion, the cellulose to ethanol route involves little or no contribution to the greenhouse effect and has a clearly positive net energy balance (five times better). Biomass contains lignin as well as cellulose and hemicellulose. During chemical pretreatment, the lignin would be liberated and used as a boiler fuel to provide heat and electricial energy.

In the conversion of biomass to fuels, the rate-limiting step is cellulose hydrolysis, especially the initial attack on the highly-ordered, insoluble structure of crystalline cellulose. The products of this attack are soluble oligosaccharides and sugars, which can be taken up by some ethanol producing cells. A great deal of effort has gone into the development of methods for conversion of cellulose to sugars. Most of this work has emphasized the biochemistry, genetics and process development of cellulase-producing fungi (especially *Trichoderma reesei*) coupled to conversion of the sugars produced to ethanol by yeast (*Saccharomyces cerevisiae*) or recombinant bacteria such as *Escherichia coli, Klebsiella oxytoca* or *Zymomonas mobilis*. However, an alternate solution is the use of anaerobic bacteria, especially the thermophilic species. These include species of *Clostridium, Ruminococcus*, and *Thermoanaerobacterium.* A co-culture of a cellulolytic strain and a saccharolytic strain of anaerobes will attack cellulose and hemicellulose, and convert the resulting oligosaccharides and sugars to ethanol. Such a process is known as consolidated bioprocessing (CBP).

By 1990, the cost of ethanol production from cellulose by the simultaneous saccharification and fermentation (SSF) process using *T. reesei* cellulase plus yeast was thought to be approaching that for ethanol production from corn (by yeast) on an unsubsidized basis. However, in a comparison made in the early 1990's [11,26], conversion of mixed hardwood flour to ethanol in a continuous fermentor was 2.5 times better using thermophilic anaerobes in a CBP process than with the SSF process using *Trichoderma* cellulase, β-glucosidase and *S. cerevisiae*; furthermore, the rate of conversion was four times higher. It was calculated then that CBP could save 50 cents per gallon as compared to the state-of-the-art SSF process since the former process combined cellulase production, hydrolysis and fermentation in a single bioreactor. Between 1997 and 2000, estimates of cost of an SSF-type process using acid hydrolysis of biomass before attack by *Trichoderma* cellulase and fermentation varied from $0.43 and $1.29 per gallon [23]. Although the price of producing fungal cellulase has dropped markedly in the last few years and the SSF process is in a demonstration stage, the potential of CBP seems greater. SSF is currently estimated to yield bioethanol at at $0.87 per gallon wholesale whereas CBP is estimated at $0.63 per gallon [21]. After many years of study, it is becoming apparent that the SSF process is not the only potential solution. In CBP, the enzymes cost very little because they are produced by the fermenting organism in the course of ethanol production. Thus, the primary advantages of a direct anaerobic bacterial conversion include elimination of capital or operating costs for enzyme production, greatly reduced diversion of substrate for enzyme production, and compatible enzyme and fermentation systems.

Bacteria that metabolize cellulose have received much interest in recent years since they can utilize lignocellulosic waste and generate ethanol, a rare property among living organisms. Indeed, their cellulase systems mediate hydrolysis of lignin-containing materials such as hardwood pretreated with dilute acid as well as model substrates not containing lignin. In principle, the concept of an anaerobic ethanol fermentation is a very simple one, i.e., a single fermentation with five biologically-mediated events involved in ethanol production: (i) cellulase and hemicellulase formation, (ii) cellulose and hemicellulose hydrolysis, (iii) uptake of sugars and oligosaccharides, (iv) hexose fermentation, and (v) pentose fermentation, all consolidated in a single process step. An advantage of such an anaerobic process is the reduced need for power-consuming agitation/aeration of reaction vessels for making biomass-degrading enzymes. Also, anaerobes generally have a low cellular growth yield, hence more of the substrate can be converted to ethanol, and in situ cellulase production is much more economical than addition of cellulase; Although mesophilic anaerobes might be usable, anaerobic thermophiles are favored as "ethanologens" for the following reasons: (i) thermophiles appear to be robust and contain stable enzymes; (ii) thermophilic fermentations run at 60°C are less prone to detrimental effects of contamination; (iii) growth at higher temperatures may facilitate the removal and recovery of volatile products such as ethanol; (iv) the rates of metabolism of cellulose and hemicellulose are high; and (v) there is a reduced need for power involved in cooling of fermentors.

Anaerobic cellulolytic bacteria break down cellulose with the formation of cellobiose and cellodextrins as main products. The cellodextrins and the cellobiose are further utilized by the organism; indeed, the cellodextrins are taken up more easily by *Clostridium thermocellum* than the disaccharide. The final end products are ethanol, acetic acid, lactic acid, hydrogen, and carbon dioxide. The major economic obstacle is the production of acetate and lactate, which limits conversion yield. A combination of recombinant DNA technology and metabolic engineering research is beginning to overcome this bottleneck. Gene transfer systems utilizing electrotransformation are now available for the anaerobes [28] and ethanol-tolerant strains have been developed [27]. The selectivity problem is being addressed by inactivation of the genes encoding acetate- and lactate-producing enzymes of both members of the co-culture system. Significant progress on this point has already been made by Lynd and coworkers [8,21].

The workhorse of the cellulolytic anaerobes is the cellulosome. In 1983, important events occurred. Eric Johnson [13] in his Ph.D. work at M.I.T. noted that the cellulase activity of *C. thermocellum* was part of a large structure with a molecular weight of over 1.5×10^6. In the same year, Bayer and Lamed and coworkers [2,19] in Israel purified a multisubunit complex from the culture supernatant which they named the cellulosome [17]. Cellulosomes are crucially important for the efficient breakdown and utilization of crystalline cellulose. The cellulosome is a macromolecular machine (multienzyme complex) which, like a ribosome, is dedicated to organized, concerted, synergistic, and efficient catalysis of cellular activities. Cellulosomes are unique in that no other extracellular protein complexes with the size and complexity of cellulosomes have been reported. They have molecular masses of 2×10^6 to 6×10^6, diameters of about 18 nm and contain 14 to 50 polypeptides ranging in size from 37 to 210 kDa. Over 95% of the endoglucanase activity of *C. thermocellum* is associated with the cellulosome.

The highly-ordered arrangement of the cellulosome gives it stability but made purification of individual proteins extremely difficult. However, purification of this complex aggregate of cellulolytic proteins was accomplished [32,33]. Using Avicel (crystalline cellulose) breakdown as a turbidimetric assay for true cellulase activity and carboxymethyl cellulose (CMC) hydrolysis as an assay for endoglucanases, the aggregate was found to contain at least four endoglucanases of different molecular weights accompanying true cellulase activity. It was dissociated by mild sodium dodecylsulfate (SDS) treatment plus EDTA and DTT but the resulting individual fractions exhibited only endoglucanase activity, the true cellulase activity being lost. Reconstitution of true cellulase activity was accomplished by combining two of the major components which were called S_S (M_r = 82,000) and S_L (M_r = 250,000). They were purified by gel filtration chromatography and by elution from an SDS-polyacrylamide gel, respectively. S_S alone acted on CMC but S_L alone had little to no enzymatic activity [18,33]. The enzymatic activity of S_S on CMC was not enhanced by S_L but its adsorption to crystalline cellulose was [33]. It was hypothesized that the cooperative degradation of crystalline cellulose involves an interaction between S_S (and presumably other cellulases), S_L and the insoluble substrate. S_L (an anchorage subunit) would function to bind S_S (and other catalytic proteins of the complex) to the cellulose surface in a manner optimal for hydrolysis [31,32]. The DNA sequence of the S_L gene (*cipA*) revealed that S_L indeed contains a cellulose-binding domain (CBD) and multiple enzyme receptor domains, consistent with the "anchor-enzyme" hypothesis [10]. The anchor-enzyme model was further confirmed using recombinant forms of S_S and S_L [15]. The anchorage function of S_L is the basis of the current understanding of cellulosome structure. Of further interest was the sequencing of the genes encoding S_S [30].

Protein S_L, which is equivalent to component S1 of Lamed et al. [18], is now called the cellulosome-integrating protein (CipA), the scaffolding protein, or 'scaffoldin' [3]. It contains approximately 1850 amino acid residues and is the most important protein of the cellulosome. In addition to its function of binding other other members of the cellulose complex to itself, it also binds to cellulose [5]. Its nucleotide sequence revealed a deduced protein size of 196,800 Da, a cellulose binding domain (CBD) [4] and nine domains of about 150-166 amino acid residues each. The nine repeated sequences, called 'cohesins' by Bayer et al. [3], are quite similar to each other, i.e., exhibiting between 60 and 100% identity, with six of the nine domains being at least 90% identical. They are the receptors that bind the individual cellulases, xylanases and other enzymes to CipA.

Protein S_S, an exoglucanase also called CelS [30], is the major catalytic subunit of the cellulosome, and is equivalent to component S8 of the Israeli group [24]. Sequencing of *celS* revealed an open reading frame of 2241bp encoding 741 amino acid residues with a predicted molecular weight of 80,670 [29,30]. The sequence revealed that CelS belongs to a new cellulase family [30] and was later classified as a member of family 48 glycosyl hydrolases. Although it is the most abundant catalytic subunit of the cellulosome, its low or complete lack of activities on CMC and other synthetic substrates explains why it had been difficult to clone its gene. Family 48 enzymes are found mostly in bacterial cellulase systems and are now considered a key component in the bacterial scheme for breaking down cellulosic materials [12]. CelS contains a duplicated 24 amino residue dockerin, the site of binding to scaffoldin. Like the cellulosome itself, recombinant CelS is competitively inhibited by cellobiose and only marginally so by glucose [14]. Thus, an association is formed by a synergistic cassette of catalytic proteins, which is optimal for hydrolysis of insoluble polymers to the level of soluble oligosaccharides.

Researchers at the University of Georgia [9,11] found even larger aggregates, ca 108 x 10^6 Da ("polycellulosomes"). Protuberances covering the surface of the cell are packed with (poly)cellulosomes, each protuberance containing several hundred cellulosomes [16]. When cells are grown on cellobiose, cellulosome complexes are packed into such discrete exocellular structures. When grown on cellulose, these polycellulosome-containing organelles (protuberancesof diameter 60-200 nm) undergo extensive structural modification [1]. After attachment to the insoluble substrate, the protuberances rapidly aggregate into "contact corridors" which physically mediate between the cellulosome, which is attached to the cellulose, and the bacterial cell surface. The proteins of the cellulosome are arranged in a highly-ordered chain-like array [22]. The cellulose-bound cellulosome clusters appear to be the sites of active cellulolysis and the products may be channeled down the fibrous structures to the cell.

The most important component of the cellulosome is the non-enzymatic scaffoldin. It is a unique scaffolding protein subunit, which assembles cellulases and related enzyme subunits. The catalytic subunits, on the other hand, contain different modules (dockerins) which are responsible for their attachment to the scaffold. Important in this relationship are (i) cohesin domains on scaffoldin, (ii) dockerin domains on the enzymes and (iii) a CBD on the scaffoldin, binding the complex to cellulose. Cellulosomes and scaffoldin have been found in many bacteria, such as *Clostridium cellulovorans, Clostridium cellulolyticum, Clostridium josui, Clostridium acetobutylicum, Acetovibrio cellulolyticus, Bacteroides cellulosolvens, Ruminococcus albus, Ruminococcus flavefaciens*, *Vibrio* sp., and the anaerobic fungal genera *Neocallimastix, Piromyces and Orpinomyces*.

The pages which follow discuss the cellulosomes of clostridia and related microorganisms such as *Ruminococcus*, their history, the interactions between their cohesins and dockerins, their catalytic makeup, and genomics. Both thermophilic and mesophilic anaerobes are discussed. Included is a chapter on cellulosome chimeras. The book also addresses the cellulases of clostridia, their regulation and ability to hydrolyze cellulose, and the biotechnological significance of all of this. As a bonus, you will also be able to read an interesting contribution on cellulosomes from anaerobic fungi.

ACKNOWLEDGEMENT

I thank Lee R. Lynd, Professor at the Thayer School of Engineering of Dartmouth College, for critically reading the manuscript and advice on bioethanol.

REFERENCES

Bayer, E. A., and Lamed, R. 1986. Ultrastructure of the cell surface cellulosome of *Clostridium thermocellum* and its interaction with cellulose. *J. Bacteriol. 167*:828-836.

Bayer, E. A., Kenig, R., and Lamed, R. 1983. Adherence of *Clostridium thermocellum* to cellulose. *J. Bacteriol. 163*:552-559.

Bayer, E. A., Morag, E., and Lamed, R. 1994. The cellulosome--a treasure-trove for biotechnology. *Trends Biotechnol. 12*:379-386.

Beguin, P., and Aubert, J. P. 1994. The biological degradation of cellulose. *FEMS Microbiol. Rev. 13*:25-58.

Beguin, P., Millet, J., and Aubert, J.–P. 1992. Cellulose degradation by *Clostridium thermocellum*: From manure to molecular biology. *FEMS Microbiol. Lett. 100*:523-528.

Bothast, R.J. and Schlicher, M. A. 2005. Biotechnological processes for conversion of corn into ethanol. *Appl. Microbiol. Biotechnol. 67*:19-25.

Demain, A. L., Newcomb, M., and Wu, J. H. D. 2005. Cellulase, clostridia and ethanol. *Microbiol. Mol. Biol. Rev. 69*:124-154.

Desai, S. G., Guerinot, M. L. and Lynd, L. R. 2004. Cloning of the L-lactate dehydrogenase gen and elimination of lactic acid production via gene knockout in *Thermoanaerobacterium saccharolyticum* JW/SL-YS485. *Appl. Microbiol. Biotechnol. 65*:600-605.

Felix, C. R., and Ljungdahl, L. G. 1993. The cellulosome: the extracellular organelle of *Clostridium*. *Annu. Rev. Microbiol. 47*:791-819.

Gerngross, U. T., Romaniec, M. P. M., Huskisson, N. S. and Demain, A. L. 1993. Sequencing of a *Clostridium thermocellum* gene (*cipA*) encoding the cellulosomal S_L-protein reveals an unusual degree of internal homology. *Mol. Microbiol. 8*:325-334.

Hon-nami, K., Coughlan, M. P., Hon-nami, H., and Ljungdahl, L. G. 1986. Separation and characterization of the complexes constituting the cellulolytic enzyme system of *Clostridium thermocellum*. *Arch. Microbiol. 145*:13-19.

Irwin, D. C., Zhang, S., and Wilson, D. B. 2000. Cloning, expression and characterization of a family 48 exocellulase, Cel48A, from *Thermobifida fusca*. *Eur. J. Biochem. 267*:4988-4997.

Johnson, E. A. 1983. Regulation of cellulase activity and synthesis in *Clostridium thermocellum*. Ph.D. thesis. Massachusetts Institute of Technology, Cambridge, MA.

Kruus, K., Andreacchi, A., Wang, W. K., and Wu, J. H. D. 1995. Product inhibition of the recombinant CelS, an exoglucanase component of the *Clostridium thermocellum* cellulosome. *Appl. Microbiol. Biotechnol. 44*:399-404.

Kruus, K., Lua, A. C., Demain, A. L., and Wu, J. H. D. 1995. The anchorage function of CipA (CelL), a scaffolding protein of the *Clostridium thermocellum* cellulosome. *Proc. Natl. Acad. Sci. USA 92*:9254-9258.

Lamed, R., and Bayer, E. A. 1986. Contact and cellulolysis in *Clostridium thermocellum* via extensile surface organelles. *Experientia 42*:72-73.

Lamed, R., and Bayer, E. A. 1988. The cellulosome of *Clostridium thermocellum. Adv. Appl. Microbiol. 33*:1-46.

Lamed, R., Setter, E., and Bayer, E. A. 1983. Characterization of a cellulose-binding, cellulase containing complex in *Clostridium thermocellum. J. Bacteriol. 156*:828-836.

Lamed, R., Setter, E., Kenig, R., and Bayer, E. A. 1983. The cellulosome-a discrete cell surface organelle of *Clostridium thermocellum* which exhibits separate antigenic, cellulose-binding and various cellulolytic activities. *Biotechnol. Bioeng. Symp. 13*:163-181.

Lynd, L. R., Weimer, P. J., van Zyl, W. H., and Pretorius, I. S. 2002. Microbial cellulose utilization: fundamentals and biotechnology. *Microbiol. Mol. Biol. Rev. 66*:506-577.

Lynd, L.R., van Zyl, W.H., McBride, J.E., and Laser, M., 2005. Consolidated bioprocessing of cellulosic biomass: an update. *Curr. Opin. Biotechnol. 16:*577-583.

Mayer, F., Coughlan, K. P., Mori, Y., and Ljungdahl, L. G. 1987. Macromolecular organization of the cellulolytic complex of *Clostridium thermocellum* as revealed by electron microscopy. *Appl. Environ. Microbiol. 53*:2785-2792.

Millet, J., Petre, D., Beguin, P., Raynaud, O., and Aubert, J. P. 1985. Cloning of ten distinct DNA fragments of *Clostridium thermocellum* coding for cellulases. *FEMS Microbiol. Lett*. 29:145-149.

Morag, E., Bayer, E. A., Hazlewood, G. P., Gilbert, H. J., and Lamed, R. 1993. Cellulase Ss (CelS) is synonymous with the major cellobiohydrolase (subunit S8) from the cellulosome of *Clostridium thermocellum. Appl. Biochem. Biotechnol. 43*:147-151.

Pessoa, A. Jr, Roberto, I. C., Menossi, M., Dos Santos, R. R., Filho, S. O., and Penna, T. C. V. 2005. Perspectives on bioenergy and biotechnology in Brazil. *Appl. Biochem. Biotechnol. 121*-124:59-70.

Soutschek-Bauer, E., Hartl, L., and Staudenbauer, W. L. 1985. Transformation of *Clostridium thermohydrosulfuricum* DSM 586 with plasmid DNA. *Biotechnol. Lett.* 7:705-710.

Strobel, H. P. and Lynn B. 2004. Proteomic analysis of ethanol sensitivity in *Clostridium thermocellum.* National meeting of the American Society for Microbiology, New Orleans, LA.

Tyurin, M. V., Desai, S. G., and Lynd, L. R. 2004. Electrotransformation of *Clostridium thermocellum*. *Appl. Environ. Microbiol. 70*:883-890.

Wang, W. K., and Wu, J. H. D. 1993. Structural features of the *Clostridium thermocellum* cellulase S_S gene. *Appl. Biochem. Biotechnol. 39*-40:149-158.

Wang, W. K., Kruus, K., and Wu, J. H. D. 1993. Cloning and DNA sequence of the gene coding for *Clostridium thermocellum* cellulase S_S (CelS), a major cellulosome component. *J. Bacteriol. 175*:1293-1302.

Wu, J. H. D. 1993. *Clostridium thermocellum* cellulosome: new mechanistic concept for cellulose degradation. In *Biocatalyst Design for Stability and Specificity*. pp. 251-264. M. E. Himmel and G. Georgiou, eds, American Chemical Society, Washington, DC.

Wu, J. H. D., and Demain, A. L. 1988. Proteins of the *Clostridium thermocellum* complex responsible for degradation of crystalline cellulose. *In* Biochemistry and Genetics of Cellulose Degradation. pp. 117-131. J. P. Aubert, P. Beguin, and J. Millet, eds., Academic Press, New York.

Wu, J. H. D., Orme-Johnson, W. H. and Demain, A. L. 1988. Two components of an extracellular protein aggregate of *Clostridium thermocellum* together degrade crystalline cellulose. *Biochemistry 27*:1703-1709.

Wyman, C. E. 2001. Twenty years of trials, tribulations, and research progress in bioethanol technology: selected key events along the way. *Appl. Biochem. Biotechnol. 91*-93:5-21.

In: Cellulosome
Eds: V. Uversky and I. A. Kataeva, pp. 11-46
ISBN 1-59454-950-8

Chapter II

THE CELLULOSOME SAGA: EARLY HISTORY

***Edward A. Bayer*[1*] *and Raphael Lamed*[2]**

[1]Department of Biological Chemistry, The Weizmann Institute of Science, Rehovot 76100 Israel,

[2]Department of Molecular Microbiology and Biotechnology, Tel-Aviv University, Ramat Aviv 69978 Israel.

ABSTRACT

The discovery of the cellulosome is a story of serendipity, a story of imagination, persistence, and a triumph over dogma. We simply sought a "cellulose-binding factor" on the cell surface of the anaerobic thermophilic bacterium, *Clostridium thermocellum,* which we inferred would be responsible for the observation that the bacterium attaches strongly to the insoluble cellulose substrate prior to its degradation. Using a then unconventional experimental approach, we unexpectedly isolated a very large multi-subunit supramolecular complex, instead of a small protein. Rather than discarding the uninvited material, we were alert enough to follow up this surprising finding experimentally. A combination of biochemical, biophysical, immunochemical and ultrastructural techniques, followed by molecular biological verification, led to the definition and proof of the cellulosome concept. The birth of the discrete, multi-enzyme cellulosome complex was thus documented. Today, cellulosomes have been confirmed in numerous but not all cellulolytic bacteria. The cellulosome-producing strains exhibit surprising diversity in the composition and architecture of the component parts.

Keywords: cohesin, dockerin, scaffoldin, protein nomenclature, ultrastructure, sequencing, cartoons

---

* Correspondence concerning this article should be addressed to Edward A. Bayer, Department of Biological Chemistry, The Weizmann Institute of Science, Rehovot 76100 Israel. Email: ed.bayer@weizmann.ac.il.

Introduction

In retrospect, several signs implied the existence of cellulosomes preceding their initial discovery and explicit description. For example, in the late 1960s, Leatherwood [1] suggested, on the basis of cellulolysis by colonies of *Ruminococcus albus*, that a combination of a separate "affinity factor" with hydrolytic components was required for effective cellulolysis by this organism. A decade later, Ait and coworkers [2] proposed that the components of the cellulase system in *Clostridium thermocellum* may form aggregates which interfere with the purification of individual enzymes. Indeed, unlike their fungal counterparts, the isolation of individual cellulases from this and related anaerobic bacterium proved difficult. It was clear that *C. thermocellum* was a particularly efficient cellulose degrader, but the enzymes from this organism that performed the degradation of the substrate remained a mystery for many years. Not so, for their fungal brethren, whose individual cellulases could be easily purified in large quantities. For this reason, perhaps, the study of fungal cellulolytic systems, such as *Trichoderma reesei*, flourished, while the study of bacterial systems lagged far behind. Indeed, the established potency of the *C. thermocellum* cellulase system called for its molecular elucidation, and the difficulties encountered in isolating the anticipated free cellulases may have frustrated its use as a bacterial model. As we now know, the basis for intense cellulolysis by this organism stems from its characteristic production of cellulosomes. Had the initial studies on bacterial cellulases centered on bacteria that fail to produce cellulosomes (e.g., *Cellulomonas fimi, Thermobifida fusca, Cellvibrio japonicus* or *Clostridium stearcorarium*), the discovery of cellulosomes might have been delayed considerably.

The efficient degradation of *C. thermocellum* was brought to the attention to the scientific community through a series of pivotal articles by Demain and coworkers [3-8] who compared the unique specific activity of its cellulase system with the best known enzyme system of the fungus, *Trichoderma viride* (later renamed *T. reesei* after its founder Elwyn T. Reese). Much lower (about fifty-fold) levels of secreted proteins of *C. thermocellum* were found to exhibit the same activity for degrading microcrystalline cellulose (Avicel) as the fungal system [5]. The term "true cellulase activity" was introduced to emphasize the capacity to degrade recalcitrant forms of cellulose relative to less crystalline or amorphous forms [5]. Several other basic properties of the *C. thermocellum* cellulase system were established in early work by this group. These include the dependence upon the presence of calcium ions and thiols [8] and the strong end-product inhibition (by cellobiose) when acting on microcrystalline cellulose [3,6], as opposed to more tractable substrates like amorphous cellulose (acid-treated) and cellodextrins. These studies were also substantiated by the work of Zeikus and colleagues [9-13].

Despite the major difficulties experienced by many research groups in isolating bacterial cellulases, some isolated reports of success were published. Persistent efforts by different research groups [11,14] eventually led to the isolation of a pair of different endoglucanases from *C. thermocellum*. These were accompanied by the use of denaturing agents, e.g., urea, to dissociate aggregates. Again, the most interesting aspect of such bacterial cellulases – their association into complexes – was overlooked! Nevertheless, even in light of the success in isolating single bacterial cellulases, it was clear that the meager examples and quantities thereof could never account for the extremely high efficiency of this bacterium in degrading cellulosic substrates. The major breakthrough in the study of anaerobic bacterial cellulase systems emanated from a closer look at the enzyme aggregates produced by *C. thermocellum*. These aggregates had been considered by most investigators to be an irrelevant nuisance, and the material generally found its way to the sink at early stages in the purification schemes.

In this review article, we reminisce on the early days of discovery of the cellulosome in *C. thermocellum*.

THE DISCOVERY

The CBF – The "Cellulose-Binding Factor"

Actually, when we commenced our initial experiments in the early 1980s that eventually led to the discovery of the cellulosome, we were not even looking for cellulases. We were much less ambitious in our objectives. Indeed, we wanted at the time to isolate a "cellulose-binding factor" (then termed CBF) from *C. thermocellum*. Our premise that such a factor existed was based on the early observation that cells of this bacterium adhered strongly and selectively to cellulose prior to and during the degradation of the substrate. As an illustration of this phenomenon, a scanning electron micrograph of *C. thermocellum* cells adhering to microcrystalline cellulose chunks is shown in Figure 1. Our early work was carried out on strain YS — an anaerobic thermophile, recently isolated at the time from a hot spring at Yellowstone National Park and characterized as a particularly potent cellulolytic strain of *C. thermocellum*.

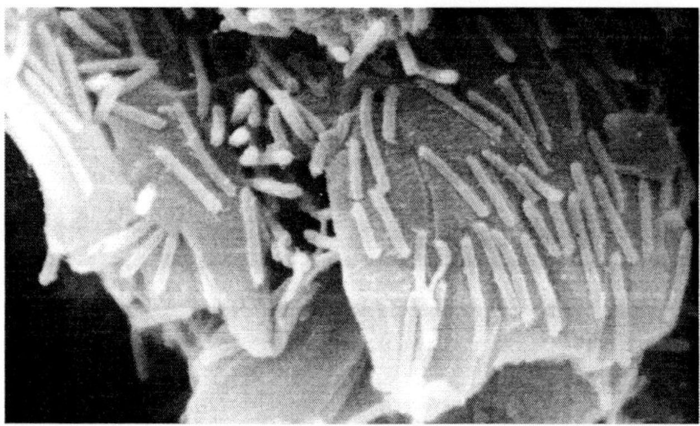

Figure 1. Scanning electron micrograph showing cells of *C. thermocellum* attached to microcrystalline cellulose (Avicel).

An Alternative Strategy: Isolation of an Adherence-Defective Mutant

In order to further study this interaction, a combined genetic-immunochemical approach was undertaken, whereby we first isolated an adherence-defective mutant [15]. This was accomplished by an enrichment procedure, which included repetitive cycles of growth on cellobiose and selective removal of adhering bacteria (Figure 2A). A spontaneous mutant was thus isolated from the wild-type strain YS. A single colony isolated by a roll-bottle technique was defined as adherence-defective mutant AD2. As shown in Figure 2B, cellobiose-grown mutant cells failed to adhere to cellulose, in contrast to the wild-type cells.

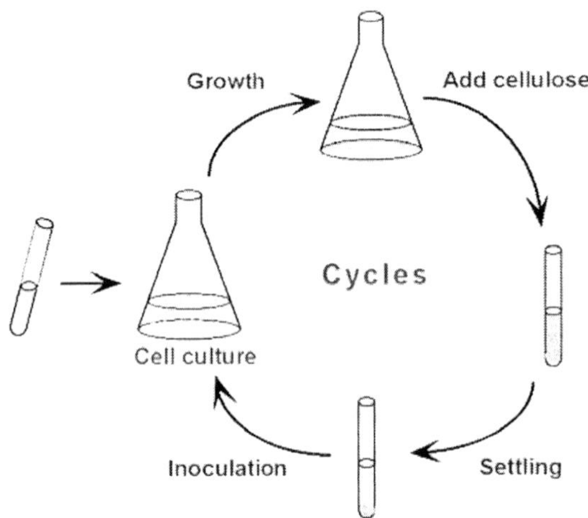

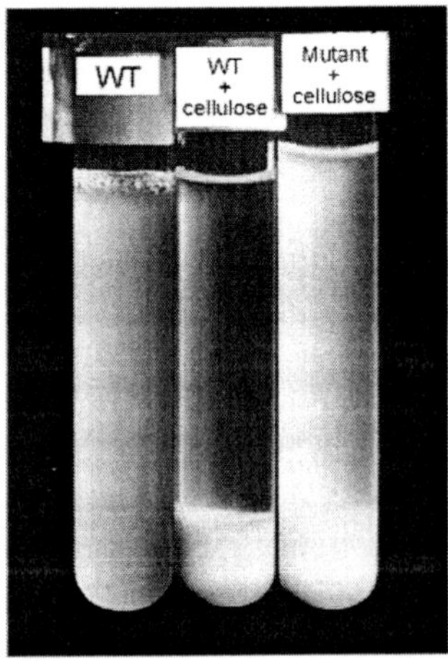

Figure 2. Selection of spontaneous adherence-defective mutant of *C. thermocellum*. A. Scheme illustrating selection procedure. An overnight starter culture (tube at left) is used to inoculate growth medium containing cellobiose as the sole source of carbon and energy. After a 24-h growth period, cellulose was added, and the adherent cells were allowed to settle. The upper portion of the clear supernatant suspension was then used to inoculate fresh growth medium. The process was repeated six times, after which the upper cell suspension remained visibly turbid. The latter suspension was cloned on solid medium using a roll-bottle technique, and single clones were isolated and examined for adherence to cellulose. An adherence-defective mutant, termed AD2, was thus isolated and essentially failed to bind to microcrystalline cellulose. B. Adherence properties of wild-type cells versus mutant AD2. *Test tube at left:* A suspension of wild-type cells in the absence of cellulose. *Middle tube:* Wild-type cells bound to cellulose after settling (note the clear cell suspension indicating more than 99% of the cells adhered to the cellulose substrate). *Right tube:* Mutant AD2 cells fail to adhere to the cellulose substrate. Bayer et al. [15], with permission.

Next, rabbits were immunized intravenously with washed intact wild-type cells, and the polyclonal antiserum preparation was rendered specific for the putative adherence factor by its adsorption onto mutant cells (Figure 3). Using this method, antigens common to both the mutant and wild-type cell surfaces were removed from the preparation leaving only the subpopulation specific to the wild-type cell. Theoretically, the residual antigen should be related to the cell surface-associated function that is present in the wild type but defective in the mutant (i.e., in this case, the attachment of the cells to cellulose). This premise was indeed borne out experimentally; the adsorbed antibody agglutinated wild-type cells but not the mutant. Armed with this function-linked (cellulose-binding) antibody preparation, we proceeded to isolate the CBF in both the extracellular and the cell-associated forms.

The procedure used to isolate the CBF involved an affinity-based procedure, by which binding to cellulose was the important defining step [16]. Microcrystalline cellulose was introduced to extracellular medium or cell sonicate (the supernatant fluids following centrifugation), the cellulose resin was washed, and adsorbed CBF was released by alkaline treatment using 1% triethylamine. The resultant solution was neutralized, the protein was concentrated and separated according to molecular size by gel filtration chromatography on a Sepharose 4B column. The major high-molecular-mass peak was then collected and pooled.

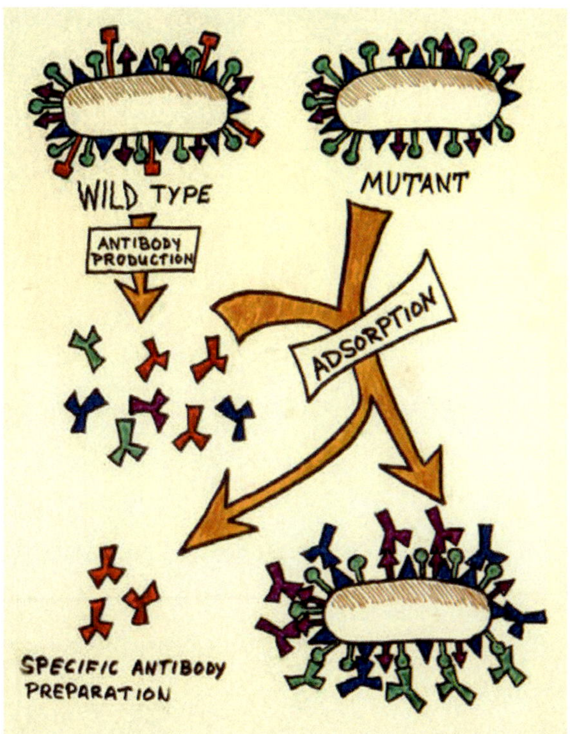

Figure 3. Scheme illustrating strategy for preparation of mono-specific antibodies against the "cellulose-binding factor" (CBF). The mutant cell surface lacks one of the components of the wild-type cell (the red-colored structures). Antibodies, elicited against the wild-type cell, are comprised of numerous subspecies (color coded), each specific for different epitopes on the cell-surface components. Adsorption of the antibody preparation onto the mutant cell purges the species common to both mutant and wild-type cells, thus leaving only the antibody species unique to the wild type. The scheme shown in the figure is the original hand-drawn diagram, heretofore unpublished, rendered by the authors over 20 years ago and used in lectures to describe the approach.

Characteristics of the CBF

The extracellular CBF preparation was demonstrated to be remarkably homogeneous [16]. Rocket immunoelectrophoresis revealed a single precipitin peak generated by the CBF-specific antibodies (Figure 4). Analytical ultracentrifugation (Figure 5) yielded a single symmetrical peak, exhibiting an intrinsic sedimentation coefficient of $s^{\circ}_{20,w}$ = 22.0 and intrinsic diffusion coefficient of $D^{\circ}_{20,w} = 1.02 \times 10^{-7}$ cm^2/sec, consistent with a molecular mass of 2.1 MDa. The relative homogeneity of the isolated CBF preparation was further indicated by transmission electron microscopy of negatively stained preparations. Particulate structures, comparatively uniform in size (~18 nm), were observed (not shown). The particles appeared to be composed of multiple subunits. Varying shapes of the molecules seemed to reflect different faces of the complex and/or flexibility in its structure. Larger structures were also observed, perhaps indicative of higher-order aggregates of the standard particles.

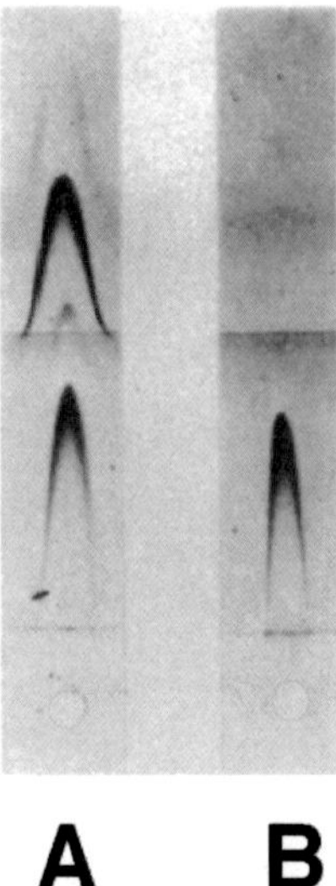

Figure 4. Rocket immunoelectrophoresis pattern of the extracellular medium (lane A) and the isolated CBF (lane B). The intermediate gel contained the CBF-specific antibody (prepared as illustrated in Figure 3), and the upper gel contained the reference whole-cell antibody preparation (without adsorption onto mutant cells). Lamed et al., [16], with permission.

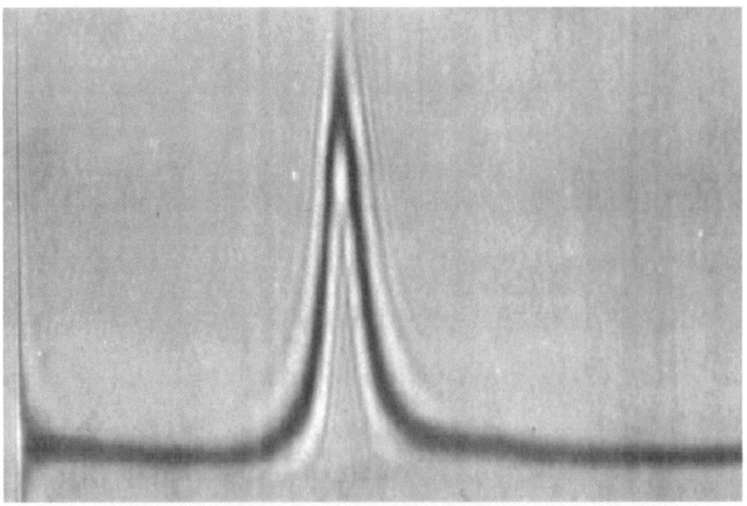

Figure 5. Analytical ultracentrifugation of the purified CBF. Note the single symmetrical peak displayed by the sedimentation pattern. Lamed et al., [16], with permission.

SDS-PAGE of the purified CBF showed a consistent, reproducible and characteristic profile of at least 14 distinct polypeptide bands, ranging in apparent molecular masses of between 48 – 210 kDa (Figure 6, lane A). However, the major surprise in the initial description of the CBF was the discovery that many of the separated polypeptides exhibited cellulase activity (Figure 6, lane B).

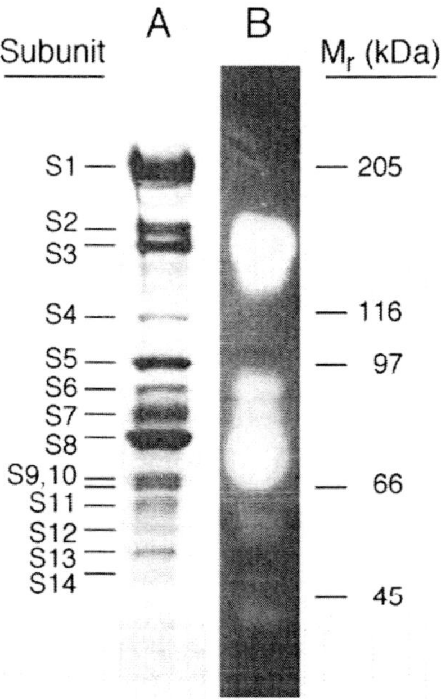

Figure 6. SDS-PAGE of the purified CBF stained for protein (lane A) or processed for cellulolytic activity (lane B). The sample in lane A was stained using Coomassie Brilliant Blue; the sample in lane B was subjected to the CMC-overlay technique, revealing numerous enzymatically active protein bands. Lamed et al., [16], with permission.

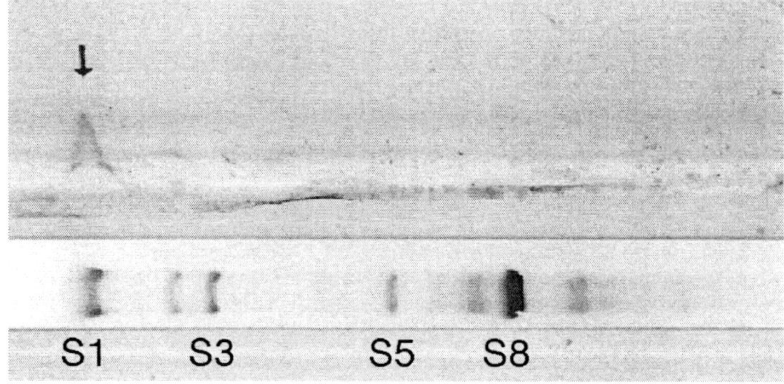

Figure 7. Crossed immunoelectrophoresis of CBF subunits separated by SDS-PAGE. The profile of the separated polypeptides (with some of the subunits identified) is superimposed at the original emplacement of the separating gel. Note the single immunprecipitate peak (arrow) coincident with the 210-kDa S1 component. Lamed et al., [16], with permission.

It was also of interest at the time to determine the antigenicity of the various subunits, and, in order to do so, the SDS-PAGE-separated proteins were subjected to crossed immunoelectrophoresis. This procedure established that only the largest subunit, the 210-kDa S1 subunit, crossreacted extensively with the CBF-specific antibodies (Figure 7). This finding indicated the intense antigenicity of S1 but did not prove that this particular subunit was responsible for cellulose-binding capacity, since the isolated material represented a cohesive complex. Ironically, however, the S1 subunit was indeed found in later work to contain the decisive cellulose-binding module that conferred this activity to the subunit, to the entire complex and even to the intact wild-type cell. It is also noteworthy that enzymatic activity was not associated with the S1 subunit (Figure 6).

Naming the Cellulosome

On the basis of the aforementioned evidence, we came to the following conclusions [16]: "Our current perception of the CBF in *C. thermocellum* can thus be viewed as a large, discrete, multisubunit complex(es) which exhibits both antigenic and cellulolytic activities... The complex apparently comprises various different forms of cellulases, each of which may bear separate specificities toward different quaternary structures on the complex cellulose substrate. The major organizational role of this complex might be designed both for effective delivery to the substrate as well as to bring into proximity the various complementary enzymes (e.g., exo- and endocellulases). In addition, the complex may be structured in such a way as to enable the protection of various product intermediates and to facilitate their transfer to other cellulase components for further hydrolysis. In any event, the cellulase subunits seem to be arranged within the CBF complex in a defined supramolecular fashion designed for highly efficient cellulose degradation."

In an extension of the original results, presented soon thereafter at the Fifth Symposium on Biotechnology for Fuels and Chemicals, held in Gatlinburg Tennessee and published in the Conference Proceedings [17], we first proposed the term "cellulosome" to describe the highly organized multi-cellulase structure of the complex, which appeared to be responsible both for cell adherence as well as economy in enzyme secretion.

The Aftermath

In the ensuing years, several groups contributed confirmatory evidence that a defined cellulase-containing multi-subunit complex indeed exists in *C. thermocellum* and other bacteria. Nevertheless, the debate remained as to whether such a complex were discrete or simply a protein aggregate. These arguments were eventually resolved by the advent of recombinant molecular biology techniques and the accumulated sequences of various cellulosomal components.

Characteristics of the Cellulosomes

Early on, we compared the properties of the purified cellulosome to those reported for the crude cellulase system of *C. thermocellum* [18]. Like the crude enzyme system, the purified cellulosome exhibited the decisive "true cellulase activity" and caused extensive solubilization of crystalline cellulose substrates. This is the first instance that a single entity was shown to effect such activity. Like the crude cellulase system, cellulosome-induced degradation of cellulose was enhanced by both calcium ions and thiols, and was subject to feedback inhibition by cellobiose. In short, the data suggested that the efficient degradation of crystalline cellulose by *C. thermocellum* essentially reflected the major contribution of its cellulosome.

The S1 and S8 components were particularly intriguing. These two subunits appeared to be the two major cellulosomal components at least according to the staining pattern in SDS-PAGE [16,17]. S1 and S8 were termed S_L and S_S, respectively, by Wu and Demain [19,20], thus designating the two dominant 'large' and 'small' cellulosomal subunits. The distinction between the two designations can perhaps be attributed to the two strains involved – S1 and S8 of strain YS versus S_L and S_S of the type strain ATCC 27405.

The S1 component was especially fascinating for several reasons: S1 exhibited most, if not all, of the antigenic activity associated with the cellulosome (Fig. 7), indicating either its high immunogenicity and/or its prominent position on the cell surface. Moreover, the S1 subunit was not observed in the free state but only associated with the cellulosome and appeared to represent a common denominator of the complex. The S1-specific antibodies served to precipitate the entire spectrum of cellulosomal subunits, further indicating their organization into a defined complex. For these and other reasons documented in the first several articles and reviews, we suggested 3 possible functions of the S1 subunit: (i) organization of the other subunits into the complex, (ii) anchoring of the complex to the cell surface and (iii) binding of the complex to the cellulosic substrate [21]. Eventually, all 3 functions were indeed found to be associated with subcomponents of the S1 subunit.

In addition to the latter features, the *C. thermocellum* S1 subunit was recognized early on as a glycoprotein [22]. The major oligosaccharide turned out to be a methylated tetrasaccharide [23,24]:

3-*O*-Me-D-Glc*p*NAc-α(l-2)-[D-Gal*p*-α(l-3)]-D-Gal*f*-α(1-2)-D-Gal

where the intriguing component, galactofuranose, had been known previously to occur in bacterial and fungal polysaccharides, in plant extensins and in a few glycolipids and glycoproteins. Eventually, a very similar but distinctive pentasaccharide was isolated and sequenced from *Bacteroides cellulosolvens*, another cellulosome-producing organism [25]. Again, the same unique trisaccharide backbone – Glc*p*NAc-α(l-2)-Gal*f*-α(1-2)-Gal*p* – was found. In this case, however, the 3-*O*-methyl group was replaced by Gal*f* in an α(l-3) linkage to Glc*p*NAc, and another Gal*f* was attached, also in an α(l-3) linkage, to the protein-linked Gal*p*. In both species of bacterial cellulosomes, the trisaccharide backbone appears to be *O*-linked to residues of Thr/Pro-rich linker segments [26]. These results suggest that such distinctive oligosaccharides may be definitive substituents of this type of cellulosomal subunit and that a similar set of glycosyl transferases would be required for their synthesis. Unfortunately, the subject of cellulosomal saccharides has been all but abandoned since publication of this original series of articles. Their presence, purpose, similarity and/or diversity among the different cellulosomes remain to be explored further.

In contrast to S1, the other dominant cellulosomal component — the S8 subunit — was recognized as an enzyme [16-18,27,28]. The major cellulosomal subunit was eventually isolated by controlled proteolytic digestion of the intact complex, followed by high-resolution ion-exchange chromatography of the liberated material [16-18,27-29]. The major peak was purified and yielded a protein of about 68 kDa. The isolated protein exhibited cellobiohydrolase activity as well as low levels of endoglucanase activity. This was the first report of a cellobiohydrolase from *C. thermocellum*. Several distinctive cellulosome-related features were also associated with S8, including inhibition by cellobiose as well as stabilization by thiols and calcium ions. The results supported the contention that subunit S8 plays a key role in the solubilization of crystalline cellulose by the cellulosome complex. Later, the gene coding for subunit S_S of the type strain (the S8 equivalent) was sequenced and named *celS* [30,31]. Upon publication of the sequence, we noted that its N-terminus was identical to the unpublished N-terminal sequence of S8 from strain YS, thus enabling the connection of the published biochemical characteristics of subunit S8 with CelS [32]. This was indeed confirmed by biochemical characterization of the expressed CelS protein [33].

Soon after its initial description, attempts were conducted to reconstitute the dissociated cellulosome or selected subunits thereof into an intact complex [28]. One of the initial problems in this approach was the detrimental set of treatments initially used to dissociate the native cellulosome, i.e., heating in the presence of SDS. Subsequent efforts to reconstitute the complex were incapable of counteracting the effects of these treatments. In time, single enzyme components of the *C. thermocellum* cellulosome were incorporated into an isolated form of the S1 subunit to produce a cellulosome-like complex. The resultant complexes were demonstrated to exhibit enhanced activity on insoluble cellulose substrates upon. In an early study [20], S_S and S_L were isolated from SDS-dissociated cellulosome, and complexes were produced by mixing the two purified subunits. The resultant preparation displayed an increase in hydrolytic activity on crystalline cellulose substrates. Subsequently, individual recombinant forms of cellulosomal enzymes were incorporated into similar types of complexes, and the observed enhancement of activity was mainly attributed to targeting of the enzyme to the solid substrate [34-36]. In an extension of this approach, the native cellulosome was dissociated using mild treatment consisting of low concentrations of SDS in the presence of EDTA and thiols at relatively low temperatures (30°C). The resulting polypeptides were reassociated, and the activity on crystalline cellulose substrates was recovered [37-40]. In subsequent work [41], the requirement for using SDS as a denaturant was eventually precluded by incubating the purified cellulosome under nondenaturing conditions (in the presence of EDTA and cellulose at 60°C). The resultant reconstituted complexes exhibited enhanced synergy on cellulose compared to the activity of the mixture of free enzymes.

In a separate work, the cellulase and xylanase activities of *C. thermocellum* have been separated into cellulosomal and noncellulosomal fractions [42]. Each of these fractions contained numerous major polypeptide bands that contributed to the different enzymatic activities. In both cases, pH and temperature optima, product pattern, and other features were almost identical. The main difference between the cellulosomal and noncellulosomal enzymes was in the average molecular weights of the respective polypeptides, which appeared responsible for the activity. The cellulosomal enzymes exhibited much larger molecular masses than their noncellulosomal counterparts. The cellulosome-associated enzymes appeared to correspond to known cellulosomal subunits. On the other hand, β-Glucosidase and β-xylosidase activities were shown to be the action of different enzymes; both were noncellulosomal components associated exclusively with the cell. Despite the lack of growth on, and utilization of xylan or its degradation products, *C. thermocellum* produces a highly developed xylanolytic apparatus, which is interlinked with its cellulase system.

Cellulosome Ultrastructure

A vital component in the initial formulation of the cellulosome concept involved the electron microscopic examination of purified samples. The first ultrastructural description by transmission electron microscopy (TEM) of the cellulosome of *C. thermocellum* involved negative staining, which revealed "bagel"-shaped particulate structures of relatively uniform size (~18 nm), composed of irregular recurring shapes that seemed to represent different types of subunits attached to a connecting entity [16].

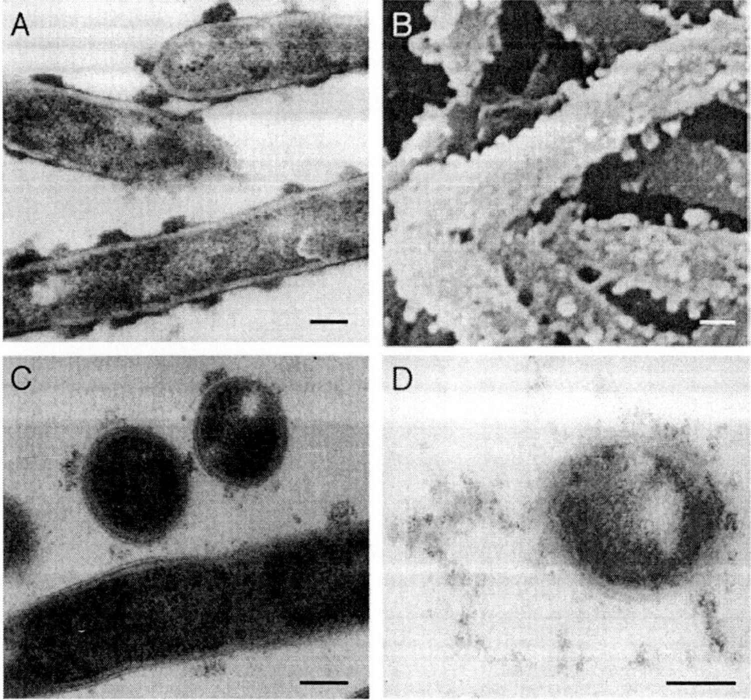

Figure 8. Electron micrographs of labelled cell-surface cellulosomes. A and B. TEM and SEM, respectively, of cationized ferritin-stained, cellobiose-grown cells of *C. thermocellum*. C and D. TEM of anti-cellulosome antibody-labelled cells grown on cellobiose and cellulose, respectively. Bars, 200 nm. Bayer et al., [21]; Bayer and Lamed, [43], with permission.

The disposition of the cellulosome on the bacterial cell surface was revealed by both TEM and scanning electron microscopy (SEM), using a cationic electron-dense probe, cationized ferritin, designed as a general stain for anionic material in thin sections [21,22,41-45]. The results of the cationized-ferritin-labeled cells were further corroborated by specific immunochemical labeling of *C. thermocellum* cells (Figure 8). Protuberance-like structures were observed lining, at intervals, on a layer of exocellular anionic material on the cell surface of cellobiose-grown cells. Immunolabeling indicated that the outer surface of the protuberances appeared to contain multiple cellulosome entities [43]. Upon interaction with the cellulose substrate, the protuberances appear to contract from the cell surface, and the cellulosomes — still attached to the cell surface by means of extended fibrous material — become intimately associated with the surface of the substrate. In cellulose-growing samples, cellulosome complexes, apparently detached from the cells, have been observed inundated on the cellulose surface.

The latter ultrastructural studies [43] served as a basis for proposing a mechanism of interaction between the cells and the cellulose substrate [27]. In the absence of cellulose (e.g., cells grown on cellobiose) the cellulosome-borne protuberance-like organelles are arranged in a resting state on the cell surface (Figure 9). When such cells bind to cellulose, these organelles appear to undergo a dramatic conformational transformation in which elongated fibers are formed between the substrate and the cell surface. The protuberances protract rapidly to form fibrous contact corridors such that cellulosome clusters are attached to the surface of the cellulose and the fibrous material structurally connect the cellulose-bound cellulosomes to the cell surface. These fibers may also function in directing the soluble products from the insoluble substrate to the cell permeases. After extended periods of growth on cellulose, cells of *C. thermocellum* desorb from the substrate, leaving cell-free cellulosome clusters which coat the entire surface of the residual cellulose matrix and presumably continue the degradation process.

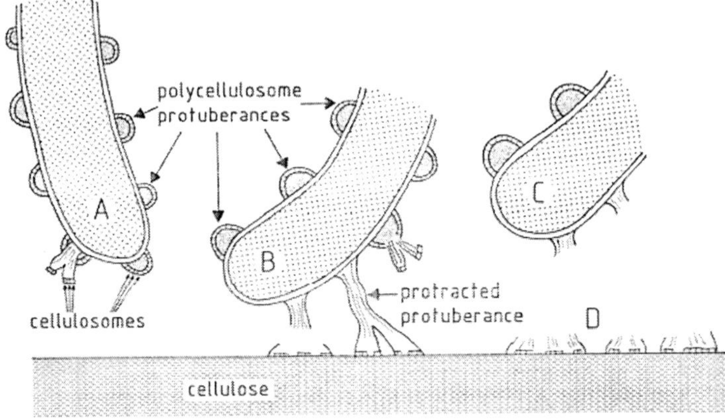

Figure 9. Schematic representation of the interaction of cell-surface cellulosome-laden protuberance-like organelles of *C. thermocellum* with the cellulose surface. The diagram is based mainly on TEM studies, and illustrates the major ultrastructural events, which accompany the adsorption/desorption processes. Cell A represents a cell immediately prior to contact with the cellulosic substrate; cell B, a cell following contact; and cell C, a cell following detachment from the substrate, leaving cellulosome clusters on the cellulose surface (D).

More sophisticated electron microscopy has subsequently been used to describe the high-level organization of negatively stained cellulosomes in polycellulosomes with different packaging of the individual cellulosome particles exhibiting the tight and more relaxed forms [46,47]. During the early stages of growth, clusters of tightly packed cellulosomes, corresponding to the previously described polycellulosome protuberances, were located on the cell surface. In the later stages, loosely packed complexes were observed. Dramatic micrographs, which revealed cellulosome subunits attached to a central fiber, were achieved (Figure 10). The notion of a multiple-cleavage mechanism and targeting was suggested on the basis of the observed array of subunits having a presumed center-to-center distance corresponding to 4 cellobiose units.

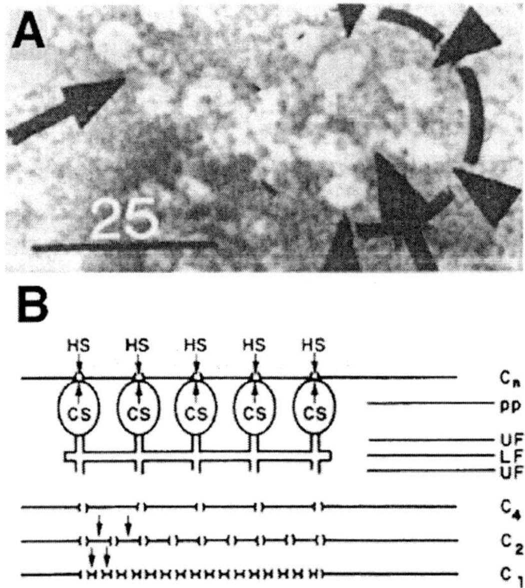

Figure 10. Ultrastructural organization of the cellulosome. A. Negatively stained image of an intact cellulosome complex. The cellulosomal subunits (arrowheads) appear to be attached to a central fiber. Bar, 25 nm. B. Digrammatic view showing the structural interpretation of the cellulosome complex. The cellulose chain (Cn) is cleaved by "multi-cutting events" occurring along the fibril. CS, catalytic site; HS, hydrolysis site; LF, central string; UF, ultrathin fibrils; C4, C2, C1, cellodextrins of various length. Mayer, et al, 1987 [46], with permission.

After a long pause during which cellulosome ultrastructure was virtually abandoned, isolated studies were carried out in the mid-1990s using a different bacterium, *C. papyrosolvens*, which presumably produced a cellulosome akin to those of other mesophilic Clostridia (e.g., *C. cellulolyticum* and *C. cellulovorans*). Fractionation of the extracellular cellulase system resulted in separate fractions, which contained distinct polypeptide composition and activity profiles on cellulosic and xylan substrates [48]. A 125-kDa glycoprotein showed no enzymatic activity and seemed to suggest a smaller version of the noncatalytic S1 subunit of *C. thermocellum* that would be characteristic of *C. papyrosolvens*. TEM studies of the separated multiprotein complexes revealed ultrastructurally distinct particles that appeared to represent the biochemically distinct fractions [49]. The particles were similar in type to the tight and loose structures observed previously for the cellulosome of *C. thermocellum*. Extraordinary micrographs of the loosely aggregated particles clearly indicated six apparent subunits attached to a thin fiber. This timely contribution also emphasized the fact that individual cellulosome complexes in a given bacterium are heterogeneous in their composition and enzymatic profile.

One final ultrastructural investigation was carried out recently [50], in which the authors concentrated on the fibrillar structures resulting from hydromechanical manipulation of intact cellulosomes. An alternating series of coiled and fibrillar stretches along the presumed S1 protein could be visualized, together with detached enzymatic subunits. The observed stretches could conceivably represent fibrillar anchoring protein(s) responsible for connecting the S1 to the cell surface.

Unfortunately, cellulosome ultrastructure was rather neglected after the original work of the 1980s. Retrospectively, the very few negative staining studies performed during the past 2 decades or so have clearly served to portend and complement subsequent structural and molecular biological studies. The advent of contemporary microscopic techniques may reopen this field for more advanced structural studies using STEM, tomography, atomic force microscopy, and other techniques.

The Proof

Recombinant DNA Technology: What the Cellulosomal Sequences Told Us

Coinciding with the initial biochemical definition of the cellulosome concept and during subsequent years, numerous cellulosomal components were systematically sequenced, first from *C. thermocellum* [51-62] and then from other bacteria [63-76]. The standard practice, at the time, was to identify a given enzymatic activity (e.g., cellulase or xylanase) in cloned fragments in a bacterial host cell and to proceed with their sequencing. As such sequences accumulated, it was noticed that many of the resultant enzymes, derived from cellulosome-producing organisms, tended to exhibit a curious reiterated or duplicated segment [30,77,78]. The function of this segment remained elusive for many years [79]. However, it eventually became apparent that the reiterated segment was exclusively associated with the cellulosome – cellulosomal enzymes had it and non-cellulosomal enzymes did not. Finally, the enigma was clarified, and its role in the incorporation of the enzyme subunits into the cellulosome complex was eventually determined [80,81].

On the other side of the coin, the equivalent of the S1 subunit was sequenced almost concurrently from both *C. cellulovorans* and *C. thermocellum* [62,67]. The first definitive outline of an S1-like cellulosome-organizing subunit, came from the sequencing of a 170-kDa protein from *C. cellulovorans* [67]. The sequence consisted of an N-terminal domain, correctly identified as a cellulose-binding domain (CBD) – now, more frequently and more generally referred to as a carbohydrate-binding module (CBM). Hence, the protein was termed CbpA (cellulose-binding protein A). In addition to the CBM, repetitive hydrophilic and hydrophobic domains were evident, the nature of which was not clear at the time. After the sequencing of the *C. cellulovorans* protein, the sequence of the *C. thermocellum* equivalent was also achieved [62]. The domain composition of the latter was quite similar, although their relative positions were different. In addition, the *C. thermocellum* protein bore a duplicated segment at its terminus, similar to that of the cellulosomal enzymes, whereas the *C. cellulovorans* protein lacked such a segment. In this case, the proposed function of the equivalent hydrophobic domains, termed internally repeated elements (IRE), was correctly recognized as potential binding components that interact with the duplicated segments of the enzyme subunits. The name of the protein – CipA (cellulosome-integrating protein A), reflects this characteristic.

These two lines of evidence laid the groundwork for proving the cellulosome concept. Final proof required experimental confirmation. The cellulosome indeed proved to be a discrete complex and not a simple protein aggregate. On the one hand, all of the sequenced cellulosomal enzymes from *C. thermocellum* contained a duplicated segment, and enzymes from other cellulosome-producing bacteria also carried such a segment, whereas noncellulosomal cellulases (xylanases, etc.) usually carried a CBM sequence instead. On the other hand, the sequencing of the two very similar S1-like proteins from two different cellulosome-producing bacteria seemed to provide the second important component. These two proteins appeared to contribute the molecular basis for at least two of the functions, originally associated with the cellulosome concept – i.e., cellulose-binding and complex formation. Indeed, the cellulose-binding function was corroborated biochemically in both cases by subcloning and expressing the relevant CBM, which bound strongly to cellulose surfaces [82,83]. The resultant CBM of *C. thermocellum* was subsequently crystallized and a plausible mechanism for its tenacious binding to the cellulose surface was proposed [84].

The incorporation of cellulosomal subunits into the complex was also demonstrated experimentally: by grafting a duplicated segment onto a noncellulosomal cellulase [80], the resultant enzymatically active fusion protein was able to bind the S1 subunit and was thus integrated into the cellulosome complex. In later publications [81,85,86], the contribution of the "repeated domains" of the S1 subunit was also demonstrated biochemically, thus completing the train of evidence. The third function, originally postulated for the S1 subunit (i.e., attachment to the cell surface), was also resolved by the sequence of CipA. Its C-terminal duplicated segment was soon shown to interact with another "anchoring" protein, which was proposed to attach the cellulosome to the cell surface [87]. Thus, at least in *C. thermocellum*, the cellulosome was proved to be a cell-surface complex whose resident CBM mediates attachment of the entire cell to the cell surface.

One critical issue remained: How is the anchoring protein attached to the cell surface. Once again, the sequence of such proteins revealed the presence of a putative module, which, upon sequence similarity searches in the database, showed homology with S-layer domains that appeared to function in mediating attachment of the parent protein to the cell surface [88]. The observation was confirmed experimentally [89]. The S-layer homology (SLH) module has since become a signature module that indicates that a given protein is anchored to the cell surface.

Table 1. Cellulosome Nomenclature.

Term	Description	Reference	
A. CBD	Cellulose-binding domain	Gilkes et al., 1991	[90]
CBM	Carbohydrate-binding module	Charnock et al., 2000	[91]
B. Scaffoldin	**Scaffolding subunit**	Bayer et al., 1994	[92]
S1	Subunit 1	Lamed et al., 1983	[16]
S_L,	Large subunit	Wu et al., 1988	[20]
CbpA	Cellulose-binding protein A (in *C. cellulovorans*)	Shoseyov, et al. 1992	[67]
CipA	Cellulosome-integrating protein A (in *C. thermocellum)*	Gerngross, et al. 1993	[62]
	Scaffolding subunit	Fujino et al., 1993	[96]
J_1		Ali et al., 1995	[139]
CipC	Cellulosome-integrating protein C (in *C. cellulolyticum)*	Pagès et al., 1996	[140]
C. Cohesin	**Cohesive binding module**	Bayer et al., 1994	[92]
SBD	Subunit-binding domain	Bayer and Lamed, 1992	[93]
HBD	Hydrophobic repeated domain	Takagi et al., 1993	[141]
IRE	Internally repeated elements	Gerngross et al., 1993	[62]
RD	Receptor domain	Salamitou et al., 1994	[81]
R	Repeat	Kruus et al., 1995	[142]
D. Dockerin	**Docking domain**	Bayer et al., 1994	[92]
RS	Reiterated sequence (stretch, segment)	Grépinet and Béguin, 1986	[143]
RR	Reiterated region	Grépinet and Béguin, 1986	[143]
RD	Reiterated domain	Joliff et al., 1986	[54]
DR	Duplicated region	Chauvaux et al., 1990	[144]
DS	Duplicated segment	Salamitou et al., 1994	[81]
DS	Duplicated ligand sequence	Lytle et al., 1996	[145]
DD	Docking domain	Ahsan et al., 1996	[146]
CDR	Conserved duplicated region	Choi and Ljungdahl, 1996	[147]

Cellulosome Nomenclature: How the Cohesin and Dockerin Modules Got their Names

As the different laboratories contributed new evidence in favor of the cellulosome concept, laboratory jargon ran amuck in the literature. In the spirit of the original abbreviation for cellulose-binding domain = CBD [90] more recently referred to more generally as carbohydrate-binding module = CBM [91], and the 3-letter convention for naming genes, 2- and 3-letter abbreviations became rampant, and in our eyes represented an obvious source of confusion (Table 1). Each laboratory had its own "in-house" jargon that spilled over into the primary literature, without regard to previously published precedence. The ambiguity would be further aggravated when expressing the interaction between the repeated domains of the S1 subunit and the duplicated segment of a cellulosomal subunit. In at least in one case, the abbreviation for both cohesin and dockerin modules was the same, published by the same laboratory in different articles!

In order to alleviate the confusion, we searched for generic terms to describe the S1 subunit, its repeated domain, and the duplicated segment of the enzyme [92]. For the S1 subunit, we chose the term "scaffoldin" to represent its functioning as a scaffolding protein. For its repeated domain or module, we chose "cohesin" in view of its cohesive interaction with the duplicated segment. For the duplicated segment itself, we chose "dockerin" in honor of its docking function.

It may be of interest to the reader to relate the following historical anecdote: In our original 1994 TIBTECH manuscript [92], we had termed the docking domain "docker" in the text and the figures. Prior to publication, we sent a copy of the manuscript to our colleague Pierre Béguin of the Pasteur Institute. Pierre sent a message back to us saying: "I thought a docker is someone who works on a wharf!" So we changed the term "docker" to "dockerin" to coincide with the term "cohesin"!

Hence, the biochemical interaction between the S1 "internally repeated element" and the duplicated segment of the enzyme subunits became the "cohesin-dockerin interaction", and much progress into the biochemistry and structure of the cohesin-dockerin interaction has been accomplished in the past decade (for details see Chapter III by Nakar, et al., this volume).

Putting the Puzzle Together

The organization of the genes encoding for the different cellulosomal components provided a modular blueprint of their structure. Thus, for a given scaffoldin, the number and arrangement of cohesins, the presence of a CBM and/or dockerin could be documented. Likewise, the modular makeup of the enzymatic subunits – the presence, position and sequence of the dockerin, the type and family of enzyme (glycoside hydrolase, polysaccharide lyase, carbohydrate esterase), the presence, position and family of ancillary modules (e.g., CBM, etc.) – could also be documented. Some of the cellulosome components are in fact very large molecules, owing to their multi-modular character. However, the determination of the basic cellulosome architecture in a given bacterium is relatively simple – almost "child's play", since the "Lego-like" arrangement of subunits and their incorporation into the complex boils down to a single type of interaction: the cohesin-dockerin interaction. Consequently, the intricate nature of cellulosomal architecture can and has been expressed pictorially in cartoon form (Figures 11 and 12). Such cartoons serve to vastly simplify the description of how the various cellulosomal components interact with each other, how they interact with the substrate, and how they interact with the cell surface.

In fact, our first pictorial description of the basic cellulosome architecture that included the major modular components was published in a review [93] before the actual sequence of the *C. thermocellum* scaffoldin was known (Fig. 11-A). Although, the structure shown in that publication contained an incorrect number and arrangement of cohesins (then termed subunit-binding domains, SBD), the major modular components and overall arrangement of the subunits into the complex were surprisingly accurate. At the time, however, we were unaware that scaffoldin sequences were in the process of being elucidated. The drawing included a CBD (CBM, according to the current terminology) and linkers between the different modules. We were indeed aware that CBDs were present as modular components of free cellulases and that the dockerins (the reiterated segments) were invariably associated with cellulosomal enzyme subunits. Nevertheless, the presence of a matching subunit-binding module on the scaffoldin (S1) subunit was the figment of long-term fantasizing on the cellulosome and its component structure – an educated guess of sorts, peppered with some imagination. The presence of linkers in the structure was prompted by our studies of the amino-acid composition of the cellulosome and its high content of prolines and threonines, which are prevalent in many types of intermodular linking segments of other proteins. The early discovery that the oligosaccharides were linked to Pro/Thr-rich regions of the presumed scaffoldin [26], also contributed to this early scheme. In any case, the structure that we initially proposed was remarkably consistent with the corrected versions (Fig. 11, B-D) that were recognized soon thereafter by us and others [92,94,95] – having benefited by the publication of the *C. thermocellum* scaffoldin sequence [62]. With very little imagination required, one can easily see how close the diagrams in Figure 11 resemble the actual micrograph shown in Figure 10 and published many years earlier.

The Cellulosome and the Cell Surface

In a similar manner, the function of the SLH module in mediating the attachment of the cellulosome complex to the cell surface has also been promoted by appropriate cartoons (Fig. 12). The first description to this effect accompanied the early report of a scaffoldin gene cluster on the *C. thermocellum* genome [96]. Whereas the genes encoding the enzyme components appear to be distributed throughout the chromosome [97], four consecutive genes encode for scaffoldin-like sequences in a multi-scaffoldin gene cluster (Fig. 13). One of these, *cipA*, is the gene that encodes the primary scaffoldin (the enzyme-integrating scaffoldin). The resultant scaffoldin bears a single CBM, 9 type-I cohesins, an X domain of unknown function and a C-terminal (type-II) dockerin. The other three genes, *olpB, orf2 and olpA* (formerly termed *orf3*), encode for proteins that each carry a C-terminal SLH module. The first two also harbor 4 and 2 cohesins, respectively, but their sequences (and binding characteristics) indicate that they are of a different type – type II. The third scaffoldin gene encodes for a protein that contains a single cohesin and SLH module; in this case, the cohesin is a type-I cohesin. In *C. thermocellum*, the type-I cohesins bind to the type-I dockerins borne by the enzyme subunits, whereas the type-II cohesins bind to the type-II *dockerin* of the CipA scaffoldin subunit. Thus, the first cartoon (shown in Fig. 12A) that attempted to describe the architectural arrangement of cellulosome-attachment to the cell surface was technically erred, in that the specificity of the OlpA type-I cohesin is for an enzyme-borne dockerin and not for the type-II dockerin of the CipA scaffoldin. On the other hand, the cartoon in Fig. 12A unequivocally showed the correct function of the SLH module in mediating attachment of its parent protein to the cell surface. In this context, OlpA would be expected to bind a single dockerin-containing enzyme to the cell surface. The reasons for this initial oversight was due to the fact that the diagram in Fig. 12A was an initial speculation that preceded the basic experiments that challenged the specificities of the different cohesins. This was only accomplished later, and the picture was thus modified in subsequent publications as shown in Figures 12B and 12C [98,99].

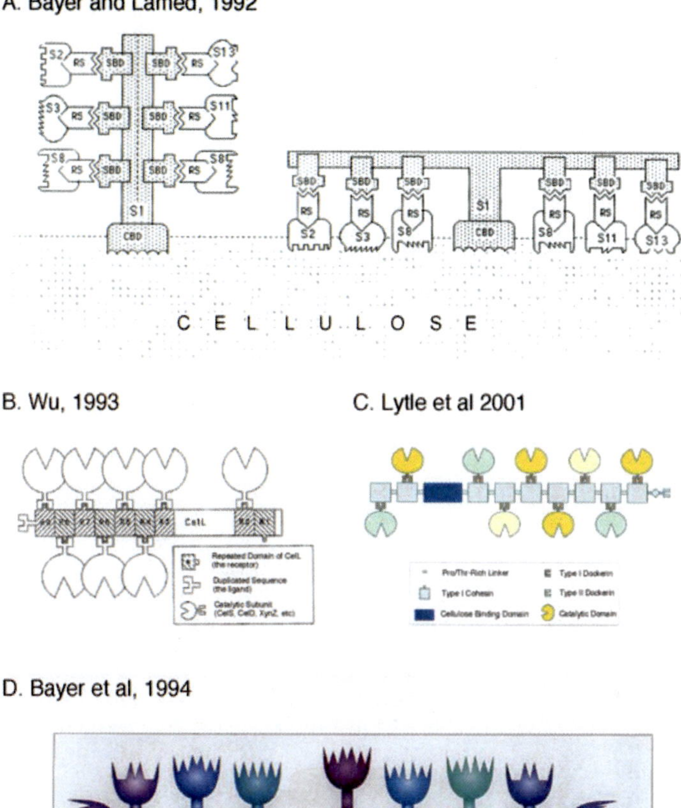

Figure 11. Development of the cellulosome principle through cartoons. A. Preliminary schematic portrayal of the cellulosome, published prior to knowledge of the actual sequence and modular organization of the S1 subunit (the CipA scaffoldin). The different enzymatic subunits (S2, S3, etc.) containing the attached reiterated segment (RS) are bound to the S1 subunit via subunit-binding domains (SBD); Bayer and Lamed, 1992 [93]. B. Cellulosome scheme, reported soon after publication of the CipA scaffoldin sequence; Wu, 1993 [95]. C. More recent rendition of B by the same group; Lytle et al., 2001 [94]. D. Early simplified scheme of the cellulosome, emphasizing the contribution of the cohesin and dockerin interaction to the integrity of the complex; Bayer and Lamed, 1994 [92].

A. Fujino et al., 1993

B. Karita et al., 1997

C. Bayer et al., 1998

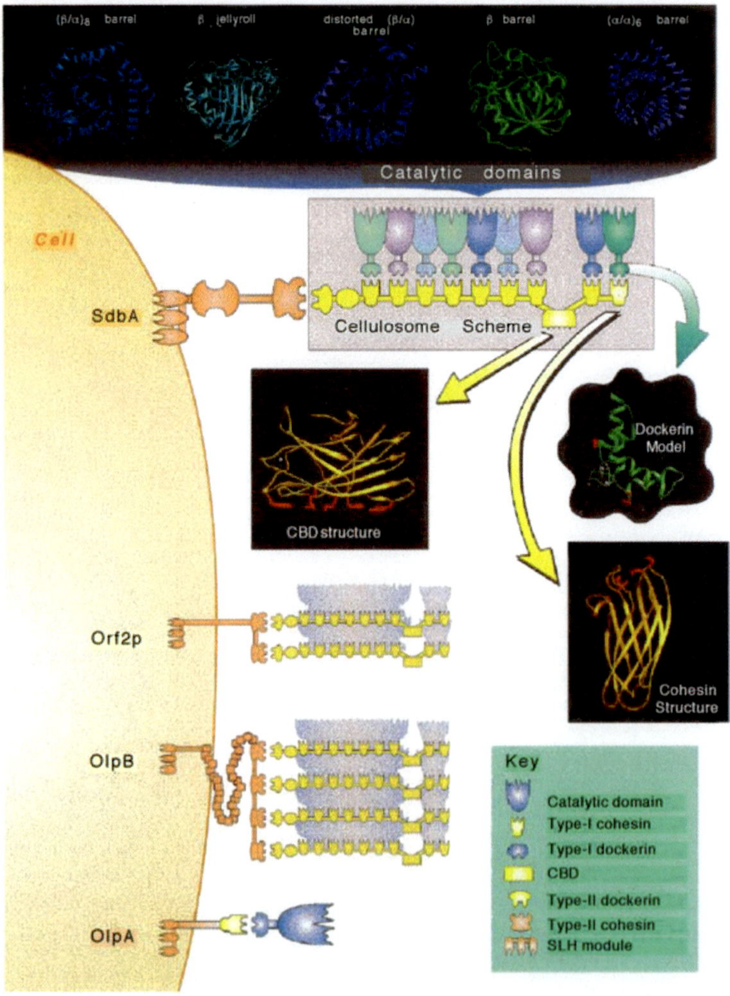

Figure 12. Cartoons illustrating attachment of the *C. thermocellum* cellulosome to the cell surface. A. Preliminary scheme illustrating basic interactions of the cellulosomal components. The scheme is the first to show the proposed attachment of the scaffoldin subunit to the cell via SLH domains; Fujino et al., 1993 [96]. B. Scheme showing different specificities of anchoring proteins in attachment of cellulosome and single enzyme to the cell surface; Karita et al., 1997 [113]. C. Scheme showing structures of the different cellulosome modules, modular architecture of the cellulosome complex and its attachment to the cell wall; Bayer et al. 1998 [99].

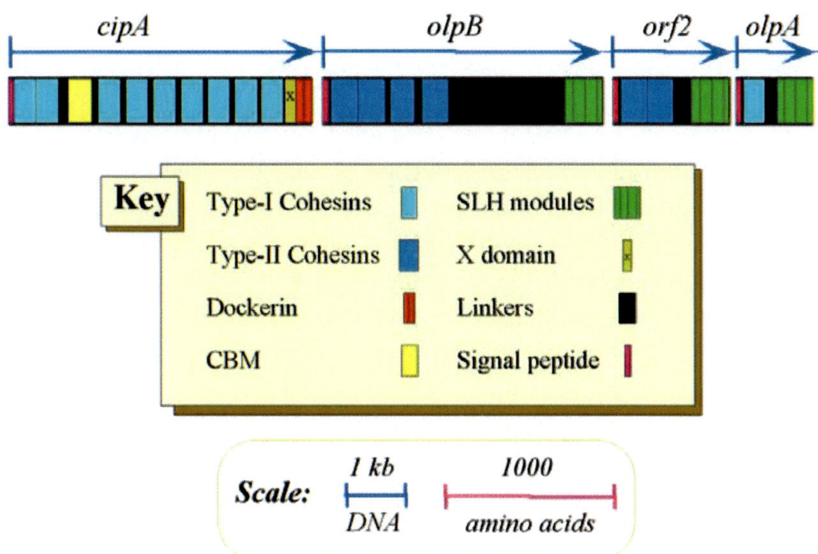

Figure 13. Multiple-scaffoldin gene cluster in *C. thermocellum.* The gene cluster comprises four genes in tandem that encode for scaffoldins, beginning with the gene that codes the primary enzyme-integrating scaffoldin *cipA*. The other three genes code for SLH-containing scaffoldins that mediate their binding to the Gram-positive bacterial cell surface.

A General But not Universal Theme - Cellulosome Diversity

Since the original elucidation of the complex structure and architecture of the cellulosome in *C. thermocellum*, cellulosomes from several other bacterial species have been described. Indeed, the presence of cellulosomes in a wide variety of cellulolytic bacteria was first reported on the basis of biochemical and ultrastructural evidence [22]. These observations were eventually verified for the most part by sequencing various dockerin- and cohesin-containing genes from the designated bactreria.

The cellulosomes of *C. cellulovorans*, *C. cellulolyticum* and *C. josui* were investigated in great detail [100,101]. Though similar in the basics – each comprised dockerin-bearing enzymes incorporated into a multiple-cohesin-, single-CBM-containing scaffoldin [67,70,73] – the cellulosomes of the latter species were found to be fundamentally different from the *C. thermocellum* standard. The observed differences served to silence the critics who seemed to view the originally described cellulosome as a definitive prototype of all cellulosomes, and who considered the broadening of the concept to include other species as an unnecessary overlapping of experimental resources. This viewpoint soon proved to be categorically flawed. A diverse and wondrous world awaited us. (Separate chapters on the *C. cellulovorans* and *C. cellulolyticum* cellulosomes are provided in this treatise by R. Doi and C. Tardif, respectively).

The basic difference between the *C. thermocellum* cellulosome and those of three mesophilic clostridia – *C. cellulovorans*, *C. cellulolyticum* and *C. josui* – was first intimated from the structure of their scaffoldins. Although all of the three latter scaffoldins sequenced contained a single N-terminal CBM and multiple (6 to 9) scaffoldins, none of them contained a dockerin. All three contained one or more X domains (of unknown function). The lack of a dockerin and (to date) SLH-containing scaffoldins in these bacteria render obscure the exact mechanism for attachment to the cell surface, if indeed this cellulosome is attached. In *C. cellulovorans,* cell-surface attachment has been attributed alternatively to the X domains of the scaffoldin or of one of the major cellulosomal enzymes. The difference between the cellulosome systems of the mesophilic clostridia and *C. thermocellum* is also reflected on the genome. Unlike the multiple-scaffoldin gene cluster of the thermophile [96], all three cellulolytic mesophiles contain a very similar type of enzyme-linked gene cluster that commences with the gene encoding for the scaffoldin, followed downstream by numerous genes encoding for cellulosomal dockerin-bearing enzymes [66,102-104]. An additional mesophile, *C. acetobutylicum*, the well-known solvent producer, also contains a similar type of cellulosome gene cluster in its genome [105,106]. The extended cluster contains a scaffoldin gene, followed downstream by a series of genes encoding for the major enzyme components. In this case, however, it is well established that *C. acetobutylicum* is not a cellulolytic organism, and the reasons for the presence and cryptic nature of a cellulosome in this bacterium are not understood. Although the bacterium appears to be incapable of degrading cellulose, low but significant levels of induction of cellulase activity occurs during growth on hemicellulose degradation products, e.g., xylose or lichenan [107].

In contrast to the arrangement in the latter mesophilc clostridia, the cellulosome systems of two other mesophilic bacteria – *Acetivibrio cellulolyticus* and *Bacteroides cellulosolvens* – are, like *C. thermocellum,* characterized by multi-scaffoldin gene clusters [104]. Also like *C. thermocellum,* the cellulosomes of both bacteria are attached to the cell surface via SLH-containing scaffoldins, but, unlike *C. thermocellum,* the modular architecture and composition of their components is distinctly different. Thus, in the *A. cellulolyticus* system, the products of three successive genes – the primary scaffoldin ScaA [108], an "adaptor" scaffoldin ScaB [109] and an anchoring scaffoldin ScaC [109] – presumably self-assemble to generate a supramolecular construction capable of containing 96 individual dockerin-containing enzyme components. The multiple cohesins on the different interconnecting scaffoldins serve to amplify the number of components contained within the cellulosome assembly: the ScaA type-I cohesins bind to the enzyme-borne dockerins, the dockerin of ScaA fits the ScaB type-II cohesins, the ScaB dockerin fits those of ScaC, and an SLH module on ScaC serves to attach the entire assembly to the cell surface. Moreover, an additional unique scaffoldin, ScaD, encoded by a fourth gene of the cluster [110], harbors three cohesins – two of which are type-II cohesins that recognize the ScaA dockerin and a third type-I cohesin that can incorporate an extra, perhaps specialized enzyme or non-enzymatic component. ScaD is the first scaffoldin to be described that contains two different types of cohesin with divergent dockerin specificities.

The cellulosome system of *B. cellulosolvens* shows a completely different but equally intriguing set of variations on the *C. thermocellum* theme [11,112]. Only two scaffoldins have been sequenced from a very short gene cluster, but they are unusually large: the gene products carry 11 and 10 cohesins, respectively. Moreover, the cohesin types are switched – those of the primary ScaA scaffoldin are of type II instead of type I, and those of the anchoring ScaB scaffoldin are of type I. *B. cellulosolvens* ScaA contains a single CBM and

ScaB contains an SLH module. Although very simple in architecture, the reconstructed cellulosome would have the capacity of carrying 110 enzyme components.

Some rumen bacteria, specifically *Ruminococcus albus* and *R. flavefaciens*, are also known to produce a cellulosome system [113-116]. In the case of *R. albus*, its cellulosome seems to be rather similar to that of *C. thermocellum*. The *R. albus* genome is currently in the process of being sequenced by The Institute for Genomic Research (TIGR, http://www.tigr.org). Thus far, over 30 dockerin-containing proteins (enzymatic – both glycoside hydrolases and other types of enzymes, e.g. proteases – as well as nonenzymatic components) have been sequenced. However, a *bona fide* scaffoldin has yet to be detected, the implications of which are unclear, and we eagerly await completion of the genome. A much more intricate and extensive cellulosome system appears to characterize *R. flavefaciens*. The cellulosome system of this rumen bacterium has been partially characterized. Numerous scaffoldins and enzymes have been sequenced [72,116-121], and the cellulosome appears to exhibit very specialized properties. The current view of the organization of the *R. flavefaciens* cellulosome is provided in a separate chapter of this volume (H.J. Flint and M.T. Rincon).

Designer Cellulosomes

The joys of gene sequencing and the information divulged from the sequences of the scaffoldins and cellulosomal enzymes have spawned a new view of the cellulosome and the biotechnological potential of its component parts. Various artificial types of cellulosomes have been proposed in an early review [92], including the concept of producing chimaeric fusion proteins, composed of cellulosomal CBMs, cohesins, dockerins and linker segments. However, at the time, these proposals were somewhat premature, since many of the basic properties of the cellulosomal modules were still uncharacterized – and, in fact, many still are. Nevertheless, in time, information regarding the binding specificity and stability of the CBMs, cohesins and dockerins has been accumulated. This eventually paved the way for the production of designer cellulosomes, consisting of chimaeric scaffoldins and dockerin-containing enzymes.

Since the cohesins of native scaffoldins usually recognize selectively the general spectrum of enzyme-borne dockerins, their mixture in solution would result in random incorporation of the components. For this reason, chimaeric scaffoldins, comprising a CBM and cohesins of divergent specificity must be produced in order to control the incorporation of selected enzymes that bear dockerins of matching specificity. By using this approach, relatively small scaffoldins, with the capacity of incorporating up to three different enzymes, have been produced [122-125]. The enzymes were incorporated stoichiometrically into the scaffoldins by virtue of the potent cohesin-dockerin interaction. By examining a variety of enzyme components and their combined action on different cellulosic substrates, high levels of synergism were observed (see chapter by H.-P. Fierobe). The synergistic effect was found to depend on two basic characteristics of the cellulosome: *targeting* of the complex to the substrate and the *proximity* of the resident cellulosomal enzymes – thus completing the cycle and proving our original suggestions over 20 years ago [16,17].

THE FUTURE

Although much has been accomplished during the past couple of decades since the original discovery of the cellulosome with respect to its composition, the concerted action of its enzymes, its structure, and diversity in nature, there are still many outstanding questions. One major void in our understanding of cellulosome biosynthesis and function is a general lack of knowledge regarding the process of cellulosome assembly at the cell exterior. Cellulosome biosynthesis and assembly has been more or less an avoided topic of study, and remains virtually untilled territory. The regulatory aspects in the biosynthesis of cellulosomal components have only recently been subjected to systematic analysis [126-132]. Although in *C. thermocellum*, the cohesin-dockerin interaction between the CipA scaffoldin and the complement of enzymes seems to be generally nonselective, an intricate molecular apparatus likely exists for the regulated biosynthesis and coordinated incorporation of components into the cellulosome. Such an apparatus begs discovery.

Another area pending extensive development is the continued application of designer cellulosomes – the construction of novel and innovative assemblies, using the component parts of cellulosomes. In addition to its utility as a tool for elucidating the secrets of cellulosome action, designer cellulosomes may eventually be used to improve the degradation of authentic cellulosic substrates with commercial prospects for waste management and/or for the production of useful products. The *in vitro* approach for production of cellulosomal components can also be extended for *in vivo* production of cellulosomes [133-137]. In addition, the strong and versatile cohesin-dockerin interaction may also be used for nanotechnological application.

Thus far, the discovery of cellulosomes has essentially been limited to the realm of anaerobic cellulose-degrading anaerobic bacteria belonging to the clostridial cluster. The highly effective and versatile nature of the cohesin-dockerin interaction, however, represents a protein-protein association that may be much more widely employed in various other types of microorganisms [138]. The need for enzyme organization on the bacterial and archaeal cell surface may also be realized by other types of protein-protein interaction, which are largely unknown or undefined at this time. The general strategy in nature for performance of various types of enzymes has now become a much more acknowledged topic than it was over 20 years ago when the cellulosome was discovered. Microbes are extremely clever, versatile, opportunistic organisms and can easily employ whatever strategy that seems to work. The cellulosome concept, as we define it today, may in fact be part of a greater scheme – not only destined for enhancing the enzymatic degradation of the plant cell wall polysaccharides. The future will undoubtedly bring additional discoveries, breakthroughs and subsequent extensions of the cellulosome concept, whose components may be employed extensively in Nature by widely different types of bacteria in complex environments.

CONCLUSION

Today, the involvement of multi-protein complexes in cellular mechanisms is almost self-evident. Not so, when we first discovered the cellulosome in *Clostridium thermocellum*.

In fact, several years of debate followed as to whether the cellulosome formed a discrete complex or whether it simply represented a nondescript aggregate of proteins. The formation of a discrete cellulosome complex initially gained credence through extensive biochemical and ultrastructural experimentation, but final proof awaited molecular biological validation.

Despite the progress regarding cellulosomal composition, architecture and enzymology, many questions linger: Is the Lego-like architecture of the cellulosome really restricted to anaerobic bacteria that degrade plant cell wall polysaccharides? Are the cohesins and dockerins simply ephemeral components of an eccentric line of evolution? Or is there a more general scheme that has thus far eluded us? Have the bacterial and archaeal dockerins evolved into the eukaryotic calcium-binding proteins? If so, what has become of the cohesins? Have they also evolved? If they have, what are their descendants? Or are the cohesins just one class of dockerin-binding proteins.

These and many other questions remain unanswered and subject to future address. In doing so, the same measures of novel approach, serendipity and astuteness, employed during the original discovery of the cellulosome concept, would indeed be desirable today. We are rather fortunate that one such approach – the sequencing of microbial genomes – is an excellent new and developing resource for our future studies on how widespread is the cellulosome concept in Nature. The advent of genomics and proteomics will certainly serve us in the future. Nevertheless, we have to ensure that our scientific vigilance, imagination and response to serendipity will meet future challenges.

Acknowledgments

This research has been supported by Research Grants 394/03 and 442/05 from the Israel Science Foundation (Jerusalem) and by grants from the United States-Israel Binational Science Foundation (BSF), Jerusalem, Israel.

References

[1] Leatherwood, J. M. (1969) Cellulase complex of Ruminococcus and a new mechanism for cellulose degradation. *Adv. Chem. Ser. 95*.

[2] Ait, N., Creuzet, N., and Forget, P. (1979) Partial purification of cellulase from Clostridium thermocellum. *J. Gen. Microbiol. 113*, 399-402.

[3] Shinmyo, A., Garcia-Martinez, D. V., and Demain, A. L. (1979) Studies on the extracellular cellulolytic enzyme complex produced by Clostridium thermocellum. *J. Appl. Biochem. 1*, 202-209.

[4] Garcia-Martinez, D. V., Shinmyo, A., Madia, A., and Demain, A. L. (1980) Studies on cellulase production by Clostridium thermocellum. *Eur. J. Appl. Microbiol. Biotechnol. 9*, 189-197.

[5] Johnson, E. A., Sakojoh, M., Halliwell, G., Madia, A., and Demain, A. L. (1982) Saccharification of complex cellulosic substrates by the cellulase system from Clostridium thermocellum. *Appl. Environ. Microbiol. 43*, 1125-1132.

[6] Johnson, E. A., Reese, E. T., and Demain, A. L. (1982) Inhibition of Clostridium thermocellum cellulase by end products of cellulolysis. *J. Appl. Biochem. 4*, 64-71.

[7] Duong, C. T. V., Johnson, E. A., and Demain, A. L. (1983) Thermophilic, anaerobic and cellulolytic bacteria. *Enzyme Ferm. Biotechnol. 7*, 156-195.

[8] Johnson, E. A., and Demain, A. L. (1984) Probable involvement of sulfhydryl groups and a metal as essential components of the cellulase of Clostridium thermocellum. *Arch. Microbiol. 137*, 135-138.

[9] Weimer, P. J., and Zeikus, J. G. (1977) Fermentation of cellulose and cellobiose by Clostridium thermocellum in the absence of Methanobacterium thermoautotrophicum. *Appl Environ Microbiol 33*, 289-297.

[10] Ng, T. K., and Zeikus, J. G. (1982) Differential metabolism of cellobiose and glucose by Clostridium thermocellum and Clostridium thermohydrosulfuricum. *J. Bacteriol. 150*, 1391-1399.

[11] Ng, T. K., and Zeikus, J. G. (1981) Purification and characterization of an endoglucanase (1,4-β-D-glucan glucanohydrolase) from Clostridium thermocellum. *Biochem. J. 199*, 341-350.

[12] Ng, T. K., Weimer, T. K., and Zeikus, J. G. (1977) Cellulolytic and physiological properties of Clostridium thermocellum. *Arch. Microbiol. 114*, 1-7.

[13] Lamed, R., and Zeikus, J. G. (1980) Ethanol production by thermophilic bacteria: relationship between fermentation product yields of and catabolic enzyme activities in Clostridium thermocellum and Thermoanaerobium brockii. *J. Bacteriol. 144*, 569-578.

[14] Petre, D., Millet, J., Longin, R., Béguin, P., Girard, H., and Aubert, J. P. (1986) Purification and properties of the endoglucanase C of Clostridium thermocellum produced in Escherichia coli. *Biochimie 68*, 687-695.

[15] Bayer, E. A., Kenig, R., and Lamed, R. (1983) Adherence of Clostridium thermocellum to cellulose. *J. Bacteriol. 156*, 818-827.

[16] Lamed, R., Setter, E., and Bayer, E. A. (1983) Characterization of a cellulose-binding, cellulase-containing complex in Clostridium thermocellum. *J. Bacteriol. 156*, 828-836.

[17] Lamed, R., Setter, E., Kenig, R., and Bayer, E. A. (1983) The cellulosome — a discrete cell surface organelle of Clostridium thermocellum which exhibits separate antigenic, cellulose-binding and various cellulolytic activities. *Biotechnol. Bioeng. Symp. 13*, 163-181.

[18] Lamed, R., Kenig, R., Setter, E., and Bayer, E. A. (1985) Major characteristics of the cellulolytic system of Clostridium thermocellum coincide with those of the purified cellulosome. *Enzyme Microb. Technol. 7*, 37-41.

[19] Wu, J. H. D., and Demain, A. L. (1988) Proteins of the Clostridium thermocellum cellulase complex responsible for degradation of crystalline cellulose, p. 117-131. In Aubert, J.-P., Béguin, P., and Millet, J. (ed.), *Biochemistry and Genetics of Cellulose Degradation*. Academic Press, London.

[20] Wu, J. H. D., Orme-Johnson, W. H., and Demain, A. L. (1988) Two components of an extracellular protein aggregate of Clostridium thermocellum together degrade crystaline cellulose. *Biochemistry 27*, 1703-1709.

[21] Bayer, E. A., Setter, E., and Lamed, R. (1985) Organization and distribution of the cellulosome in Clostridium thermocellum. *J. Bacteriol. 163*, 552-559.

[22] Lamed, R., Naimark, J., Morgenstern, E., and Bayer, E. A. (1987) Specialized cell surface structures in cellulolytic bacteria. *J. Bacteriol. 169*, 3792-3800.

[23] Gerwig, G., de Waard, P., Kamerling, J. P., Vliegenthart, J. F. G., Morgenstern, E., Lamed, R., and Bayer, E. A. (1989) Novel O-linked carbohydrate chains in the cellulase complex (cellulosome) of Clostridium thermocellum. *J. Biol. Chem. 264*, 1027-1035.

[24] Gerwig, G., Kamerling, J. P., Vliegenthart, J. F. G., Morag (Morgenstern), E., Lamed, R., and Bayer, E. A. (1991) Primary structure of O-linked carbohydrate chains in the cellulosome of different Clostridium thermocellum strains. *Eur. J. Biochem. 196*, 115-122.

[25] Gerwig, G., Kamerling, J. P., Vliegenthart, J. F. G., Morag, E., Lamed, R., and Bayer, E. A. (1992) Novel oligosaccharide constituents of the cellulase complex of Bacteroides cellulosolvens. *Eur. J. Biochem. 205*, 799-808.

[26] Gerwig, G., Kamerling, J. P., Vliegenthart, J. F. G., Morag, E., Lamed, R., and Bayer, E. A. (1993) The nature of the carbohydrate-peptide linkage region in glycoproteins from the cellulosomes of Clostridium thermocellum and Bacteroides cellulosolvens. *J. Biol. Chem. 268*, 26956-26960.

[27] Lamed, R., and Bayer, E. A. (1988) The cellulosome concept: exocellular/extracellular enzyme reactor centers for efficient binding and cellulolysis, p. 101-116. In Aubert, J.-P., Beguin, P., and Millet, J. (ed.), *Biochemistry and Genetics of Cellulose Degradation*. Academic Press, London.

[28] Lamed, R., and Bayer, E. A. (1988) The cellulosome of Clostridium thermocellum. *Adv. Appl. Microbiol. 33*, 1-46.

[29] Morag, E., Halevy, I., Bayer, E. A., and Lamed, R. (1991) Isolation and properties of a major cellobiohydrolase from the cellulosome of Clostridium thermocellum. *J. Bacteriol. 173*, 4155-4162.

[30] Wang, W. K., and Wu, J. H. (1993) Structural features of the Clostridium thermocellum cellulase SS gene. *Appl Biochem Biotechnol 39-40*, 149-158.

[31] Wang, W. K., Kruus, K., and Wu, J. H. D. (1993) Cloning and DNA sequence of the gene coding for Clostridium thermocellum cellulase SS (CelS), a major cellulosome component. *J. Bacteriol. 175*, 1293-1302.

[32] Morag, E., Bayer, E. A., Hazlewood, G. P., Gilbert, H. J., and Lamed, R. (1993) Cellulase SS (CelS) is synonymous with the major cellobiohydrolase (subunit S8) from the cellulosome of Clostridium thermocellum. *Appl. Biochem. Biotechnol. 43*, 147-151.

[33] Wang, W. K., Kruus, K., and Wu, J. H. D. (1994) Cloning and expression of the Clostridium thermocellum cellulase celS gene in Escherichia coli. *Appl. Microbiol. Biotechnol. 42*, 346-352.

[34] Kataeva, I., Guglielmi, G., and Béguin, P. (1997) Interaction between Clostridium thermocellum endoglucanase CelD and polypeptides derived from the cellulosome-integrating protein CipA: stoichiometry and cellulolytic activity of the complexes. *Biochem. J. 326*, 617-624.

[35] Ciruela, A., Gilbert, H. J., Ali, B. R. S., and Hazlewood, G. P. (1998) Synergistic interaction of the cellulosome integrating protein (CipA) from Clostridium thermocellum with a cellulosomal endoglucanase. *FEBS Lett. 422*, 221-224.

[36] Garcia-Campayo, V., and Béguin, P. (1997) Synergism between the cellulosome-integrating protein CipA and endoglucanase CelD of Clostridium thermocellum. *J. Biotechnol. 57*, 39-47.

[37] Beattie, L., Bhat, K. M., and Wood, T. M. (1994) The effect of cations on reassociation of the components of the cellulosome cellulase complex synthesized by the bacterium Clostridium thermocellum. *Appl. Microbiol. Biotechnol. 40*, 740-744.

[38] Bhat, K. M., and Wood, T. M. (1992) The cellulase of the anaerobic bacterium Clostridium thermocellum: isolation, dissociation, and reassociation of the cellulosome. *Carbohydr. Res. 227*, 293-300.

[39] Bhat, S., Goodenough, P. W., Bhat, M. K., and Owen, E. (1994) Isolation of four major subunits from Clostridium thermocellum cellulosome and their synergism in the hydrolysis of crystalline cellulose. *Int. J. Biol. Macromol. 16*, 335-342.

[40] Bhat, M. K. (1998) Clostridium thermocellum cellulosome: Dissociation, isolation and characterisation of subunits and the potential biotechnological implications. *Recent Res. Devel. Biotech. Bioeng. 1*, 59-84.

[41] Morag, E., Yaron, S., Lamed, R., Kenig, R., Shoham, Y., and Bayer, E. A. (1996) Dissociation of the cellulosome of Clostridium thermocellum under nondenaturing conditions. *J. Biotechnol. 51*, 235-242.

[42] Morag, E., Bayer, E. A., and Lamed, R. (1990) Relationship of cellulosomal and noncellulosomal xylanases of Clostridium thermocellum to cellulose-degrading enzymes. *J. Bacteriol. 172*, 6098-6105.

[43] Bayer, E. A., and Lamed, R. (1986) Ultrastructure of the cell surface cellulosome of Clostridium thermocellum and its interaction with cellulose. *J. Bacteriol. 167*, 828-836.

[44] Lamed, E., and Bayer, E. A. (1986) Contact and cellulolysis in Clostridium thermocellum via extensile surface organelles. *Experientia 42*, 72-73.

[45] Lamed, R., Naimark, J., Morgenstern, E., and Bayer, E. A. (1987) Scanning electron microscopic delineation of bacterial surface topology using cationized ferritin. *J. Microbiol. Methods* 7, 233-240.

[46] Mayer, F., Coughlan, M. P., Mori, Y., and Ljungdahl, L. G. (1987) Macromolecular organization of the cellulolytic enzyme complex of Clostridium thermocellum as revealed by electron microscopy. *Appl. Environ. Microbiol. 53*, 2785-2792.

[47] Mayer, F. (1988) Cellulolysis: ultrastructural aspects of bacterial systems. *Electron Microsc. Rev. 1*, 69-85.

[48] Pohlschröder, M., Leschine, S. B., and Canale-Parola, E. (1994) Multicomplex cellulase-xylanase system of Clostridium papyrosolvens C7. *J. Bacteriol. 176*, 70-76.

[49] Pohlschröder, M., Canale-Parola, E., and Leschine, S. B. (1995) Ultrastructural diversity of the cellulase complexes of Clostridium papyrosolvens C7. *J. Bacteriol. 177*, 6625-6629.

[50] Madkour, M., and Mayer, F. (2003) Structural organization of the intact bacterial cellulosome as revealed by electron microscopy. *Cell Biol. Int. 27*, 831-836.

[51] Cornet, P., Millet, J., Béguin, P., and Aubert, J.-P. (1983) Characterization of two cel (cellulose degradation) genes of Clostridium thermocellum coding for endoglucanases. *Bio/technology 1*, 589-594.

[52] Béguin, P., Cornet, P., and Aubert, J.-P. (1985) Sequence of a cellulase gene of the thermophilic bacterium Clostridium thermocellum. *J. Bacteriol. 162*, 102-105.
[53] Millet, J., Petre, D., Béguin, P., Raynaud, O., and Aubert, J.-P. (1985) Cloning of ten distinct DNA fragments of Clostridium thermocellum coding for cellulases. *FEBS Lett. 29*, 145-149.
[54] Joliff, G., Béguin, P., and Aubert, J.-P. (1986) Nucleotide sequence of the cellulase gene celD encoding endoglucanase D of Clostridium thermocellum. *Nucl. Acids Res. 14*, 8605-8613.
[55] Béguin, P., Millet, J., Grépinet, O., Navarro, A., Juy, M., Amit, A., Poljak, R., and Aubert, J.-P. (1988) The cel (cellulose degradation) genes of Clostridium thermocellum, p. 267-282. In Aubert, J.-P., Béguin, P., and Millet, J. (ed.), *Biochemistry and Genetics of Cellulose Degradation*. Academic Press, London.
[56] Grépinet, O., Chebrou, M.-C., and Béguin, P. (1988) Nucleotide sequence and deletion analysis of the xylanase gene (xynZ) of Clostridium thermocellum. *J. Bacteriol. 170*, 4582-4588.
[57] Hazlewood, G. P., Romaniec, M. P. M., Davidson, K., Grépinet, O., Béguin, P., Millet, J., Raynaud, O., and Aubert, J.-P. (1988) A catalogue of Clostridium thermocellum endoglucanase, β-glucosidase and xylanase genes cloned in Eshcherichia coli. *FEMS Microbiol. Lett. 51*, 231-236.
[58] MacKenzie, C. R., Yang, R. C., Patel, G. B., Bilous, D., and Narang, S. A. (1989) Identification of three distinct Clostridium thermocellum xylanase genes by molecular cloning. *Arch. Microbiol. 152*, 377-381.
[59] Tuka, K., Zverlov, V. V., Bumazkin, B. K., Velikodvorskaya, G. A., and Strongin, A. Y. (1990) Cloning and expression of Clostridium thermocellum genes coding for thermostable exoglucanases (cellobiohydrolases) in Escherichia coli cells. *Biochem. Biophys. Res. Commun. 169*, 1055-1060.
[60] Yagüe, E., Béguin, P., and Aubert, J.-P. (1990) Nucleotide sequence and deletion analysis of the cellulase-encoding gene celH of Clostridium thermocellum. *Gene 89*, 61-67.
[61] Navarro, A., Chebrou, M.-C., Béguin, P., and Aubert, J.-P. (1991) Nucleotide sequence of the cellulase gene celF of Clostridium thermocellum. *Research Microbiology 142*, 927-936.
[62] Gerngross, U. T., Romaniec, M. P. M., Kobayashi, T., Huskisson, N. S., and Demain, A. L. (1993) Sequencing of a Clostridium thermocellum gene (cipA) encoding the cellulosomal SL-protein reveals an unusual degree of internal homology. *Mol. Microbiol. 8,* 325-334.
[63] Faure, E., Belaich, A., Bagnara, C., Gaudin, C., and Belaich, J.-P. (1989) Sequence analysis of the Clostridium cellulolyticum endoglucanase-A-encoding gene celCCA. *Gene 84*, 39-46.
[64] Foong, F., Hamamoto, T., Shoseyov, O., and Doi, R. H. (1991) Nucleotide sequence and characteristics of endoglucanase gene engB from Clostridium cellulovorans. *J. Gen. Microbiol. 137*, 1729-1736.

[65] Shima, S., Igararashi, Y., and Kodama, T. (1991) Nucleotide sequence analysis of the endoglucanase-encoding gene, celCCD, of Clostridium cellulolyticum. *Gene 104*, 33-38.

[66] Bagnara-Tardif, C., Gaudin, C., Belaich, A., Hoest, P., Citard, T., and Belaich, J.-P. (1992) Sequence analysis of a gene cluster encoding cellulases from Clostridium cellulolyticum. *Gene 119*, 17-28.

[67] Shoseyov, O., Takagi, M., Goldstein, M. A., and Doi, R. H. (1992) Primary sequence analysis of Clostridium cellulovorans cellulose binding protein A. *Proc. Natl. Acad. Sci. USA 89*, 3483-3487.

[68] Reverbel-Leroy, C., Belaich, A., Bernadac, A., Gaudin, C., Belaich, J.-P., and Tardif, C. (1996) Molecular study and overexpression of the Clostridium cellulolyticum celF cellulase gene in Escherichia coli. *Microbiology 142*, 1013-1023.

[69] Gal, L., Gaudin, C., Belaich, A., Pagès, S., Tardif, C., and Belaich, J.-P. (1997) CelG from Clostridium cellulolyticum: a multidomain endoglucanase acting efficiently on crystalline cellulose. *J. Bacteriol. 179*, 6595-6601.

[70] Pagès, S., Belaich, A., Fierobe, H.-P., Tardif, C., Gaudin, C., and Belaich, J.-P. (1999) Sequence analysis of scaffolding protein CipC and ORFXp, a new cohesin-containing protein in Clostridium cellulolyticum: comparison of various cohesin domains and subcellular localization of ORFXp. *J. Bacteriol. 181*, 1801-1810.

[71] Flint, H. J., McPherson, C. A., and Bisset, J. (1989) Molecular cloning of genes from Ruminococcus flavefaciens encoding xylanase and β(1-3,1-4)glucanase activities. *Appl Environ Microbiol 55*, 1230-1233.

[72] Kirby, J., Martin, J. C., Daniel, A. S., and Flint, H. J. (1997) Dockerin-like sequences in cellulases and xylanases from the rumen cellulolytic bacterium Ruminococcus flavefaciens. *FEMS Microbiol. Lett. 149*, 213-219.

[73] Kakiuchi, M., Isui, A., Suzuki, K., Fujino, T., Fujino, E., Kimura, T., Karita, S., Sakka, K., and Ohmiya, K. (1998) Cloning and DNA sequencing of the genes encoding Clostridium josui scaffolding protein CipA and cellulase CelD and identification of their gene products as major components of the cellulosome. *J. Bacteriol. 180*, 4303-4308.

[74] Ohara, H., Noguchi, J., Karita, S., Kimura, T., Sakka, K., and Ohmiya, K. (2000) Sequence of egV and properties of EgV, a Ruminococcus albus endoglucanase containing a dockerin domain. *Biosci. Biotechnol. Biochem. 64*, 80-88.

[75] Taguchi, H., Hagiwara, D., Genma, T., Karita, S., Kimura, T., Sakka, K., and Ohmiya, K. (2004) Cloning of the Ruminococcus albus cel5D and cel9A genes encoding dockerin module-containing endoglucanases and expression of cel5D in Escherichia coli. *Biosci. Biotechnol. Biochem. 68*, 1557-1564.

[76] Jindou, S., Karita, S., Fujino, E., Fujino, T., Hayashi, H., Kimura, T., Sakka, K., and Ohmiya, K. (2002) alpha-Galactosidase Aga27A, an enzymatic component of the Clostridium josui cellulosome. *J Bacteriol 184*, 600-604.

[77] Schimming, S., Schwarz, W., and Staudenbauer, W. (1992) Structure of the Clostridium thermocellum gene licB and the encoded β-1,3-1,4-glucanase. A catalytic region homologous to Bacillus lichenases joined to the reiterated domain of clostridial cellulases. *Eur. J. Biochem. 204*, 13-19.

[78] Tokatlidis, K., Salamitou, S., Béguin, P., Dhurjati, P., and Aubert, J.-P. (1991) Interaction of the duplicated segment carried by Clostridium thermocellum cellulases with cellulosome components. *FEBS Lett. 291*, 185-188.

[79] Hall, J., Hazlewood, G. P., Barker, P. J., and Gilbert, H. J. (1988) Conserved reiterated domains in Clostridium thermocellum endoglucanases are not essential for activity. *Gene 69*, 29-38.

[80] Tokatlidis, K., Dhurjati, P., and Béguin, P. (1993) Properties conferred on Clostridium thermocellum endoglucanase CelC by grafting the duplicated segment of endoglucanase CelD. *Protein Eng. 6*, 947-952.

[81] Salamitou, S., Raynaud, O., Lemaire, M., Coughlan, M., Béguin, P., and Aubert, J.-P. (1994) Recognition specificity of the duplicated segments present in Clostridium thermocellum endoglucanase CelD and in the cellulosome-integrating protein CipA. J. *Bacteriol. 176*, 2822-2827.

[82] Poole, D. M., Morag, E., Lamed, R., Bayer, E. A., Hazlewood, G. P., and Gilbert, H. J. (1992) Identification of the cellulose binding domain of the cellulosome subunit S1 from Clostridium thermocellum. *FEMS Microbiol. Lett. 99*, 181-186.

[83] Goldstein, M. A., Takagi, M., Hashida, S., Shoseyov, O., Doi, R. H., and Segel, I. H. (1993) Characterization of the cellulose binding domain of the Clostridium cellulovorans cellulose-binding protein A. *J. Bacteriol. 175*, 5762-5768.

[84] Tormo, J., Lamed, R., Chirino, A. J., Morag, E., Bayer, E. A., Shoham, Y., and Steitz, T. A. (1996) Crystal structure of a bacterial family-III cellulose-binding domain: A general mechanism for attachment to cellulose. *EMBO J. 15*, 5739-5751.

[85] Leibovitz, E., and Béguin, P. (1996) A new type of cohesin domain that specifically binds the dockerin domain of the Clostridium thermocellum cellulosome-integrating protein CipA. *J. Bacteriol. 178*, 3077-3084.

[86] Leibovitz, E., Ohayon, H., Gounon, P., and Béguin, P. (1997) Characterization and subcellular localization of the Clostridium thermocellum scaffoldin dockerin binding protein SdbA. *J. Bacteriol. 179*, 2519-2523.

[87] Béguin, P., Chauvaux, S., Guglielmi, G., Matuschek, M., Leibovitz, E., Chaveroche, M.-K., Miras, I., Alzari, P., and Gounon, P. (1999) The Clostridium thermocellum cellulosome: Organization and mode of attachment to the cell, p. 437-443. In Ohmiya, K., Hayashi, K., Sakka, K., Kobayashi, Y., Karita, S., and Kimura, T. (ed.), *Genetics, biochemistry and ecology of cellulose degradation*. Uni Publishers Co., Tokyo.

[88] Lemaire, M., Miras, I., Gounon, P., and Béguin, P. (1998) Identification of a region responsible for binding to the cell wall within the S-layer protein of Clostridium thermocellum. *Microbiology 144*, 211-217.

[89] Chauvaux, S., Matuschek, M., and Béguin, P. (1999) Distinct affinity of binding sites for S-layer homologous domains in Clostridium thermocellum and Bacillus anthracis cell envelopes. *J Bacteriol 181*, 2455-2458.

[90] Gilkes, N. R., Henrissat, B., Kilburn, D. G., Miller, R. C. J., and Warren, R. A. J. (1991) Domains in microbial β-1,4-glycanases: sequence conservation, function, and enzyme families. *Microbiol. Rev. 55*, 303-315.

[91] Charnock, S. J., Bolam, D. N., Turkenburg, J. P., Gilbert, H. J., Ferreira, L. M., Davies, G. J., and Fontes, C. M. (2000) The X6 "thermostabilizing" domains of xylanases are

carbohydrate-binding modules: structure and biochemistry of the Clostridium thermocellum X6b domain. *Biochemistry 39*, 5013-5021.

[92] Bayer, E. A., Morag, E., and Lamed, R. (1994) The cellulosome — A treasure-trove for biotechnology. *Trends Biotechnol. 12*, 378-386.

[93] Bayer, E. A., and Lamed, R. (1992) The cellulose paradox: pollutant par excellence and/or a reclaimable natural resource? *Biodegradation 3*, 171-188.

[94] Lytle, B. L., Volkman, B. F., Westler, W. M., Heckman, M. P., and Wu, J. H. D. (2001) Solution structure of a type I dockerin domain, a novel prokaryotic, extracellular calcium-binding domain. *J. Mol. Biol. 307*, 745-753.

[95] Wu, J. H. D. (1993) Clostridium thermocellum cellulosome — New mechanistic concept for cellulose degradation, p. 251-264. In Himmel, M. E. and Georgiou, G. (ed.), *Biocatalyst design for stability and specificity*. ACS Symposium Series No. 516.

[96] Fujino, T., Béguin, P., and Aubert, J.-P. (1993) Organization of a Clostridium thermocellum gene cluster encoding the cellulosomal scaffolding protein CipA and a protein possibly involved in attachment of the cellulosome to the cell surface. *J. Bacteriol. 175*, 1891-1899.

[97] Guglielmi, G., and Béguin, P. (1998) Cellulase and hemicellulase genes of Clostridium thermocellum from five independent collections contain few overlaps and are widely scattered across the chromosome. *FEMS Microbiol. Lett. 161*, 209-215.

[98] Karita, S., Sakka, K., and Ohmiya, K. (1997) Cellulosomes, cellulase complexes, of anaerobic microbes: their structure models and functions, p. 47-57. In Onodera, R., Itabashi, H., Ushida, K., Yano, H., and Sasaki, Y. (ed.), *Rumen microbes and digestive physiology in ruminants, Vol. 14*. Japan Sci. Soc. Press, Tokyo/S.Karger, Basel.

[99] Bayer, E. A., Shimon, L. J. W., Lamed, R., and Shoham, Y. (1998) Cellulosomes: structure and ultrastructure. *J. Struct. Biol. 124*, 221-234.

[100] Belaich, J.-P., Belaich, A., Fierobe, H.-P., Gal, L., Gaudin, C., Pagès, S., Reverbel-Leroy, C., and Tardif, C. (1999) The cellulolytic system of Clostridium cellulolyticum, p. 479-487. In Ohmiya, K., Hayashi, K., Sakka, K., Kobayashi, Y., Karita, S., and Kimura, T. (ed.), *Genetics, biochemistry and ecology of cellulose degradation*. Uni Publishers Co., Tokyo.

[101] Doi, R. H., and Tamura, Y. (2001) The Clostridium cellulovorans cellulosome: An enzyme complex with plant cell wall degrading activity. *Chem. Rec. 1*, 24-32.

[102] Belaich, J. P., Tardif, C., Belaich, A., and Gaudin, C. (1997) The cellulolytic system of Clostridium cellulolyticum. *J Biotechnol 57*, 3-14.

[103] Tamaru, Y., Karita, S., Ibrahim, A., Chan, H., and Doi, R. H. (2000) A large gene cluster for the Clostridium cellulovorans cellulosome. *J Bacteriol 182*, 5906-5910.

[104] Bayer, E. A., Belaich, J.-P., Shoham, Y., and Lamed, R. (2004) The cellulosomes: Multi-enzyme machines for degradation of plant cell wall polysaccharides. *Annu. Rev. Microbiol. 58*, 521-554.

[105] Nolling, J., Breton, G., Omelchenko, M. V., Makarova, K. S., Zeng, Q., Gibson, R., Lee, H. M., Dubois, J., Qiu, D., Hitti, J., Wolf, Y. I., Tatusov, R. L., Sabathe, F., Doucette_Stamm, L., Soucaille, P., Daly, M. J., Bennett, G. N., Koonin, E. V., and Smith, D. R. (2001) Genome sequence and comparative analysis of the solvent-producing bacterium Clostridium acetobutylicum. *J. Bacteriol. 183*, 4823-4838.

[106] Sabathe, F., Belaich, A., and Soucaille, P. (2002) Characterization of the cellulolytic complex (cellulosome) of Clostridium acetobutylicum. *FEMS Microbiol. Lett. 217*, 15-22.

[107] Lopez-Contreras, A. M., Gabor, K., Martens, A. A., Renckens, B. A., Claassen, P. A., Van Der Oost, J., and De Vos, W. M. (2004) Substrate-induced production and secretion of cellulases by Clostridium acetobutylicum. *Appl. Environ. Microbiol. 70*, 5238-5243.

[108] Ding, S.-Y., Bayer, E. A., Steiner, D., Shoham, Y., and Lamed, R. (1999) A novel cellulosomal scaffoldin from Acetivibrio cellulolyticus that contains a family-9 glycosyl hydrolase. *J. Bacteriol. 181*, 6720-6729.

[109] Xu, Q., Kenig, R., Shoham, Y., Bayer, E. A., and Lamed, R. (2003) The cellulosome system of Acetivibrio cellulolyticus includes a novel type of adaptor protein and a cell-surface anchoring protein. *J. Bacteriol. 185*, 4548-4557.

[110] Xu, Q., Barak, Y., Kenig, R., Shoham, Y., Bayer, E. A., and Lamed, R. (2004) A novel Acetivibrio cellulolyticus anchoring scaffoldin that bears divergent cohesins. *J. Bacteriol. 186*, 5782-5789.

[111] Ding, S.-Y., Bayer, E. A., Steiner, D., Shoham, Y., and Lamed, R. (2000) A scaffoldin of the Bacteroides cellulosolvens cellulosome that contains 11 type II cohesins. *J. Bacteriol. 182*, 4915-4925.

[112] Xu, Q., Bayer, E. A., Goldman, M., Kenig, R., Shoham, Y., and Lamed, R. (2004) Architecture of the Bacteroides cellulosolvens cellulosome: description of a cell-surface anchoring scaffoldin and a family-48 cellulase. *J. Bacteriol. 186*, 968-977.

[113] Karita, S., Kimura, T., Sakka, K., and Ohmiya, K. (1997) Purification of the Ruminococcus albus endoglucanase IV using a cellulose-binding domain as an affinity tag. *J. Ferment. Bioeng. 84*, 354-357.

[114] Morrison, M., and Miron, J. (2000) Adhesion to cellulose by Ruminococcus albus: a combination of cellulosomes and Pil-proteins? *FEMS Microbiol. Lett. 185*, 109-115.

[115] Ohara, H., Karita, S., Kimura, T., Sakka, K., and Ohmiya, K. (2000) Characterization of the cellulolytic complex (cellulosome) from Ruminococcus albus. *Biosci. Biotechnol. Biochem. 64*, 254-260.

[116] Ding, S.-Y., Rincon, M. T., Lamed, R., Martin, J. C., McCrae, S. I., Aurilia, V., Shoham, Y., Bayer, E. A., and Flint, H. J. (2001) Cellulosomal scaffoldin-like proteins from Ruminococcus flavefaciens. *J. Bacteriol. 183*, 1945-1953.

[117] Aurilia, V., Martin, J. C., McCrae, S. I., Scott, K. P., Rincon, M. T., and Flint, H. J. (2000) Three multidomain esterases from the cellulolytic rumen anaerobe Ruminococcus flavefaciens 17 that carry divergent dockerin sequences. *Microbiology 146*, 1391-1397.

[118] Rincon, M. T., McCrae, S. I., Kirby, J., Scott, K. P., and Flint, H. J. (2001) EndB, a multidomain family 44 cellulase from Ruminococcus flavefaciens 17, binds to cellulose via a novel cellulose-binding module and to another R. flavefaciens protein via a dockerin domain. *Appl. Environ. Microbiol. 67*, 4426-4431.

[119] Rincon, M. T., Ding, S.-Y., McCrae, S. I., Martin, J. C., Aurilia, V., Lamed, R., Shoham, Y., Bayer, E. A., and Flint, H. J. (2003) Novel organization and divergent

dockerin specificities in the cellulosome system of Ruminococcus flavefaciens. *J. Bacteriol. 185*, 703-713.

[120] Rincon, M. T., Martin, J. C., Aurilia, V., McCrae, S. I., Rucklidge, G., Reid, M., Bayer, E. A., Lamed, R., and Flint, H. J. (2004) ScaC, an adaptor protein carrying a novel cohesin that expands the dockerin-binding repertoire of the Ruminococcus flavefaciens 17 cellulosome. *J. Bacteriol. 186*, 2576-2585.

[121] Rincon, M. T., Ding, S.-Y., Martin, J. C., McCrae, S. I., Aurilia, V., Lamed, R., Shoham, Y., Bayer, E. A., and Flint, H. J. (2004) Protein:protein interactions involved in the assembly of lignocellulolytic enzyme complexes in Ruminococcus flavefaciens, p. 241-248. In Ohmiya, K., Sakka, K., Karita, S., Kimura, T., and Sakka, M. (ed.), *Genetics, biotechnology of lignocellulose degradation and biomass utilization.* Uni Publishers Co., Ltd., Tokyo, Japan.

[122] Leibovitz, E., and Béguin, P. (1998) Comparison of two scaffolding polypeptides for the integration of different proteins in synthetic complexes derived from the Clostridium thermocellum cellulosome. *Enzyme Microb. Technol. 22*, 588-593.

[123] Fierobe, H.-P., Mechaly, A., Tardif, C., Belaich, A., Lamed, R., Shoham, Y., Belaich, J.-P., and Bayer, E. A. (2001) Design and production of active cellulosome chimeras: Selective incorporation of dockerin-containing enzymes into defined functional complexes. *J. Biol. Chem. 276*, 21257-21261.

[124] Fierobe, H.-P., Bayer, E. A., Tardif, C., Czyzek, M., Mechaly, A., Belaich, A., Lamed, R., Shoham, Y., and Belaich, J.-P. (2002) Degradation of cellulose substrates by cellulosome chimeras: Substrate targeting versus proximity of enzyme components. *J. Biol. Chem. 277*, 49621-49630.

[125] Fierobe, H.-P., Mingardon, F., Mechaly, A., Belaich, A., Rincon, M. T., Lamed, R., Tardif, C., Belaich, J.-P., and Bayer, E. A. (2005) Action of designer cellulosomes on homogeneous versus complex substrates: Controlled incorporation of three distinct enzymes into a defined tri-functional scaffoldin. *J. Biol. Chem. 280*, 16325-16334.

[126] Attwood, G. T., Blaschek, H. P., and White, B. A. (1994) Transcriptional analysis of the Clostridium cellulovorans endoglucanase gene, engB. *FEMS Microbiol. Lett. 124*, 277-284.

[127] Mishra, S., Béguin, P., and Aubert, J. (1991) Transcription of Clostridium thermocellum endoglucanase genes celF and celD. *J. Bacteriol. 173*, 80-85.

[128] Han, S. O., Yukawa, H., Inui, M., and Doi, R. H. (2003) Regulation of expression of cellulosomal cellulase and hemicellulase genes in Clostridium cellulovorans. *J. Bacteriol. 185*, 6067-6075.

[129] Han, S. O., Cho, H. Y., Yukawa, H., Inui, M., and Doi, R. H. (2004) Regulation of expression of cellulosomes and noncellulosomal (hemi)cellulolytic enzymes in Clostridium cellulovorans during growth on different carbon sources. *J. Bacteriol. 186*, 4218-4227.

[130] Dror, T. W., Morag, E., Rolider, A., Bayer, E. A., Lamed, R., and Shoham, Y. (2003) Regulation of the cellulosomal celS (cel48A) gene of Clostridium thermocellum is growth-rate dependent. *J. Bacteriol. 185*, 3042-3048.

[131] Dror, T. W., Rolider, A., Bayer, E. A., Lamed, R., and Shoham, Y. (2003) Regulation of expression of scaffoldin-related genes in Clostridium thermocellum. *J. Bacteriol. 185*, 5109-5116.

[132] Dror, T. W., Rolider, A., Bayer, E. A., Lamed, R., and Shoham, Y. (2005) Regulation of major cellulosomal endoglucanases of Clostridium thermocellum differs from that of a prominent cellulosomal xylanase. *J. Bacteriol. 187*, 2261-2266.

[133] Cho, H. Y., Yukawa, H., Inui, M., Doi, R. H., and Wong, S. L. (2004) Production of minicellulosomes from Clostridium cellulovorans in Bacillus subtilis WB800. *Appl. Environ. Microbiol. 70*, 5704-5707.

[134] Murashima, K., Chen, C. L., Kosugi, A., Tamaru, Y., Doi, R. H., and Wong, S. L. (2002) Heterologous production of Clostridium cellulovorans engB, using protease-deficient Bacillus subtilis, and preparation of active recombinant cellulosomes. *J. Bacteriol. 184*, 76-81.

[135] Perret, S., Casalot, L., Fierobe, H.-P., Tardif, C., Sabathe, F., Belaich, J.-P., and Belaich, A. (2004) Production of heterologous and chimeric scaffoldins by Clostridium acetobutylicum ATCC 824. *J. Bacteriol. 186*, 253-257.

[136] Perret, S., Belaich, A., Fierobe, H. P., Belaich, J. P., and Tardif, C. (2004) Towards designer cellulosomes in Clostridia: mannanase enrichment of the cellulosomes produced by Clostridium cellulolyticum. *J. Bacteriol. 186*, 6544-6552.

[137] Sabathe, F., and Soucaille, P. (2003) Characterization of the CipA scaffolding protein and in vivo production of a minicellulosome in Clostridium acetobutylicum. *J. Bacteriol. 185*, 1092-1096.

[138] Bayer, E. A., Coutinho, P. M., and Henrissat, B. (1999) Cellulosome-like sequences in Archaeoglobus fulgidus: an enigmatic vestige of cohesin and dockerin domains. *FEBS Lett. 463*, 277-280.

[139] Ali, B. R., Romaniec, M. P., Hazlewood, G. P., and Freedman, R. B. (1995) Characterization of the subunits in an apparently homogeneous subpopulation of Clostridium thermocellum cellulosomes. *Enzyme Microbiol. Technol. 17*, 705-711.

[140] Pagès, S., Belaich, A., Tardif, C., Reverbel-Leroy, C., Gaudin, C., and Belaich, J.-P. (1996) Interaction between the endoglucanase CelA and the scaffolding protein CipC of the Clostridium cellulolyticum cellulosome. *J. Bacteriol. 178*, 2279-2286.

[141] Takagi, M., Hashida, S., Goldstein, M. A., and Doi, R. H. (1993) The hydrophobic repeated domain of the Clostridium cellulovorans cellulose-binding protein (CbpA) has specific interactions with endoglucanases. *J. Bacteriol. 175*, 7119-7122.

[142] Kruus, K., Lua, A. C., Demain, A. L., and Wu, J. H. D. (1995) The anchorage function of CipA (CelL), a scaffolding protein of the Clostridium thermocellum cellulosome. *Proc. Natl. Acad. Sci. USA 92*, 9254-9258.

[143] Grepinet, O., and Beguin, P. (1986) Sequence of the cellulase gene of Clostridium thermocellum coding for endoglucanase B. *Nucleic Acids Res 14*, 1791-1799.

[144] Chauvaux, S., Béguin, P., Aubert, J.-P., Bhat, K. M., Gow, L. A., Wood, T. M., and Bairoch, A. (1990) Calcium-binding affinity and calcium-enhanced activity of Clostridium thermocellum endoglucanase D. *Biochem. J. 265*, 261-265.

[145] Lytle, B., Myers, C., Kruus, K., and Wu, J. H. D. (1996) Interactions of the CelS binding ligand with various receptor domains of the Clostridium thermocellum cellulosomal scaffolding protein, CipA. *J. Bacteriol. 178*, 1200-1203.

[146] Ahsan, M. M., Kimura, T., Karita, S., Sakka, K., and Ohmiya, K. (1996) Cloning, DNA sequencing, and expression of the gene encoding Clostridium thermocellum cellulase CelJ, the largest catalytic component of the cellulosome. *J. Bacteriol. 178*, 5732-5740.

[147] Choi, S. K., and Ljungdahl, L. G. (1996) Structural role of calcium for the organization of the cellulosome of Clostridium thermocellum. *Biochemistry 35*, 4906-4910.

In: Cellulosome
Eds: V. Uversky and I. A. Kataeva, pp. 47-67

ISBN 1-59454-950-8

Chapter III

THE COHESIN-DOCKERIN INTERACTION

David Nakar[1], Tal Handelsman[2], Yoav Barak[1], Yuval Shoham[2], Raphael Lamed[3] and Edward A. Bayer[1,*]

[1]Department of Biological Chemistry, The Weizmann Institute of Science, Rehovot, Israel
[2]Department of Biotechnology and Food Engineering, and Institute of Catalysis Science and Technology, Technion — Israel Institute of Technology, Haifa, Israel
[3]Department of Molecular Microbiology and Biotechnology, Tel-Aviv University, Ramat Aviv,Israel.

ABSTRACT

Cellulosome architecture is defined by the intermodular interaction between cohesins and dockerins, which usually display very high affinities. Within and among the different cellulosome-producing microbial species, the sequences of both cohesins and dockerins separate into distinct types, reflecting their phylogenetic relationship.

The crystal structure of the type-I cohesin module revealed an elongated, conical molecule, consisting of a nine-stranded β-sandwich with an extensive aromatic core. Structures of type-II cohesins were also determined recently, which showed a jelly-roll topology with 3 additional secondary structures: namely an α helix and two "β-flaps" that disrupt the normal course of a β-strand. In contrast, the dockerin module comprises a duplicated sequence of about 22 amino acids (both required for binding), connected by a linker segment. The solution structure of the ~70-residue type-I dockerin from *C. thermocellum* Cel48S showed a pair of calcium-binding loop-helix motifs, related by a 2-fold pseudo-symmetry axis. Calcium induces protein folding and is required for stable tertiary structure and cohesin-dockerin interaction.

* Correspondence concerning this article should be addressed to Edward A. Bayer, Department of Biological Chemistry, The Weizmann Institute of Science, Rehovot 76100 Israel. Email: ed.bayer@weizmann.ac.il.

The recently solved three-dimensional structure of the type-I cohesin-dockerin complex from demonstrated that the overall structure of the cohesin module is essentially unchanged. The dockerin, however, appears to form 3 α-helices in a conformation defined by a loop-helix motif followed by a helix-loop-helix motif. Helices 1 and 3 are closer together in the complex compared to the free form. The cohesin interacts predominantly with α-helix 3 of the dockerin and to a lesser extent with α-helix 1. The protein-protein contact is mediated mainly by hydrophobic interactions and relatively few intermolecular hydrogen bonds.
Comparative binding analysis between the cohesins and dockerins of *C. thermocellum* versus *C. cellulolyticum* suggested that *inter*-species binding is nonselective, whereas *intra*-species interaction is specific. The dockerin residues that appear to account for the observed specificity are at positions 10, 11, 18 and 22 in both of the duplicated sequences. In the case of the cohesin, only three residues (Ala36, Asn37 and Glu-131) appear to be involved in interspecies specificity between the two bacteria, representing but 3 of the 16 contact residues that participate in the cohesin-dockerin interaction.
Our current knowledge of the sequences, the structures and interactions among the different types of cohesins and dockerins is still rather sparse. Current inroads in genome sequencing and continued progress in structure determination should provide vital information regarding the diversity of cohesin-dockerin interactions.

Keywords: protein-protein interaction, binding specificity, phylogenetic diversity, protein families, cellulosome architechture, crystal structure

INTRODUCTION

The cellulosome complex is composed of a conglomerate of subunits, each of which comprises a set of interacting functional modules [1-14]. A major cellulosomal subunit, called scaffoldin, is responsible for organizing the cellulolytic subunits into the complex. This is accomplished by the interaction of two complementary classes of modules, each of which is located on one of the two individual interacting subunits, i.e., a cohesin module on the scaffoldin and a dockerin module on each of the enzymatic subunits [15-17]. The cohesin-dockerin interaction defines cellulosomal architecture. It dictates the incorporation of individual enzymes into the complex by binding enzymes to the scaffoldins, connecting between scaffoldins in the assembly of more intricate cellulosomes, and attachment of the complexes to the cell surfaces. The early biochemical examination of the cellulosome complex of *C. thermocellum* indicated an exceptionally strong interaction [2]. Subsequent analyses confirmed this observation as the cohesin-dockerin interactions display high affinities, with dissociation constants (K_D) estimated at 10^{-9}-10^{-12} M and even less [18-22]. Consequently, the interaction between the type-I cohesin and dockerin of *C. thermocellum* is one of the most tenacious protein-protein interactions known.

The cohesin-dockerin interaction was shown to be calcium dependent [23-25], an observation later explained by the dockerin structure, which requires calcium for stable configuration [26] as predicted earlier [27]. To confirm the exclusive role of the complementary cohesin and dockerin modules in binding, it was demonstrated that neither of the non-cohesin modules of the *C. cellulolyticum* CipC scaffoldin – namely, the cellulose-binding module (CBM), the hydrophilic module, and linker sequence adjacent to the first cohesin – nor the catalytic module of a major cellulosomal cellulase (Cel5A) have any influence on the interaction. However, removal of the linker located upstream of the first conserved amino acid from the dockerin module reduced the affinity for one of the cohesin modules by ~600-fold compared with the same dockerin that contains the 11-residue linker sequence [18]. It thus appears that the incorporation of the various cellulosome subunits into the cellulosome complex is dependent solely on the cohesin-dockerin interaction.

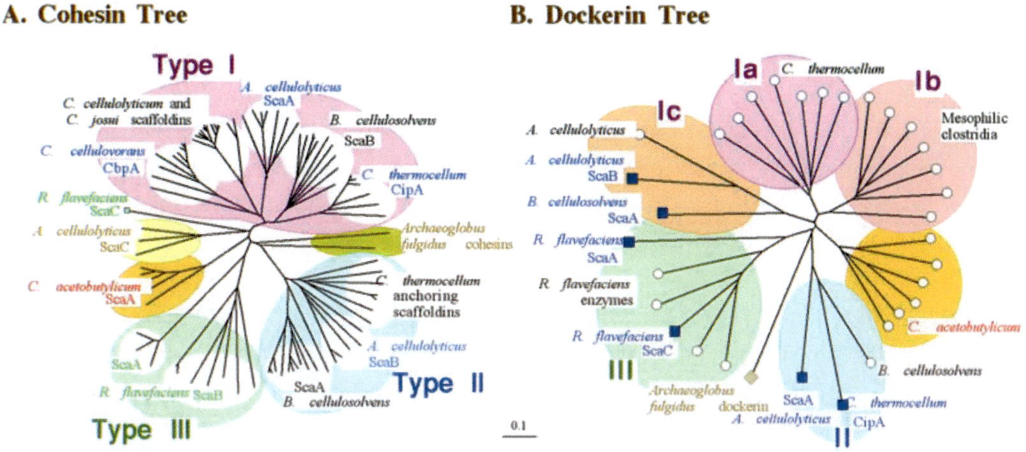

A. Cohesin Tree **B. Dockerin Tree**

Figure 1. Phylogenetic relationship of the cohesins and dockerins from different species. (*A*) The cohesin tree. The type-I cohesins include those from the anchoring scaffoldin from *B. cellulosolvens* and the primary scaffoldins derived from *A. cellulolyticus* and many of the known clostridia. Of the remaining groups, the type-II cohesins are the most distantly located and include cohesins borne by the primary, adaptor, and anchoring scaffoldins from *B. cellulosolvens*, *A. cellulolyticus*, and *C. thermocellum*, respectively. The type-III *R. flavefaciens* scaffoldins comprise the primary ScaA cohesins and those borne by the presumed anchoring ScaB. The cohesins from the primary, adaptor, and anchoring scaffoldins of *C. acetobutylicum*, *R. flavefaciens*, and *A. cellulolyticus*, respectively, radiate from intermediate points along the phylogenetic route that separates the type-I from the type-II cohesins. The two curious cohesins discovered in the archaeon *Archaeoglobus fulgidus* map along the line joining the type-II and type-III cohesins. (*B*) The dockerin tree. Enzyme-borne dockerins are represented by white circles, the scaffoldin-borne dockerins by blue squares, and the *Archaeoglobus* dockerin by a beige diamond. The dockerins form groups: those that interact with the type-I and allied cohesins (designated Ia, Ib, and Ic); those that interact with the type-II scaffoldins (II); and those that interact with the type-III cohesins (III). The *C. acetobutylicum* and *Archaeoglobus* dockerins map at positions consistent with their species-specific preference of cohesins. Bayer et al. [1], with permission.

Phylogenetic Diversity Among Cohesins and Dockerins

The elucidation of the scaffoldin sequences from *C. cellulovorans*, *C. thermocellum* and other cellulosome-producing bacteria allowed the assessment of their intra- and interspecies phylogenetic relationship [28]. The number and phylogenetic distribution of the cohesins is in flux, and the current status is presented in Figure 1*A*. Initial observations showed that cohesin sequences from the primary scaffoldins of *C. thermocellum* differed from those of its anchoring scaffoldins [17,29]. The former were classified as type-I and the latter as type II cohesins, based in addition to sequence homology on the identity of the parent protein and the binding specificity of the dockerin module. Later, the discovery of the *R. flavefaciens* cohesins, added a distinct class of cohesins designated type III [30,31]. However the connection between cohesin type and parent protein was severed when the reversed situation was discovered, wherein the primary scaffoldin of *Bacteroides cellulosolvens* was found to contain type-II cohesins as opposed to the anchoring scaffoldin that bears those of type I [32,33].

The phylogenetic map of the dockerins reflects, to a great extent, that of their cohesin counterparts (Figure 1*B*). The enzyme-borne dockerins that interact with the type-I cohesins are closely grouped and most distant from the dockerins that interact with the type-II cohesins which are also grouped. Dockerins interacting with the *R. flavefaciens* type-III cohesins, the *Archaeoglobus* dockerin [34], and dockerins of the *C. acetobutylicum* enzymes [35] branch out from separate intermediate nodes, in accordance with the distribution of their cohesin counterparts.

New groups of cohesin and dockerins are being described continuously. The main challenge remains to unfold the relationship (if indeed one exists) between the phylogenetic distributions, the identity of the parent protein, the protein sequence and the binding specificity.

THE COHESIN MODULE

The scaffoldin of the *C. thermocellum* cellulosome contains nine very similar cohesin modules, separated from each other by linker segments [36]. Cohesins 3 to 8 (numbered according to their position from the N-terminus of the scaffoldin polypeptide) show greater than 95% identity. Cohesins 4, 5, 6 and 8 show near-complete identity and can be considered 'consensus' cohesins in the scaffoldin of this species. Of the nine sequences, cohesins 1, 2 and 9 are the most variable: cohesins 2 and 9 show about 75% identity with the consensus cohesins, and cohesin-1 shows 63% identity. Cohesins 1 and 2 also have a 3-residue deletion relative to the other cohesins. Generally, the type-I cohesin modules within other scaffoldins display similar levels of homology (e.g. 42-90% identity in the CbpA scaffoldin of *C. cellulovorans*), while amino acid sequence conservation is weaker between cohesin modules from different species (typically 30-50% identity). The sequence similarity between type-I and type-II cohesins is lower, typically 15-25% identity.

The three-dimensional structure of cohesin-2 from *C. thermocellum* was solved by X-ray crystallography [37]. The 138-residue cohesin module is an elongated, conical molecule with a jelly-roll topology and approximate dimensions of 46 Å × 28 Å × 21 Å (Figure 2). The topological motif consists of a nine-stranded β-sandwich, which results from the association of a four-stranded antiparallel β-sheet, and a five-stranded mixed β–sheet. The two sheets of the sandwich are composed of strands 8, 3, 6, 5 and 9, 1, 2, 7, 4, respectively, where β–strand 9 (C terminus) and β-strand 1 (N terminus) run parallel, and the remaining strands are all antiparallel. The nine β-strands of cohesin-2 are assembled around an extensive aromatic core comprising mainly phenylalanine residues. The solvent-accessible surface of the cohesin-2 module is devoid of any obvious binding pocket or cleft. The crystal structures of two additional type-I cohesin modules, namely cohesin 7 from the same scaffoldin subunit of *C. thermocellum* and cohesin-1 from *C. cellulolyticum,* were also reported [38,39]. The three cohesin modules share similar structure and fold, and they are also somewhat similar to the fold displayed by the *C. thermocellum* cellulose-binding module, despite the absence of sequence identity [37].

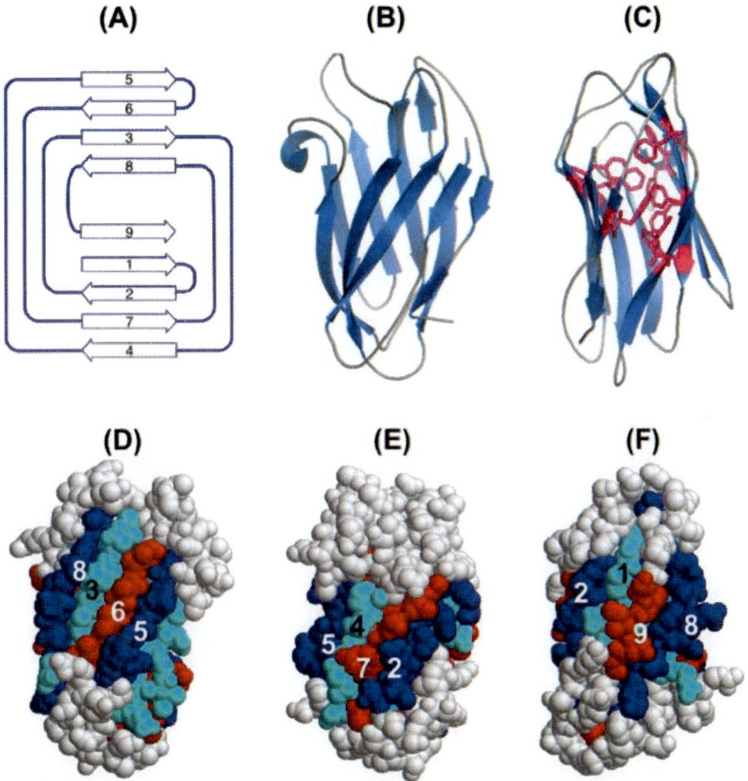

Figure 2. Three-dimensional structure of *C. thermocellum* cohesin-2 (PDB code 1ANU). (A) Topology diagram, representing the two β-sheets, composed of 4 and 5 strands. The β strands are numbered 1 to 9. (B,C) Schematic ribbon diagrams of the cohesin. Phenylalanine residues of the hydrophobic core colored in red. (D,E,F) Rotated views of the cohesin module in spacefill representation.

The solution and crystal structures of type-II cohesins were also published recently [40-42]. The crystal structure of the type-II cohesin from the ScaB cohesin of A. *cellulosolvens* revealed a jelly-roll topology similar to that of the type-I cohesins, with several remarkable and distinctive appendages (Figure 3). These include an α-helix, separating β-strands 6 and 7, another α-helix formation in the C-terminal linker region, and two extraordinary "β-flaps" that rupture the central section of β-strands 4 and 8 but revert to the original path to complete the strand. Structure-based sequence alignment of representative type-II cohesins indicated that these secondary structural elements are definitive of this type of cohesin [42].

All four cohesin modules thus far described were found to generate crystallographic dimers, which exhibit highly complementary interfaces with numerous intermolecular salt bridges, hydrogen bonds, and hydrophobic interactions. The interactions take place mainly via conserved residues located on the 8,3,6,5-face. All three type-I cohesins formed dimers in the crystal, sharing roughly the same orientation [38]. Comparison between the type-II and the type-I dimers (1TYJ versus 1ANU) reveals a clear discrepancy in their relative orientation and interacting regions. The presence of the protruding β-flaps in 1TYJ appears to cause a displacement in the zone of interaction by about 10 Å towards the crown of the molecule compared to 1ANU and a relative rotation of approximately 50° [42]. In both cases, the coincidence of the functional and crystallographic interfaces would appear to cast doubt on a possible physiological role of dimer formation among the cohesin molecules. Nevertheless, formation of crystallographic dimers may indeed signify a biological role in cellulosome regulation or assembly, whereby unoccupied cohesin sites would be protected until a prospective dockerin-borne enzyme is incorporated into the complex [37,42].

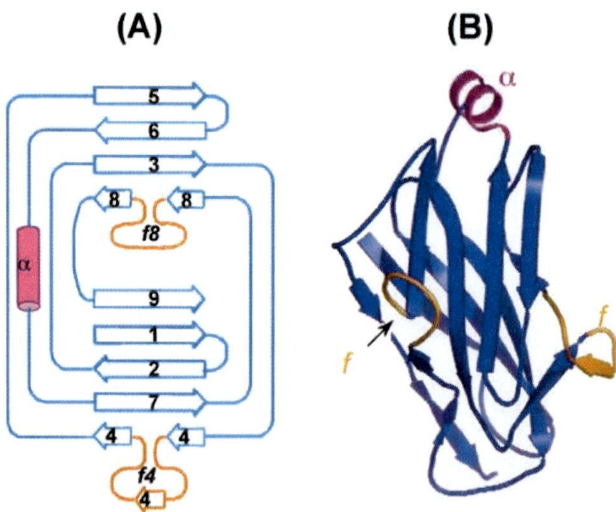

Figure 3. Three-dimensional structure of *B. cellulosolvens* cohesin-II (PDB code 1TYJ). (A) Topology diagram, representing the two β-sheets, composed of 4 and 5 strands. The β strands are numbered 1 to 9. The additional secondary structures, the two "β-flaps" (*f*4 and *f*8) and the α helix, are colored in orange and magenta, respectively. (B) Schematic ribbon diagrams of a type-II cohesin. The topology of the type-II cohesin is essentially analogous to that of the type-I cohesin, except for an α helix (colored magenta) formed within the loop that separates β strands 6 and 7 and two "β-flaps" (*f*4 and *f*8; colored orange) that originate from the midst of β strands 4 and 8, both of which return to their original course to complete the strand. Noach et al. [42], with permission.

The Dockerin Module

The catalytic subunits of the cellulosome bind to the scaffoldin via a dockerin module, usually found at their C-termini. Some scaffoldins also contain a dockerin, usually of type-II. All the dockerin modules described so far consist of a duplicated sequence of about 22 amino acids, connected by a 9-16 residue linker [27]. Experiments with recombinant truncated dockerins demonstrated that both segments of the dockerin are required for binding [18,43].

The first 12 residues of each conserved duplicated segment bear homology to the calcium-binding loop of the known EF-hand motif found in eukaryotic calcium-binding proteins, like calmodulin and troponin C [44,45]. This notion was corroborated by the demonstration that the cohesin-dockerin interaction is calcium dependent [23-25]. Moreover, biophysical and structural analyses of the type-I dockerin from *C. thermocellum* Cel48S and the type-II dockerin from *C. thermocellum* CipA revealed that calcium induces protein folding and is required for a stable tertiary structure [26,46,47].

The solution structure of the Cel48S dockerin was solved by three-dimensional NMR spectroscopy [48]. The ~70 residue dockerin module consists of a pair of calcium-binding loop-helix motifs connected by a ten-residue linker (Figure 4). The two loop-helix motifs are related by a 2-fold pseudo-symmetry axis. The first helix is a typical α-helix extending from Asp19 to Arg29, while the second helix extends from Asp51 to Ile59 and has a small kink in it. Residues of the N-terminus, which is part of the linker connecting the dockerin to the catalytic module, appear to be dynamically disordered. The structure reveals that despite sequence homology to the EF-hand motif, the dockerin module has a different structure [48]. Specifically, the E helices entering each loop of the classical EF-hand motif are absent from the dockerin module, thus confirming the early conclusions of Pagés, et al. [27].

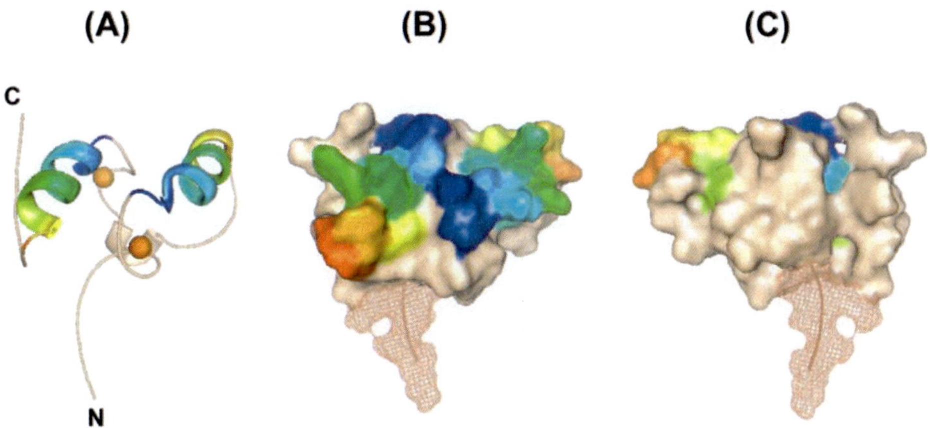

Figure 4. Solution structure of the dockerin module from C. thermocellum Cel48S (PDB code 1DAQ). (A) Ribbon diagram of the mean structure. Calcium ions are shown as orange spheres. (B,C) Molecular surface representation of the dockerin from two opposite views. Each helix is rainbow-colored from blue to red. The N-terminus linker, connecting the dockerin to the catalytic module, is shown as brown mesh.

Biophysical experiments suggested that upon calcium binding an exposed region comprising hydrophobic residues of the CipA dockerin becomes more exposed or additional hydrophobic regions become accessible [47]. In this regard, the NMR structure of the calcium-bound Cel48S dockerin revealed a Y-shaped, solvent-accessible hydrophobic patch surrounded by positions 10 and 11 of both loops (i.e., residues Ser17, Thr18, Ser49, and Thr50) [48]. Interestingly, the calcium-bound Cel48S dockerin is monomeric in solution [48], whereas the CipA dockerin in the presence of calcium forms homodimers [47].

THE STRUCTURE OF THE COHESIN-DOCKERIN COMPLEX

The three-dimensional structure of the cohesin-dockerin complex from *C. thermocellum* was recently solved by X-ray crystallography [49]. The complex, composed of cohesin-2 from CipA and the dockerin module from xylanase 10B (formerly called XynY [50]), both from *C. thermocellum*, is shown in Figure 5A. The data show that while the cohesin module remains essentially unchanged, the dockerin appears to undergo conformational adjustments upon binding.

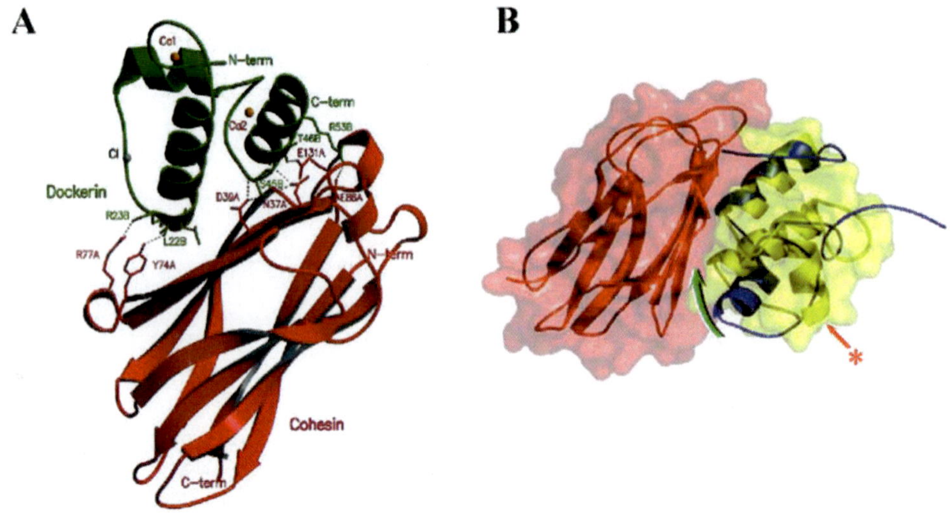

Figure 5. Crystal structure of the *C. thermocellum* type-I cohesin-dockerin complex [PDB code 1OHZ]. (A) The complex is formed between a cohesin-2 module (red) and a Ca^{2+}-bound dockerin (green). The residues involved in intermodular contacts are shown as stick models. The two Ca^{2+}-binding sites of the dockerin module are represented as orange spheres. (B) Surface complementarity of the cohesin (red) - dockerin (yellow) complex. The dockerin solution structure (PDB code 1DAQ) is overlaid in blue and reveals the apparent movement of α-helix 3 away from the cohesin. Carvalho et al. (49), with permission.

The dockerin of the complex forms 3 α-helices, in a conformation defined by a loop-helix motif followed by a helix-loop-helix motif, connected by a six-residue segment. Helices 1 and 3 correspond to the duplicated conserved segments, while helix 2 is composed of the unconserved linker between them. It seems that in the complex structure the dockerin is in a "tighter" conformation, which brings α-helices 1 and 3 closer together, compared to the free form. It is likely that the dockerin has a flexible conformation in solution, and, upon binding to the cohesin module, the protein-protein interactions impose structural constrains onto the dockerin fold such that it adopts a single conformation. Indeed, superposition of the solution structure of the dockerin with that of the complex shows that α-helix 3 of the solution structure is too distant to interact with the cohesin module without conformational change (Figure 5B). However, it is worth noting that two different dockerin modules (i.e., derived from Cel48S versus Xyn10B) were used to determine the structures of the respective free and complexed forms of the dockerin. In this context, the above observations concerning the conformational changes, notably the apparent formation of the additional helix (helix 2), must be taken with reservations, since the Xyn10B dockerin contains 2 additional amino acids compared with that of Cel48S, positioned precisely within the extra α-helix. It is thus not certain whether such a helix would be formed in a cohesin-complex with the Cel48S dockerin and whether in solution the dockerin of Xyn10B would not possess the extra helix.

The structure of the cohesin-dockerin heterodimer shows that the 8,3,6,5 β-sheet of the cohesin interacts predominantly with α-helix 3 of the dockerin, but also, to a lesser extent, with α-helix 1 [49]. Protein-protein contact is mediated mainly by hydrophobic interactions and relatively few intermolecular hydrogen bonds. The hydrophobic residues participating in complex formation are Ala36, Val41, Ala72, Ile79, Val81, Leu83, and Ala-85 from β-strands 3, 5, and 6 of the cohesin module, and residues Val21, Leu22, Val48, Leu49, and Leu50, from α-helices 1 and 3 of the dockerin module [49]. The proteins also interact via a series of hydrogen bonds (see Figure 6 for details).

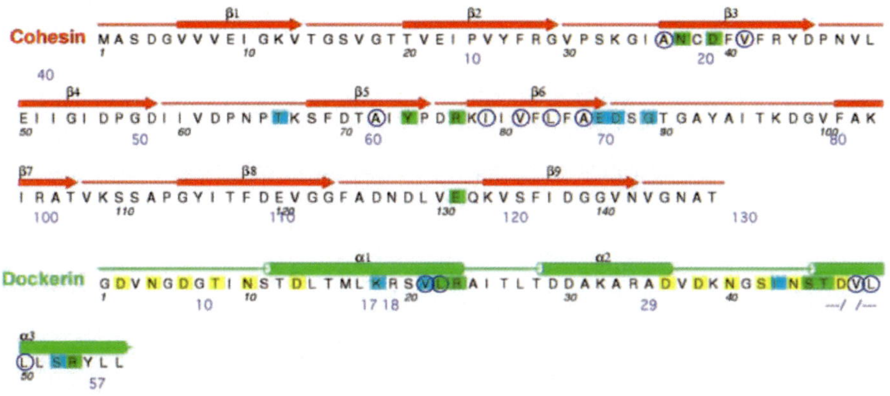

Figure 6. Amino acid sequence of the cohesin and dockerin modules from the crystal complex, with their secondary structure. Residues involved in direct contacts between the two subunits are marked with green squares. Blue squares represent contacts mediated by bridging water molecules. The blue open circles indicate hydrophobic residues in the contact surface. The residues involved in Ca^{2+} coordination are marked with yellow squares. For convenience, residues are numbered according to their position in the complex structure (black) and according to the position in the cohesin (PDB 1ANU) and dockerin (PDB 1DAQ) separate structures (blue). Carvalho et al. [49], with permission.

Recent site-directed mutagenesis studies on the seventh cohesin module of *C. thermocellum* CipA [51,52] proposed that residues Asp39, Tyr74, and Glu86 and Gly89 of this module play a key role in the formation of Coh-Doc complexes and cohesin–cohesin dimers [52]. Regarding Asp39, recent work has shown that substituting Asp34 (the Asp39 equivalent) of the second cohesin module of *C. thermocellum* CipA into Asn resulted in a 1000-fold reduction in the affinity [53]. Interestingly, other homologous substitutions of the cohesin, including the replacement of an entire loop region that contained 3 distinct contact residues, exhibited relatively little or no effect on the binding to the target dockerin [54]. The crystal structure of the Coh-Doc complex reveals that the equivalent residues in cohesin 2 (Asp39, Tyr74, and Glu86, respectively) do indeed interact directly with the dockerin, whereas Gly89 (equivalent to Gly89 of CipA) makes a water-mediated link with of Arg53 of the dockerin module.

Very recently, a type-II cohesin-dockerin complex from *C. thermocellum* was reported to be crystallized [47], and the structure of the type-II heterodimer awaits final solution.

Binding Specificity

Experimental evidence has shown that the cohesins of both *C. thermocellum* and *C. cellulolyticum* recognize nearly all of the dockerins on their own enzymatic subunits [23,55-57]. It thus seemed that within a given species the binding process of the enzymes to the various cohesins of the scaffoldin is nonselective. However, despite the relatively high sequence homology, interaction among cohesins and dockerins between the two *Clostridia* was found to be species specific, i.e., the cohesin module from *C. thermocellum* interacts with the dockerins from its own cellulosome but fails to recognize dockerins from *C. cellulolyticum* and vice versa [27].

Affinity blotting of the cohesin-dockerin interaction between *C. thermocellum* and another clostridial species, *C. josui,* showed that species specificity was indeed generally conserved, but with at least one surprising exception: the dockerin module of *C. thermocellum* Xyn11A was able to bind cohesins from *C. josui*, [19]. This finding was further confirmed by surface plasmon resonance (SPR). In an independent work [58], the dockerins of both Xyn11A and Xyn11B were shown to bind cohesins from both *C. thermocellum* and *C. cellulolyticum*. In this context, the cohesin and dockerin sequences of *C. cellulolyticum* and *C. josui* are very similar, and it is perhaps not surprising that they interact similarly with the *C. thermocellum* cohesins. It is not clear what is responsible for the interspecies crossreactivity exhibited by *C. thermocellum* Xyn11A and Xyn11B. At this point, the above-described crossreactivity can be considered atypical, although this certainly does not rule out the possibility of divergent cohesin specificities in a single scaffoldin. The more we learn about newly sequenced types of cellulosomes, the more we discover how versatile and divergent is the cellulosome architecture in nature.

The specificity of the cohesin-dockerin interaction indeed has important consequences regarding cellulosome architecture in a given bacterium. In this regard, the *C*-terminal dockerin of the CipA scaffoldin of *C. thermocellum* fails to recognize any of its own complement of "type-I" cohesins; rather, it binds specifically to a divergent type of cohesin (termed "type-II" cohesin) harbored by another type of scaffoldin located on the cell surface [29,59,60].

Based on the known species specificity of the cohesin-dockerin interaction, sequences of various dockerins were analyzed in order to conclude which positions could account for the observed specificity. At first, four suspected residues were indicated which collectively act as a possible recognition code. These are two tandem pairs at positions 10 and 11 of the duplicated dockerin sequence (these are serine and threonine at positions 17, 18 in the first segment and 49, 50 in the second segment of *C. thermocellum* Cel48S) [27]. To validate the prediction, the four suspected dockerin residues of one species were converted to imitate those of the other. The resultant dockerins gained new specificity: they recognized the cohesins from both rival species and their own [51,61]. Subsequent mutagenesis suggested that positions 18 and 22 of the duplicated sequence (which correspond to positions 25, 29 in the first segment and 57, 61 in the second segment of *C. thermocellum* CelS) also play a role in the specific interaction [21,22].

In more recent work, the differential recognition of the *C. thermocellum* and *C. cellulolyticum* cohesins was examined by progressive rounds of swapping (see Figure 7), in which corresponding homologous stretches were interchanged [62]. The information obtained from each round of swapping served to constrict the putative protein surface area where the specificity-determining residues might reside and was then used for the next round of swapping. In the final round of swapping three residues were discerned (Ala36, Asn37 and Glu-131), reputedly involved in interspecies binding specificity. These residues represent but 3 of the 16 contact residues that participate in the cohesin-dockerin interaction (designated in Figure 8). By replacing only these three residues of the *C. cellulolyticum* cohesin, the specificity of the resultant mutated cohesin was converted, which bound preferentially to the rival dockerin with approximately 20% capacity of the wild-type interaction.

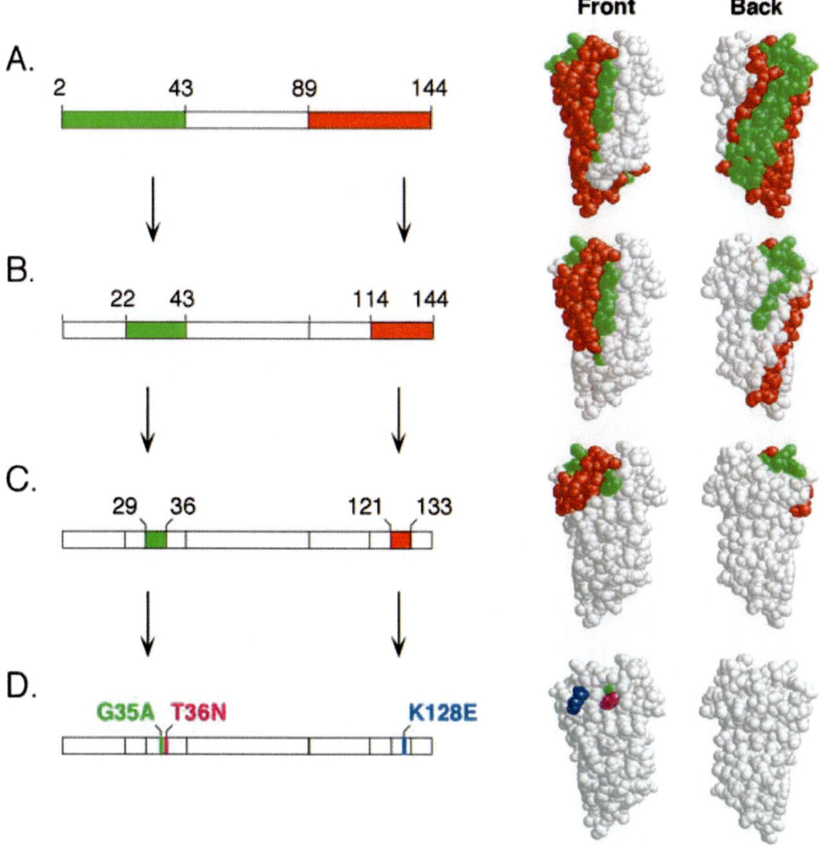

Figure 7. Mapping the recognition trail on the *C. cellulolyticum* cohesin molecule, as elucidated by the four generations of homologue swapping experiments. The scheme shows segments along the cohesin gene, during the progressive rounds of swapping (*A–D*), that appeared to influence the interspecies preference of dockerin binding between *C. thermocellum* and *C. cellulolyticum.* On the *right* of the scheme are shown the corresponding regions or amino acid residues on the cohesin structure (PDB code 1G1K), color-coded as in the respective scheme. *Front* represents the 8365-strand surface of the cohesin molecule, and *Back* represents the 91274 surface. Nakar et al. [62], with permission.

Solving the 3D structure of the cohesin-dockerin complexes with differential recognition characteristics could reveal the structural dissimilarities that drive differential recognition. In the case of the type-I interaction from *C. thermocellum* [49], the authors succeeded in crystallizing the cohesin together with the dockerin module of xylanase Xyn10B. The cohesin was the same cohesin (the second from the *N*-terminus) whose crystal structure was previously elucidated [37], but the Xyn10B dockerin module was different from that of Cel48S used to determine the solution structure [48]. The residues in the dockerin module of Xyn10B that interact with the cohesin are highly conserved in the dockerins located in other *C. thermocellum* cellulosomal enzymes [49], but, as mentioned above in a previous section, the intervening linker between the two duplicated segments has 2 extra amino acids compared to that of Cel48S. According to the structure of the heterodimer, the cohesin residues that form direct hydrogen bonds with the dockerin are completely conserved in the other 8 cohesins of the *C. thermocellum* scaffoldin. This is completely consistent with the inability of the cohesin modules of *C. thermocellum* CipA to discriminate between the different dockerins appended to cellulosomal enzymes [3,27,63]. When we compare the dockerin modules derived from different species, however, the lack of conservation of Ser45 and Thr46 and the replacement of Arg53 (for example, with a lysine in the dockerin modules of the *C. cellulolyticum* cellulosomal enzymes) provide at least a partial explanation for why cohesins or dockerin modules from one cellulosome-producing bacterium fail to recognize the orthologous counterpart from a different bacterium.

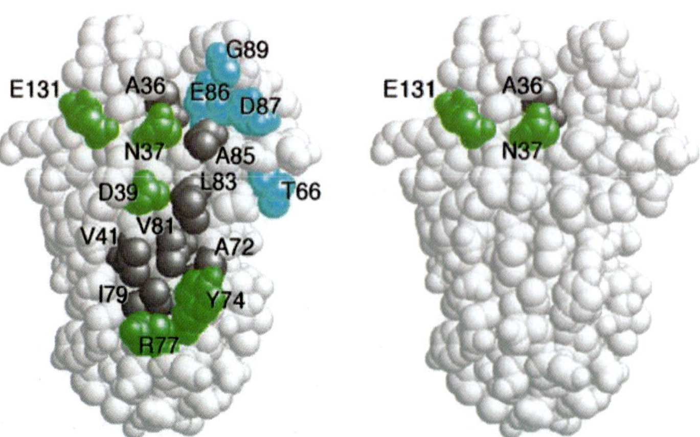

Figure 8. Comparison of recognition residues versus contact residues on the *C. thermocellum* cohesin surface (PDB code 1ANU). A. Contact residues are those singled out by Carvalho et al. [49]. B. Putative recognition residues identified by progressive homologue swapping (Nakar, et al., 2004). Amino acids involved in direct hydrogen bonding are colored green, hydrophobic residues in the cohesin-dockerin contact surface are colored gray, and amino acids which interact with the dockerin molecule via bridging water molecules are colored cyan. Nakar et al. [62], with permission.

The Cohesin-Dockerin Interaction and Cellulosome Assembly

Our knowledge regarding the assembly of the cellulosome is still in its infancy. We know from the above discussion, that on the biochemical level the incorporation of dockerin-containing enzymes into the cohesin-containing scaffoldins is essentially nonselective. The consequences of this would be that simple mixture (i.e., *in vitro*) of cellulosomal components would result in a random distribution among individual scaffoldins. It would seem that the regulation and control of incorporation would take place on the cellular level, via a defined mechanism for biosynthesis and assembly.

Early experiments have shown that the cohesins of both *C. thermocellum* and *C. cellulolyticum* recognize nearly all the dockerins on their own enzymatic subunits [23,27], which would seem to be consistent with the intrinsic similarity of their sequences. It was thus suggested that, within a given species, the incorporation of the cellulosomal enzymes into a complex is a non-selective process, and the catalytic subunits would indeed appear to interact randomly along the scaffoldin. However, comparative binding analysis of several cohesin and dockerin modules from the cellulosome of *C. josui* have recently revealed a certain selectivity with a maximal 34-fold difference in the K_D values, measured by surface plasmon resonance (SPR) [19]. It is worth noting that the maximum differences in K_D values were not necessarily correlated with lower homology in cohesin sequences. At any rate, a 34-fold difference in specificity at such high affinities (~10^{10} M^{-1}) would probably not account for selective incorporation of specific enzymes into the complex.

It is clear that the cellulosome fraction is not homogenous in composition. The observed heterogeneity has been experimentally demonstrated particularly well with *C. papyrosolvens*. The cellulosome population from this species was separated into seven distinct subpopulations by anion-exchange chromatography, whereby each subpopulation exhibited morphological differences as well as functional differences [64-66]. Electron micrographs of the isolated *C. papyrosolvens* cellulosomes indicated that the different fractions not only contained different cellulosomes but, within each fraction, the morphology of the cellulosomes seemed to be homogeneous. Anion-exchange chromatography has also been used to obtain apparent homogeneous subpopulations of *C. thermocellum* cellulosomes that differ with respect to subunit composition and enzyme activities such as avicelase (the degradation of microcrystalline cellulose or Avicel) and xylanase activity [67]. Subpopulations of cellulosomes from *C. cellulovorans,* in which the function and subunit composition of the cellulosomes were clearly different, have also been observed [68]. Although the exact process is far from being understood, the evidence of relatively distinct homogenous fractions suggests that some form of organization takes place during complex assembly [12]. Otherwise, a much greater number of compositionally different complexes would likely be observed, rather than a few homogenous fractions.

CONCLUSION

The cellulosome complex was first discovered using combined biochemical, ultrastructural and immunochemical techniques [69-71]. Later, the advent of molecular biological techniques led to the sequencing of the scaffoldins that contained a series of repeating modules [31-33,36,72-79]. The defining module, the cohesin, was demonstrated to be responsible for binding the dockerin modules borne by each cellulolytic enzyme subunit [15-17]. In time, several cohesins were crystallized, which revealed a similar jelly-roll topology that folds into a 9-stranded β-sandwich [38,39,41,42,80]. In contrast the solution structure of only one dockerin has been solved by three-dimensional NMR spectroscopy [2648]. The ~70-residue dockerin module consists of a pair of calcium-binding loop-helix motifs connected by a ca. ten-residue linker. In solution, the two loop-helix motifs are related by 2-fold pseudo-symmetry axis. More recently the three-dimensional structure of a cohesin-dockerin complex has been solved by X-ray crystallography [49]. This structure revealed the contact residues involved in the formation of this complex. To understand which residues are binding residues and which are specificity-determining residues, additional complexes need to be solved and biochemical analysis of different cohesin-dockerin pairs should be performed.

In fact, insight into the specificity characteristics of the cohesin-dockerin interaction has long been subject to experimentation, using combined bioinformatics, mutation and biochemical approach [63]. It was noted earlier that the cohesin-dockerin interaction between two closely related cellulosomal species, *C. thermocellum* and *C. cellulolyticum*, observes intraspecies fidelity whereas between the two species, cohesin-dockerin specificity diverges [27]. Comparative bioinformatics and homologue scanning among the suspected dockerin residues revealed the main specificity-determining residues [21,22,27,51,61]. Identification of these residues for the cohesin module required a more sophisticated novel approach, termed progressive homologue swapping [62]. For this purpose, the entire cohesin module was systematically divided into progressively smaller sequences until only three specificity-determining residues remained. Progressive homologue swapping supplements the more conventional site-directed mutagenesis of selected surface residues [53,54] and provides an inherent strategy for zeroing in on relevant specificity residues that would otherwise remain undetected. Whether the positions of these specificity-determining residues are conserved throughout the entire dockerin and cohesin populations in other types and/or species remains to be determined. Although comparative bioinformatics provides some indications, more specificity determinants of cohesin-dockerin pairs should be established by biochemical approaches together with additional structures of different heterodimers. These efforts will eventually enable us to decipher the specificity codes of both the cohesin and the dockerin modules among the different types of cellulosomes and their interacting components.

The ultimate research goals in this context are to understand the importance of the cohesin-dockerin interaction for cellulosome architecture and its significance regarding the endurance of cellulosome-producing organisms in nature. Moreover, if we can learn to control this interaction, we will be able to utilize these components for biotechnological and nanotechnological purposes. We have progressed greatly into the elucidation of the intricate structures of many different types of cellulosomes. Our findings have provided insight into the involvement of the cohesin-dockerin interaction in the organization of cellulosome architecture. Finally, we have started to reveal the molecular determinants of this interaction, which have allowed us to fabricate designer cellulosomes with custom-made organization and function.

ACKNOWLEDGMENTS

The authors are indebted to Felix Frolow, Linda J.W. Shimon and Ilit Noach for continued collaboration on the X-ray crystallography of cellulosomal components. Fig. 3 of this article is a product of this collaborative effort. This research was supported by the Israel Science Foundation (Grant Nos. 394/03, 442/05 and 446/01) and by grants from the United States-Israel Binational Science Foundation (BSF), Jerusalem, Israel.

REFERENCES

[1] Bayer, E. A., Belaich, J.-P., Shoham, Y., and Lamed, R. (2004) The cellulosomes: Multi-enzyme machines for degradation of plant cell wall polysaccharides. *Annu. Rev. Microbiol. 58*, 521-554.

[2] Lamed, R., and Bayer, E. A. (1988) The cellulosome of Clostridium thermocellum. *Adv. Appl. Microbiol. 33*, 1-46.

[3] Bayer, E. A., Morag, E., and Lamed, R. (1994) The cellulosome — A treasure-trove for biotechnology. *Trends Biotechnol. 12*, 378-386.

[4] Bayer, E. A., Chanzy, H., Lamed, R., and Shoham, Y. (1998) Cellulose, cellulases and cellulosomes. *Curr. Opin. Struct. Biol. 8*, 548-557.

[5] Bayer, E. A., Shimon, L. J. W., Lamed, R., and Shoham, Y. (1998) Cellulosomes: structure and ultrastructure. *J. Struct. Biol. 124*, 221-234.

[6] Bayer, E. A., Shoham, Y., and Lamed, R. (2000) Cellulose-decomposing prokaryotes and their enzyme systems. In Dworkin, M., Falkow, S., Rosenberg, E., Schleifer, K.-H., and Stackebrandt, E. (ed.), *The Prokaryotes: An Evolving Electronic Resource for the Microbiological Community, 3rd edition* (latest update release 3.7, September 2001). http://link.springer.de/link/service/books/10125/index.htm. Springer-Verlag, New York.

[7] Belaich, J.-P., Tardif, C., Belaich, A., and Gaudin, C. (1997) The cellulolytic system of Clostridium cellulolyticum. *J. Biotechnol. 57*, 3-14.

[8] Béguin, P., and Lemaire, M. (1996) The cellulosome: an exocellular, multiprotein complex specialized in cellulose degradation. *Crit. Rev. Biochem. Molec. Biol. 31*, 201-236.

[9] Felix, C. R., and Ljungdahl, L. G. (1993) The cellulosome - the exocellular organelle of Clostridium. *Annu. Rev. Microbiol. 47*, 791-819.

[10] Demain, A. L., Newcomb, M., and Wu, J. H. (2005) Cellulase, clostridia, and ethanol. *Microbiol. Mol. Biol. Rev. 69*, 124-154.

[11] Lynd, L. R., Weimer, P. J., van Zyl, W. H., and Pretorius, I. S. (2002) Microbial cellulose utilization: fundamentals and biotechnology. Microbiol. Mol. Biol. Rev. 66, 506-577.

[12] Doi, R. H., and Kosugi, A. (2004) Cellulosomes: plant-cell-wall-degrading enzyme complexes. *Nat. Rev. Microbiol. 2*, 541-551.

[13] Doi, R. H., Kosugi, A., Murashima, K., Tamaru, Y., and Han, S. O. (2003) Cellulosomes from mesophilic bacteria. *J. Bacteriol. 185*, 5907-5914.

[14] Schwarz, W. H. (2001) The cellulosome and cellulose degradation by anaerobic bacteria. *Appl. Microbiol. Biotechnol. 56*, 634-649.

[15] Tokatlidis, K., Salamitou, S., Béguin, P., Dhurjati, P., and Aubert, J.-P. (1991) Interaction of the duplicated segment carried by Clostridium thermocellum cellulases with cellulosome components. *FEBS Lett. 291*, 185-188.

[16] Tokatlidis, K., Dhurjati, P., and Béguin, P. (1993) Properties conferred on Clostridium thermocellum endoglucanase CelC by grafting the duplicated segment of endoglucanase *CelD. Protein Eng. 6*, 947-952.

[17] Salamitou, S., Tokatlidis, K., Béguin, P., and Aubert, J.-P. (1992) Involvement of separate domains of the cellulosomal protein S1 of Clostridium thermocellum in binding to cellulose and in anchoring of catalytic subunits to the cellulosome. *FEBS Lett. 304*, 89-92.

[18] Fierobe, H.-P., Pagès, S., Belaich, A., Champ, S., Lexa, D., and Belaich, J.-P. (1999) Cellulosome from Clostridium cellulolyticum: Molecular study of the dockerin/cohesin interaction. *Biochemistry 38*, 12822-12832.

[19] Jindou, S., Souda, A., Karita, S., Kajino, T., Beguin, P., Wu, D., Inagaki, M., Kimura, T., Sakka, K., and Ohmiya, K. (2004) Cohesin/dockerin interactions within and between Clostridium josui and Clostridium thermocellum: Binding selectivity between cognate dockerin and cohesin domains and species specificity. *J. Biol. Chem. 279*, 9867-9874.

[20] Fierobe, H.-P., Mechaly, A., Tardif, C., Belaich, A., Lamed, R., Shoham, Y., Belaich, J.-P., and Bayer, E. A. (2001) Design and production of active cellulosome chimeras: Selective incorporation of dockerin-containing enzymes into defined functional complexes. *J. Biol. Chem. 276*, 21257-21261.

[21] Mechaly, A., Fierobe, H.-P., Belaich, A., Belaich, J.-P., Lamed, R., Shoham, Y., and Bayer, E. A. (2001) Cohesin-dockerin interaction in cellulosome assembly: A single hydroxyl group of a dockerin domain distinguishes between non-recognition and high-affinity recognition. *J. Biol. Chem. 276*, 9883-9888.

[22] Mechaly, A., Fierobe, H.-P., Belaich, A., Belaich, J.-P., Lamed, R., Shoham, Y., and Bayer, E. A. (2001) Cohesin-dockerin interaction in cellulosome assembly: A single hydroxyl group of a dockerin domain distinguishes between non-recognition and high-affinity recognition (Erratum). *J. Biol. Chem. 276*, 19678.

[23] Yaron, S., Morag, E., Bayer, E. A., Lamed, R., and Shoham, Y. (1995) Expression, purification and subunit-binding properties of cohesins 2 and 3 of the Clostridium thermocellum cellulosome. *FEBS Lett. 360*, 121-124.

[24] Chauvaux, S., Béguin, P., Aubert, J.-P., Bhat, K. M., Gow, L. A., Wood, T. M., and Bairoch, A. (1990) Calcium-binding affinity and calcium-enhanced activity of Clostridium thermocellum endoglucanase D. *Biochem. J. 265*, 261-265.

[25] Choi, S. K., and Ljungdahl, L. G. (1996) Structural role of calcium for the organization of the cellulosome of Clostridium thermocellum. *Biochemistry 35*, 4906-4910.

[26] Lytle, B., Volkman, B. F., Westler, W. M., and Wu, J. H. D. (2000) Secondary structure and calcium-induced folding of the Clostridium thermocellum dockerin domain determined by NMR spectroscopy. *Arch. Biochem. Biophys. 379*, 237-244.

[27] Pagès, S., Belaich, A., Belaich, J.-P., Morag, E., Lamed, R., Shoham, Y., and Bayer, E. A. (1997) Species-specificity of the cohesin-dockerin interaction between Clostridium thermocellum and Clostridium cellulolyticum: Prediction of specificity determinants of the dockerin domain. *Proteins 29*, 517-527.

[28] Bayer, E. A., Ding, S.-Y., Mechaly, A., Shoham, Y., and Lamed, R. (1999) Emerging phylogenetics of cellulosome structure, p. 189-201. In Gilbert, H. J., Davies, G. J., Henrissat, B., and Svensson, B. (ed.), *Recent advances in carbohydrate bioengineering.* The Royal Society of Chemistry, Cambridge.

[29] Leibovitz, E., and Béguin, P. (1996) A new type of cohesin domain that specifically binds the dockerin domain of the Clostridium thermocellum cellulosome-integrating protein CipA. *J. Bacteriol. 178*, 3077-3084.

[30] Ding, S.-Y., Rincon, M. T., Lamed, R., Martin, J. C., McCrae, S. I., Aurilia, V., Shoham, Y., Bayer, E. A., and Flint, H. J. (2001) Cellulosomal scaffoldin-like proteins from Ruminococcus flavefaciens. *J. Bacteriol. 183*, 1945-1953.

[31] Rincon, M. T., Ding, S.-Y., McCrae, S. I., Martin, J. C., Aurilia, V., Lamed, R., Shoham, Y., Bayer, E. A., and Flint, H. J. (2003) Novel organization and divergent dockerin specificities in the cellulosome system of Ruminococcus flavefaciens. *J. Bacteriol. 185*, 703-713.

[32] Ding, S.-Y., Bayer, E. A., Steiner, D., Shoham, Y., and Lamed, R. (2000) A scaffoldin of the Bacteroides cellulosolvens cellulosome that contains 11 type II cohesins. *J. Bacteriol. 182*, 4915-4925.

[33] Xu, Q., Bayer, E. A., Goldman, M., Kenig, R., Shoham, Y., and Lamed, R. (2004) Architecture of the Bacteroides cellulosolvens cellulosome: description of a cell-surface anchoring scaffoldin and a family-48 cellulase. *J. Bacteriol. 186*, 968-977.

[34] Bayer, E. A., Coutinho, P. M., and Henrissat, B. (1999) Cellulosome-like sequences in Archaeoglobus fulgidus: an enigmatic vestige of cohesin and dockerin domains. *FEBS Lett. 463*, 277-280.

[35] Sabathe, F., Belaich, A., and Soucaille, P. (2002) Characterization of the cellulolytic complex (cellulosome) of Clostridium acetobutylicum. *FEMS Microbiol. Lett. 217*, 15-22.

[36] Gerngross, U. T., Romaniec, M. P. M., Kobayashi, T., Huskisson, N. S., and Demain, A. L. (1993) Sequencing of a Clostridium thermocellum gene (cipA) encoding the cellulosomal SL-protein reveals an unusual degree of internal homology. *Mol. Microbiol. 8*, 325-334.

[37] Shimon, L. J. W., Bayer, E. A., Morag, E., Lamed, R., Yaron, S., Shoham, Y., and Frolow, F. (1997) A cohesin domain from Clostridium thermocellum: the crystal structure provides new insights into cellulosome assembly. *Structure 5*, 381-390.

[38] Spinelli, S., Fierobe, H. P., Belaich, A., Belaich, J. P., Henrissat, B., and Cambillau, C. (2000) Crystal structure of a cohesin module from Clostridium cellulolyticum: Implications for dockerin recognition. *J. Mol. Biol. 304*, 189-200.

[39] Tavares, G. A., Béguin, P., and Alzari, P. M. (1997) The crystal structure of a type I cohesin domain at 1.7 Å resolution. *J. Mol. Biol. 273*, 701-713.

[40] Smith, S. P., Béguin, P., Alzari, P. M., and Gehring, K. (2002) 1H, 13C, 15N NMR sequence-specific resonance assignment of a Clostridium thermocellum type II cohesin module. *J. Biomolec. NMR 23*, 73-74.

[41] Noach, I., Lamed, R., Xu, Q., Rosenheck, S., Shimon, L. J. W., Kenig, R., Bayer, E. A., and Frolow, F. (2003) Preliminary X-ray characterization and phasing of a type II cohesin domain from the cellulosome of Acetivibrio cellulolyticus. *Acta Crystallogr. D D59*, 1670-1673.

[42] Noach, I., Frolow, F., Jakoby, H., Rosenheck, S., Shimon, L. J. W., Lamed, R., and Bayer, E. A. (2005) Crystal structure of a type-II cohesin module from the Bacteroides cellulosolvens cellulosome reveals novel and distinctive secondary structural elements. J. *Mol. Biol. 348*, 1-12.

[43] Lytle, B., and Wu, J. H. D. (1998) Involvement of both dockerin subdomains in assembly of the Clostridium thermocellum cellulosome. *J. Bacteriol. 180*, 6581-6585.

[44] Kawasaki, H., Nakayama, S., and Kretsinger, R. H. (1998) Classification and evolution of EF-hand proteins. *Biometals 11*, 277-295.

[45] Lewit-Bentley, A., and S., R. (2000) EF-hand calcium-binding proteins. *Curr. Opin. Struct. Biol. 10*, 637-643.

[46] Adams, J. J., Jang, C. J., Spencer, H. L., Elliott, M., and Smith, S. P. (2004) Expression, purification and structural characterization of the scaffoldin hydrophilic X-module from the cellulosome of Clostridium thermocellum. *Protein Expr. Purif. 38*, 258-263.

[47] Adams, J. J., Webb, B. A., Spencer, H. L., and Smith, S. P. (2005) Structural characterization of type II dockerin module from the cellulosome of Clostridium thermocellum: Calcium-Induced effects on conformation and target recognition. *Biochemistry 44*, 2173-2182.

[48] Lytle, B. L., Volkman, B. F., Westler, W. M., Heckman, M. P., and Wu, J. H. D. (2001) Solution structure of a type I dockerin domain, a novel prokaryotic, extracellular calcium-binding domain. *J. Mol. Biol. 307*, 745-753.

[49] Carvalho, A. L., Dias, F. M., Prates, J. A., Nagy, T., Gilbert, H. J., Davies, G. J., Ferreira, L. M., Romão, M. J., and Fontes, C. M. (2003) Cellulosome assembly revealed by the crystal structure of the cohesin-dockerin complex. *Proc. Natl. Acad. Sci. USA 100*, 13809-13814.

[50] Fontes, C. M., Hazlewood, G. P., Morag, E., Hall, J., Hirst, B. H., and Gilbert, H. J. (1995) Evidence for a general role for non-catalytic thermostabilizing domains in xylanases from thermophilic bacteria. *Biochem. J. 307*, 151-158.

[51] Schaeffer, F., Matuschek, M., Guglielmi, G., Miras, I., Alzari, P. M., and Béguin, P. (2002) Duplicated dockerin subdomains of Clostridium thermocellum endoglucanase CelD bind to a cohesin domain of the scaffolding protein CipA with distinct thermodynamic parameters and a negative cooperativity. *Biochemistry 41*, 2106-2114.

[52] Miras, I., Schaeffer, F., Béguin, P., and Alzari, P. M. (2002) Mapping by site-directed mutagenesis of the region responsible for cohesin-dockerin interaction on the surface of the seventh cohesin domain of Clostridium thermocellum CipA. *Biochemistry 41*, 2115-2119.

[53] Handelsman, T., Barak, Y., Nakar, D., Mechaly, A., Lamed, R., Shoham, Y., and Bayer, E. A. (2004) Cohesin-dockerin interaction in cellulosome assembly: A single Asp-to-Asn mutation disrupts high-affinity cohesin-dockerin binding. *FEBS Lett. 572*, 195-200.

[54] Handelsman, T., Mechaly, A., Barak, Y., Nakar, D., Lamed, R., Shoham, Y., and Bayer, E. A. (2004) Effect of rational mutagenesis of selected cohesin residues on the high-affinity cohesin-dockerin interaction, p. 194-206. In Sahah, B. C. and Hayashi, K. (ed.), ACS Series #889, *Lignocellulose Biodegradation*. American Chemical Society, Washington, DC.

[55] Pagès, S., Belaich, A., Tardif, C., Reverbel-Leroy, C., Gaudin, C., and Belaich, J.-P. (1996) Interaction between the endoglucanase CelA and the scaffolding protein CipC of the Clostridium cellulolyticum cellulosome. *J. Bacteriol. 178*, 2279-2286.

[56] Gal, L., Pagès, S., Gaudin, C., Belaich, A., Reverbel-Leroy, C., Tardif, C., and Belaich, J.-P. (1997) Characterization of the cellulolytic complex (cellulosome) produced by Clostridium cellulolyticum. *Appl. Environ. Microbiol. 63*, 903-909.

[57] Lytle, B., Myers, C., Kruus, K., and Wu, J. H. D. (1996) Interactions of the CelS binding ligand with various receptor domains of the Clostridium thermocellum cellulosomal scaffolding protein, CipA. *J. Bacteriol. 178*, 1200-1203.

[58] Barak, Y., Handelsman, T., Nakar, D., Mechaly, A., Lamed, R., Shoham, Y., and Bayer, E. A. (in press) Matching fusion-protein systems for affinity analysis of two interacting families of proteins: The cohesin-dockerin interaction. *J. Molec. Recog. 18*, 491-501.

[59] Salamitou, S., Raynaud, O., Lemaire, M., Coughlan, M., Béguin, P., and Aubert, J.-P. (1994) Recognition specificity of the duplicated segments present in Clostridium thermocellum endoglucanase CelD and in the cellulosome-integrating protein CipA. *J. Bacteriol. 176,* 2822-2827.

[60] Jindou, S., Kajino, T., Inagaki, M., Karita, S., Beguin, P., Kimura, T., Sakka, K., and Ohmiya, K. (2004) Interaction between a type-II dockerin domain and a type-II cohesin domain from Clostridium thermocellum cellulosome. Biosci. Biotechnol. *Biochem. 68*, 924-926.

[61] Mechaly, A., Yaron, S., Lamed, R., Fierobe, H.-P., Belaich, A., Belaich, J.-P., Shoham, Y., and Bayer, E. A. (2000) Cohesin-dockerin recognition in cellulosome assembly: Experiment versus hypothesis. *Proteins 39*, 170-177.

[62] Nakar, D., Handelsman, T., Shoham, Y., Fierobe, H.-P., Belaich, J. P., Morag, E., Lamed, R., and Bayer, E. A. (2004) Pinpoint mapping of recognition residues on the cohesin surface by progressive homologue swapping. *J. Biol. Chem. 279*, 42881-42888.

[63] Bayer, E. A., Morag, E., Lamed, R., Yaron, S., and Shoham, Y. (1998) Cellulosome structure: four-pronged attack using biochemistry, molecular biology, crystallography and bioinformatics, p. 39-65. In Claeyssens, M., Nerinckx, W., and Piens, K. (ed.), *Carbohydrases from Trichoderma reesei and other microorganisms*. The Royal Society of Chemistry, London.

[64] Pohlschröder, M., Canale-Parola, E., and Leschine, S. B. (1995) Ultrastructural diversity of the cellulase complexes of Clostridium papyrosolvens C7. J. Bacteriol. 177, 6625-6629.

[65] Pohlschröder, M., Leschine, S. B., and Canale-Parola, E. (1993) Regulation of the multicomplex cellulase-xylanase system of Clostridium papyrosolvens, p. 86-94. In Shimada, K., Hoshino, S., Ohmiya, K., Sakka, K., Kobayashi, Y., and Karita, S. (ed.), *Genetics, biochemistry and ecology of lignocellulose degradation*. Uni Publishers Co., Ltd., Tokyo, Japan.

[66] Pohlschröder, M., Leschine, S. B., and Canale-Parola, E. (1994) Multicomplex cellulase-xylanase system of Clostridium papyrosolvens C7. *J. Bacteriol. 176*, 70-76.

[67] Ali, B. R., Romaniec, M. P., Hazlewood, G. P., and Freedman, R. B. (1995) Characterization of the subunits in an apparently homogeneous subpopulation of Clostridium thermocellum cellulosomes. *Enzyme Microbiol. Technol. 17*, 705-711.

[68] Murashima, K., Kosugi, A., and Doi, R. H. (2002) Determination of subunit composition of Clostridium cellulovorans cellulosomes that degrade plant cell walls. *Appl Environ Microbiol 68*, 1610-1615.

[69] Bayer, E. A., Kenig, R., and Lamed, R. (1983) Adherence of Clostridium thermocellum to cellulose. *J. Bacteriol. 156*, 818-827.

[70] Lamed, R., Setter, E., and Bayer, E. A. (1983) Characterization of a cellulose-binding, cellulase-containing complex in Clostridium thermocellum. *J. Bacteriol. 156*, 828-836.

[71] Lamed, R., Setter, E., Kenig, R., and Bayer, E. A. (1983) The cellulosome — a discrete cell surface organelle of Clostridium thermocellum which exhibits separate antigenic, cellulose-binding and various cellulolytic activities. *Biotechnol. Bioeng. Symp. 13*, 163-181.

[72] Shoseyov, O., and Doi, R. H. (1990) Essential 170-kDa subunit for degradation of crystalline cellulose by Clostridium cellulovorans cellulase. *Proc. Natl. Acad. Sci. USA 87*, 2192-2195.

[73] Shoseyov, O., Takagi, M., Goldstein, M. A., and Doi, R. H. (1992) Primary sequence analysis of Clostridium cellulovorans cellulose binding protein A. *Proc. Natl. Acad. Sci. USA 89*, 3483-3487.

[74] Fujino, T., Béguin, P., and Aubert, J.-P. (1993) Organization of a Clostridium thermocellum gene cluster encoding the cellulosomal scaffolding protein CipA and a protein possibly involved in attachment of the cellulosome to the cell surface. *J. Bacteriol. 175*, 1891-1899.

[75] Kakiuchi, M., Isui, A., Suzuki, K., Fujino, T., Fujino, E., Kimura, T., Karita, S., Sakka, K., and Ohmiya, K. (1998) Cloning and DNA sequencing of the genes encoding Clostridium josui scaffolding protein CipA and cellulase CelD and identification of their gene products as major components of the cellulosome. *J. Bacteriol. 180*, 4303-4308.

[76] Pagès, S., Belaich, A., Fierobe, H.-P., Tardif, C., Gaudin, C., and Belaich, J.-P. (1999) Sequence analysis of scaffolding protein CipC and ORFXp, a new cohesin-containing protein in Clostridium cellulolyticum: comparison of various cohesin domains and subcellular localization of ORFXp. *J. Bacteriol. 181*, 1801-1810.

[77] Ding, S.-Y., Bayer, E. A., Steiner, D., Shoham, Y., and Lamed, R. (1999) A novel cellulosomal scaffoldin from Acetivibrio cellulolyticus that contains a family-9 glycosyl hydrolase. *J. Bacteriol. 181*, 6720-6729.

[78] Rincon, M. T., Martin, J. C., Aurilia, V., McCrae, S. I., Rucklidge, G., Reid, M., Bayer, E. A., Lamed, R., and Flint, H. J. (2004) ScaC, an adaptor protein carrying a novel cohesin that expands the dockerin-binding repertoire of the Ruminococcus flavefaciens 17 cellulosome. *J. Bacteriol. 186*, 2576-2585.

[79] Xu, Q., Barak, Y., Kenig, R., Shoham, Y., Bayer, E. A., and Lamed, R. (2004) A novel Acetivibrio cellulolyticus anchoring scaffoldin that bears divergent cohesins. *J. Bacteriol. 186*, 5782-5789.

[80] Shimon, L. J. W., Frolow, F., Yaron, S., Lamed, R., Morag, E., Bayer, E. A., and Shoham, Y. (1997) Crystallization and preliminary X-ray analysis of a cohesin domain of the cellulosome from Clostridium thermocellum. *Acta Crystallogr. D 53*, 114-115.

In: Cellulosome
Eds: V. Uversky and I. A. Kataeva, pp. 69-94
ISBN 1-59454-950-8

Chapter IV

CATALYTIC COMPONENTS OF THE CELLULOSOME

***Pedro M. Alzari*[1] *and Pierre Béguin*[2*]**

[1]Unité de Biochimie structurale

[2]Plate-forme 5 Production de Protéines recombinantes et d'Anticorps, Institut Pasteur, Paris, France

ABSTRACT

Based on the nearly complete sequence of the *Clostridium thermocellum* genome (http://genome.jgi-psf.org/draft_microbes/cloth/cloth.home.html), some 70 genes of this bacterium encode polypeptides that include a dockerin domain and may thus be included within the quaternary structure pf the cellulosome. In most cases, these polypeptides include readily identified modules such as catalytic domains or carbohydrate-binding modules. Most of these modules belong to widely distributed families and may serve as paradigms to investigate the structure and function of a vast variety of related enzymes. In this chapter, we review the structure and mechanism of the best studied examples of catalytic domains and carbohydrate-binding modules present in components of the cellulosome. Interestingly, it appears that glycosyl hydrolases splitting different substrates often share the same basic framework and mode of action.

Keywords: Catalytic domains, cellulose-binding modules, retaining glycosidases, inverting glycosidases

---

* Correspondence concerning this article should be addressed to Pierre Béguin Plate-forme 5 Production de Protéines recombinantes et d'Anticorps, Institut Pasteur, 25 rue du Dr. Roux, 75724 Paris cedex 15 France. E-mail: beguin@pasteur.fr

THE CELLULOSOME: A GIGANTIC LEGO® SET

The multiplicity of enzyme components in cellulase systems has been known for decades. It was E.T. Reese who originally suggested the presence in *Trichoderma reesei* of at least two components, the amorphogenic C_1 factor and the hydrolytic component C_x [76]. Since then, the number of distinct proteins involved in the process of cellulose degradation has been largely extended: most of the true cellulolytic microorganisms known to date produce at least four or five distinct endoglucanases, one or two β–glucosidases, and two complementary cellobiohydrolases, which hydrolyze cellulose chains form the reducing and the non-reducing end, respectively. In addition, most cellulolytic microorganisms also produce a set of glycoside hydrolases required to degrade and metabolize the hemicellulose components of plan cell walls: xylanases, mannanases, arabinofuranosidases, acetylxylan esterases, pectinases.

From the beginning, it was clear that the biochemical complexity of the cellulosome was considerable: the original papers by Lamed, Bayer and coworkers establishing the existence of the cellulosome as a specific biologically relevant complex also showed that it contained at least 14 components that could be separated by SDS-PAGE [46,47]. However, isolating and characterizing the subunits of the complex by classical biochemical analysis turned out to be quite difficult owing to their very tight association. Cloning and expression of the corresponding genes in *Escherichia coli* proved to be most helpful in this respect, and by the end of the 1980s, the targeted search for *Clostridium thermocellum* genes encoding cellulases and hemicellulases by several laboratories had yielded large collections of clones [63,74,81,82]. While there was some degree of duplication between these collections, by 1998 it was clear that over 30 proteins participated in the degradation of plant cell wall polysaccharides [32]. These included a vast majority of hydrolytic enzymes, most of which possessed a repeated sequence of some 24 residues. This sequence, termed dockerin [3], was shown to promote *in vitro* binding to a large, non catalytic subunit [94,95], termed scaffoldin [3], which had been shown previously to act as a cellulose-binding factor [101]. This was followed by the cloning of genes that encoded the scaffoldins CbpA of *Clostridium cellulovorans* [89], CipA of *C. thermocellum* [29] and *Clostridium cellulolyticum* CipC [68]. The latter were found to harbour 8 or 9 repetitive modules, which suggested that such modules, termed cohesins [3], might serve as anchoring sites for the dockerins borne by the catalytic subunits. Indeed, polypeptides including a single module were produced by subcloning and expressing the corresponding DNA fragments and were shown to interact specifically with dockerin-containing polypeptides [24,68,92,103]. In addition, several genes encoding bi- or multifunctional components thought to mediate the attachment of individual enzymes or whole cellulosomes to the surface of the bacterial cell were identified [25,50,84,85]. One of these components, termed OlpA, contains a cohesin domain of the same type (type I) as those borne by the scaffoldin, and binds the dockerins borne by the catalytic subunits. The other polypeptides, such as OlpB, Orf2p and SdbA, contain cohesins of a different type (type II), which are specific for the divergent dockerin borne by the scaffoldin CipA. All of these polypeptides possess SLH domains, which are known to promote the attachment of proteins to the cell surface [52,53,55,62,67] by binding to a secondary cell wall polysaccharide containing N-acetylglucosamine, N-acetylmannosamine, and pyruvate [20,57,61,78]. Indeed, cell fractionation and electron microscopy demonstrated that OlpA, OlpB, and SdbA are located in the cell envelope [51,53,54].

The *C. thermocellum* genome has now been almost entirely sequenced by the Joint Genome Institute at the U.S. Department of Energy, and the annotated data are available at their web site (http://genome.jgi-psf.org/draft_microbes/cloth/cloth.home.html). Thus, the list of putative cellulosome components has been extended considerably. Some 70 genes encode polypeptides carrying a type I dockerin and thus susceptible to bind to the scaffoldin CipA. The majority can either be identified as previously characterized components or related to known families of cellulases, hemicellulases or glycoside esterases; however, at least 13 of them show no obvious connection to known proteins. Since there is little or no binding specificity between different type I cohesins and dockerins [38,56,85,103], these figures would imply that the number of possible cellulosome combinations might exceed the matter available in the whole Universe! Of course, the number of biochemically relevant combinations is likely much more modest: there is as yet no evidence that all genes are expressed, that all enzymes are functional, nor that each enzyme fulfills a specific function. In addition, the abundance of each particular combination is highly dependent on the rate of production of its individual subunits. Nevertheless, the cellulosome concept covers a set of complexes whose diversity will likely defy exhaustive biochemical analysis for quite some time.

The paradigm established for the quaternary organization of cellulosome-type complexes produced by cellulolytic clostridia has since been further extended to other systems such as ruminococci [16,79,80] and even anaerobic rumen fungi [21,91]. The topic of cellulosomes as a multienzyme machines for the degradation of plant cell wall polysaccharides has been reviewed recently [2] and is covered by several chapters of this volume. In this chapter, we will therefore focus on the structure and function of the catalytic subunits that have been characterized from cellulosome-producing microorganisms.

STRUCTURAL PARADIGMS

The diversity of cellulosome-related components would be quite intractable were it not for the fact that they consist of identifiable modules, such as catalytic domains or cellulose-binding domains, which can be ordered into structural families based on sequence similarities. Catalytic domains are at the core of the cellulose and hemicellulose hydrolysis process. They behave as independently folding units and possess hydrolytic activity, at least against soluble model substrates, even if activity against natural, insoluble substrates such as crystalline cellulose is much reduced or undetectable. The classification introduced by B. Henrissat in 1989 [35] and periodically updated by P. Coutinho and B. Henrissat (http://afmb.cnrs-mrs:fr/CAZY) currently includes over 90 distinct enzyme families of glycosidases. At a higher level, many of these families can be ordered into clans, whose members share very few identical residues, but possess very similar three-dimensional structures and active site topologies. Glycoside hydrolases belonging to the same families as those synthesized by cellulosome-producing microorganisms also occur in a wide variety of organisms, prokaryotic and eukaryotic alike. Hence much can be learned about a large number of systems by studying a few paradigms in detail.

Catalytic Domains

A look at the annotated draft of the *C. thermocellum* genome sequence (http://genome.jgi-psf.org/draft_microbes/cloth/cloth.home.html) shows that the largest battalion of cellulolytic enzymes belongs to family 9 of glycoside hydrolases (GH-9, 16 members), followed by GH-5 (8 members). In addition, there are two genes encoding cellulases of family 48 (*celS* and an open reading frame whose corresponding polypeptide lacks a dockerin) and one gene encoding an endoglucanase of family 8 (*celA*). GH-6 and GH-7 cellulases are not represented. While this is to be expected for family 7, which so far has been found exclusively in fungi, the absence of family 6 is more intriguing, since cellulases of this type have been found in a wide variety of microorganisms ranging from fungi to Gram positive bacteria. As regards hemicellulases, there are numerous genes encoding enzymes with putative or demonstrated activity against the various polysaccharides present in plant cell walls: xylanases, lichenases, mannanases, arabinofuranosidases, pectin esterase and pectate lyase etc. The most conspicuous groups are GH-10 xylanases (5 genes), GH-43 α-arabinofuranosidases (4 genes), and feruloyl esterases (3 genes, two of which are fused to xylanase genes). Although data are not yet as complete, the glycoside hydrolase genes cloned from *C. cellulolyticum*, *C. cellulovorans* and *Clostridium josui* seem to follow a similar pattern.

The general principles underpinning catalysis are now well understood and substantiated by biochemical, structural and mutagenesis studies [4,12,13]. Part of the energy required to rupture the glycosidic bond is provided by subsites that bind sugar protomers on either side of the reaction site and force the polysaccharide chain into a strained conformation. As a result, distortion of the sugar lying on the -1 side of the bond to be cleaved (which usually adopts a half-chair or boat conformation) favors the acid-base-catalyzed hydrolysis of the glycosidic bond [17,31,104]. In terms of their catalytic mechanisms, glycosidases can be generally classified in two large groups according to the stereochemical outcome of the hydrolytic reaction, namely inversion or retention of the anomeric conformation. Koshland first proposed fifty years ago that retaining glycosidases worked via a double-displacement mechanism involving an enzyme nucleophile, whereas inverting glycosidases worked by a single chemical step, the displacement of the aglycon by a nucleophilic water molecule [43]. This hypothesis has now been largely substantiated by extensive biochemical, structural and mutagenesis studies. Two catalytic groups of the protein participate in the reaction. One is a glutamic acid residue, which protonates the glycosidic oxygen, thereby promoting cleavage of the glycosidic bond and release of the leaving group. The other one is a nucleophile, generally an aspartate residue, which promotes the nucleophilic substitution at the level of the C_1 carbon of the glycosidic bond. Two cases are possible depending on the distance of the carboxylate group from the reaction center [12]. If the carboxylate is sufficiently close to the proton donor (about 5.5 Å), it acts as a nucleophile and forms a glycosyl enzyme intermediate. The latter is then hydrolyzed in a second step. An incoming water molecule releases a proton, which restores the acidic form of the proton donor, and the hydroxyl ion substitutes for the carboxylate, releasing the glycosyl moiety. If instead the nucleophilic carboxylate is at a greater distance form the proton donor (typically 9-10 Å), it acts as a general base to promote the deprotonation of a nucleophilic water molecule, and the reaction proceeds in a single step. The two mechanisms have a different stereochemical outcome. In the first case, a double nucleophilic substitution occurs at the anomeric carbon, leading to retention of the β configuration; in the second case, a single substitution occurs, and the configuration of the anomeric carbon is inverted from β to α. In all cases tested, enzymes belonging to the same structural family were shown to share the same type of mechanism, reflecting conservation of the topology of the active site [28].

Retaining Glycosidases: GH-5 Endoglucanases and GH-10 Xylanases

Glycosidase families GH-5 and GH-10 are among those most largely represented in the *C. thermocellum* genome, with 8 and 5 genes respectively (http://genome.jgi-psf.org/draft_microbes/cloth/cloth.home.html). Dominguez *et al.* [19] determined and compared the 3-D structures of two *C. thermocellum* enzymes from these families, namely GH-5 endoglucanase Cel5B (formerly CelC) and GH-10 xylanase Xyn10A (formerly XynZ). The two enzymes fold into the classical $(\beta/\alpha)_8$ barrel topology (Fig. 1A), first observed for triose phosphate isomerase (TIM) and characteristic of a large number of functionally unrelated enzymes. An insertion of 54 amino acids in Cel5B between the sixth β-strand and the sixth α-helix forms a distinct subdomain adjacent to the active site composed of four helices and a short two-stranded β-structure. This subdomain is present in some, but not all, family 5 glycosidases, and may help to modulate substrate-binding specificity. As it was found for GH-9 glycosidases (see below), it is conceivable that similar insertions in connecting loops close to the active site might reshape the substrate-binding groove, so that homologous GH-5 enzymes could display distinct endo/exo modes of action. A possible example may be *C. thermocellum* cellobiohydrolase Cel5F (encoded by the gene *celO*) [105], whose catalytic domain belongs to the GH-5 family (including mostly endoglucanases) but displays a predominant exoglucanase mode of action.

Despite the lack of sequence similarity and the different substrate specificities, the structure of GH-5 Cel5B and GH-10 Xyn10A resemble each other in overall shape and orientation of the substrate-binding groove (Fig. 1A), as well as in specific structural features such as the presence of two β-bulges at strands 3 and 7, the conformation of some loops at the bottom of the barrel and the conservation of various intramolecular salt bridges. Equivalent regions of the two enzymes can be superimposed with an rms deviation of 1.9 Å for 159 corresponding Cα positions. The resulting alignment reveals a better match towards the bottom of the barrel (C-terminal end of helices and N-terminal end of strands), opposite to the catalytic center and consistent with structural changes being required to evolve a distinct specificity.

Substrate hydrolysis by both GH-5 Cel5B and GH-10 Xyn10A proceeds by a two-step acid-base mechanism leading to retention of configuration at the anomeric carbon. As shown in Fig. 1B, the two glycosidases have a similar active site architecture. The nucleophilic residue involved in catalysis (Cel5B Glu280 [99], Xyn10A Glu754) lies at the bottom of the active site crevice, partially stacked between two conserved residues in each family (His and Trp in GH-5, Gln and Trp in GH-10). The acid catalyst (Cel5B Glu140 [65], Xyn10A Glu645) is positioned at the end of β-strand 4, in a less restrained environment. The structures of wild-type Cel5B and the Glu140Gln mutant in complexed and uncomplexed forms revealed conformational changes of the loop containing the acid catalyst upon substrate binding, suggesting an induced fit activation mechanism [18].

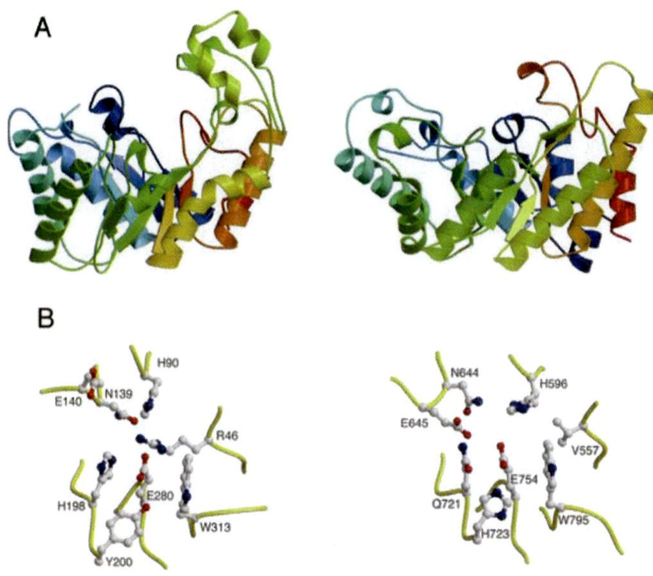

Figure 1. Clan GH-A retaining glycosidases. (A) Backbone structures of *C. thermocellum* endoglucanase Cel5B (left, PDB code 1cec) and xylanase Xyn10A (right, code 1xyz) displaying the classical $(\beta/\alpha)_8$ fold. The rainbow coloring starts at the NH_2-terminal end (dark blue). The substrate-binding cleft faces the top of the Figure. (B) Schematic view of the respective catalytic centers showing conserved residues in family 5 (left) and family 10 (right) glycosidases.

The structural and mechanistic similarity between Cel5B and Xyn10A illustrates how different substrate specificities may be derived from the same fold while preserving a similar disposition of identical catalytic residues. Indeed, the same $(\beta/\alpha)_8$ fold and catalytic machinery shared by GH-5 and GH-10 enzymes is characteristic of glycosidases with quite distinct substrate specificities from at least seventeen different GH families [34,37], (http://afmb.cnrs-mrs.fr/CAZY/). These are now grouped in a vast superfamily, or clan, termed clan GH-A. This remarkable versatility of an ancestral protein fold illustrates how evolution dealt with the problem of glycoside hydrolase specificity, in order to face the extensive diversity of carbohydrate substrates, which are composed of sugar moieties with many different stereochemistries linked together in a large number of possible combinations.

Inverting Glycosidases: GH-8 Endoglucanases and GH-48 Cellobiohydrolases

An inverting cellobiohydrolase belonging to family 48 is found as a major catalytic component in the cellulosomes of *C. thermocellum* [64], *C. cellulolyticum* [77], *C. josui* [40], *C. cellulovorans* [54] and of the anaerobic fungus *Piromyces* [90]. *C. thermocellum* Cel48A was the first catalytic subunit shown to act synergistically with the scaffolding subunit CipA to degrade crystalline cellulose [102], indicating that this enzyme is a key component of the cellulosome. The two most extensively studied examples of cellulosomal enzymes belonging to family 48 are Cel48A (formerly CelS) from *C. thermocellum* and Cel48F from *C. cellulolyticum* (formerly CelF). Both enzymes display typical characteristics of exoglucanases [44,45,100] and processive enzymes [77]. The 3-D structures of Cel48F [71] and Cel48A [33] have been determined and are substantially equivalent. The catalytic domain folds into an $(\alpha/\alpha)_6$ barrel, with long helical connecting loops at the N-terminal side of the inner α-helices (Fig. 2). The substrate-binding region consists of a 25-Å 1 tunnel, which covers about two-thirds of the barrel side and can accommodate at least six glucosyl residues [33,71], followed by an open cleft that can bind two or three additional sugar residues. Linear sliding of the uncleaved cellulose chain within the tunnel, favored by stacking interactions of the glucosyl rings with aromatic protein residues, accounts for the processive mode of action of these enzymes. The pattern of observed protein-carbohydrate interactions suggests that the substrate chain occupies discrete positions within the tunnel and that chain threading might thus occur by two sugar units (the structural repeating unit of the polymer) at a time. Cleavage and release of cellobiose is predicted to occur at the reducing end of the chain.

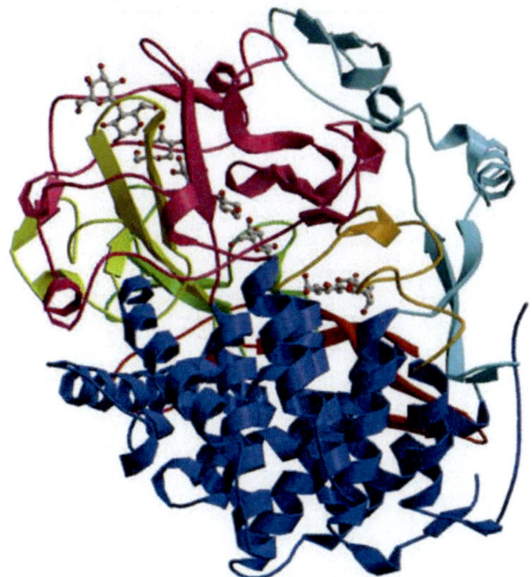

Figure 2. Backbone structure of inverting cellobiohydrolase Cel48A, the major enzymatic component of the *C. thermocellum* cellulosome. The helices forming the (a_6/a_6) core are colored in blue. At the N-terminus of the inner helices, the helix-connecting segments forming the substrate-binding site are shown in different colours. The position of glucosyl residues in the Cel48A-cellohexaose complex is shown in ball-and-stick mode. One ligand binds inside the tunnel (positions -7 to -2, upper left), and two additional glucosyl residues are seen in the open cleft at positions +1 and +2 (right).

Despite sharing only 12 % identity with Cel48A, *C. thermocellum* endoglucanase Cel8A (formerly CelA) displays the same $(\alpha/\alpha)_6$ barrel [1]. When these structures are superimposed, the $(\alpha/\alpha)_6$ barrels of Cel8A and Cel48A are found to display a similar helical packing (Fig. 3A), with an rms deviation of 1.8 Å for 202 equivalent Cα positions in their core regions. Furthermore, critical amino acid residues in the active center are conserved and the glucosyl residues bind in a similar way to equivalent subsites (Fig. 3B), strongly supporting a conserved catalytic machinery in GH families 8 and 48. The two families are thus grouped into the same clan (clan GH-M) (http://afmb.cnrs-mrs.fr/CAZY/). However, the substrate-binding region is strikingly different in these two enzyme families. The long tunnel that characterizes processive exoglucanase Cel48A is converted into an acidic open groove in endoglucanase Cel8A (Fig. 3A), thus accounting for the different exo/endo mode of action of the two enzymes.

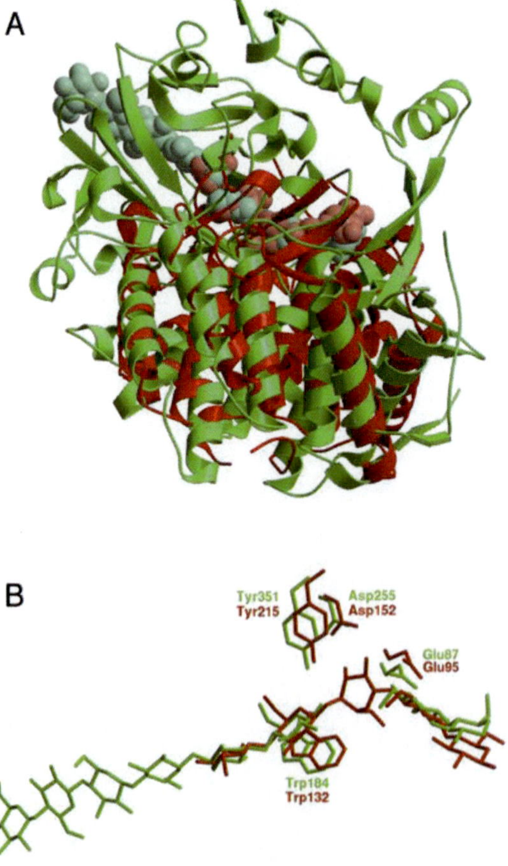

Figure 3. (A) Structural superposition of family 8 endoglucanase Cel8A (red, code 1cem) and family 48 cellobiohydrolase Cel48A (green, code 1l2a). (B) Detailed view showing bound substrate at the catalytic center of Cel8A and Cel48A. The two enzymes bind substrate in a similar way and have conserved amino acid residues at critical positions for catalysis.

The catalytic mechanism of GH-8 and GH-48 glycosidases can be described in terms of the available crystal structures of *C. cellulolyticum* Cel48A (formerly CelF) [71], *C.*

thermocellum Cel48A [33] and endoglucanase Cel8A [1] in complex with different oligosaccharides. In particular, the structure of endoglucanase Cel8A in complex with substrate was recently determined for a point mutant (Glu555Gln) affecting the proton donor of the enzyme [31]. The atomic resolution of the data (0,94 Å) provides a precise description of the catalytic center and protein-carbohydrate interactions. The substrate binds to a deep acidic cleft, occupying six D-glucosyl-binding subsites, three on either side of the scissile glycosidic linkage. Protein-carbohydrate interactions are characterized by several stacking interactions between sugar rings and aromatic protein residues (Fig. 4A) and by an extended network of direct and water-mediated hydrogen-bonding interactions. These interactions impose a torsional stress upon the substrate chain, which forces the glucosyl ring at subsite –1 into a distorted boat conformation (Fig. 4B), further stabilized by a bidentate hydrogen-bond of its OH-3 and OH-4 hydroxyls with Asp152. This strained conformation facilitates the formation of an oxocarbenium ion intermediate upon the concerted attack on the glycosidic linkage by the proton donor Glu95 and the nucleophilic water (assisted by Asp278 acting as a general base, see Fig. 4). Cel48A is thought to operate through a similar catalytic mechanism, although an important difference is that GH-48 enzymes lack a carboxylate group equivalent to Cel8A Asp278, suggesting that a carboxylic group would not be strictly required for catalysis. In contrast, a conserved tyrosine residue (Cel48A Tyr351, equivalent to Cel8A Tyr215) must play an important role in catalysis, since its hydroxyl group was found to form a strong hydrogen bonding interaction with the water nucleophile in Cel8A (Fig. 4B).

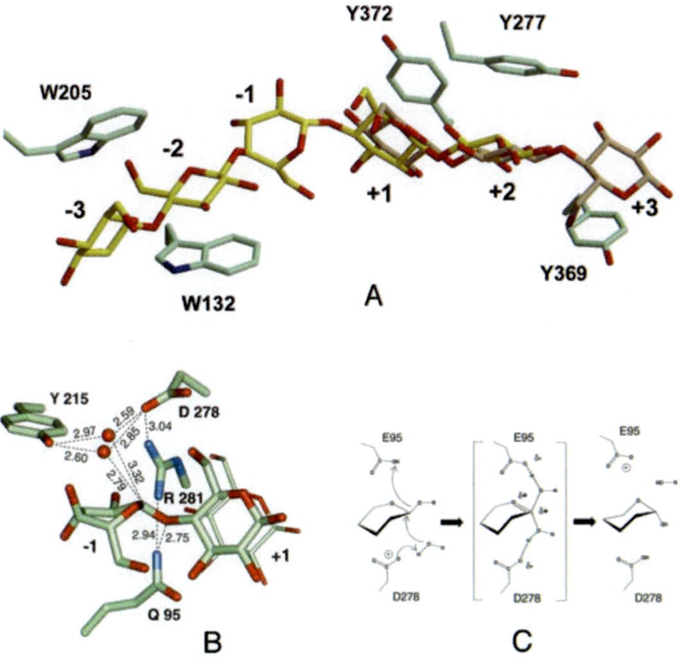

Figure 4. The active center of Cel8A$_{E95Q}$ (code 1kwf). (A) Protein-sugar stacking interactions along the substrate-binding cleft. (B) Local geometry of the catalytic center showing the two catalytic residues and the putative nucleophilic water molecule. (C) Proposed catalytic mechanism, in which the boat conformation of the glucosyl ring favors a transition state with substantial oxocarbenium ion-like character.

Inverting Glycosidases: GH-9 endo/exo Glucanases

Inverting cellulases belonging to glycosidase family 9 are the best represented in the *C. thermocellum* genome, with as many as 16 different members (http://genome.jgi-psf.org/draft_microbes/cloth/cloth.home.html). Enzymes from this family may behave as strict endoglucanases, mixed endo/exo-glucanases or strict exoglucanases. A few crystal structures of GH-9 enzymes are currently available, including those of two *C. thermocellum* glycosidases with a distinct endo/exo mode of action: endoglucanase Cel9A (formerly CelD) [39] and cellobiohydrolase CbhA [87] (Fig. 5). These enzymes share the same $(\alpha/\alpha)_6$ topology that was described above for *C. thermocellum* Cel8A and Cel48A. However, GH-9 glycoside hydrolases differ considerably from GH-8 and GH-48 and are therefore not classified in the same clan. In particular, the architecture and orientation of the corresponding active-site clefts are quite different. When the structures Cel9A and Cel8A are superimposed with the α-helices in the same topological order, the long substrate-binding grooves in CelA and CelD run in approximately perpendicular directions [1].

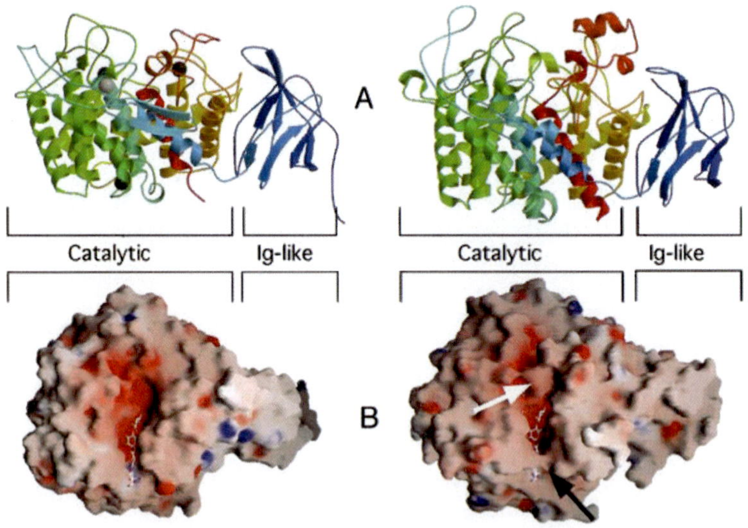

Figure 5. (A) Ribbon representation of family 9 endoglucanase Cel9A (left, code 1clc) and cellobiohydrolase CbhA (right, code 1ut9), with the substrate-binding cleft facing the top of the Figure. Color coding as in Fig. 1. (B) Surface representation of the same enzymes, colored according to charge, with the substrate-binding cleft viewed from the top. For CbhA, the bulge blocking the cleft beyond site -2 and the bridge above site +2 are indicated by a white and a black arrow, respectively.

The $(\alpha/\alpha)_6$ barrels forming the catalytic domains of Cel9A and CbhA are both tightly associated with an N-terminal immunoglobulin domain of unknown function (Fig. 5). Again, the active site cleft is found at the N-terminal end of the inner helices; several aromatic residues stabilize the carbohydrate moiety through stacking interactions and the sugar residue bound at subsite –1 is clearly distorted [87]. In both cases, a conserved glutamate residue (Glu795 in CbhA, Glu555 in Cel9A) is favorably positioned to act as a proton donor in the reaction, while two conserved aspartic acid residues (Asp383/386 in CbhA, Asp198/201 in Cel9A) are thought to deprotonate the water molecule that carries out the nucleophilic attack on the anomeric carbon of the glucose residue. Mutagenesis of these residues in Cel9A resulted in nearly total loss of activity [10].

The shape of the active sites explains why Cel9A and CbhA behave like an endoglucanase and a cellobiohydrolase, respectively. The substrate-binding site of Cel9A is a long open groove, specifically adapted to bind the cellulose substrate at random places along the chain and therefore accounting for its endo mode of action. In contrast, the active site cleft of CbhA is narrower and forms a bridge above subsite +2 upon substrate binding (Fig. 5B). In addition, the key structural element that distinguishes CbhA as an exoglucanase is a steric blockage of the substrate-binding cleft beyond the –2 subsite, which indicates that the enzyme attacks cellulose chains from the non-reducing end. A longer helix-connecting loop (at CbhA residues 590-620) adopts a conformation differing from the equivalent loop in Cel9A, making the –2 subsite specific for a terminal glucosyl residue. As described above for Cel8A and Cel48A, these structural differences underline the general principle that endo- and exo-glycosidases can be supported by a same catalytic mechanism, whose endo/exo mode of action can be modulated through conformational modifications in loop regions.

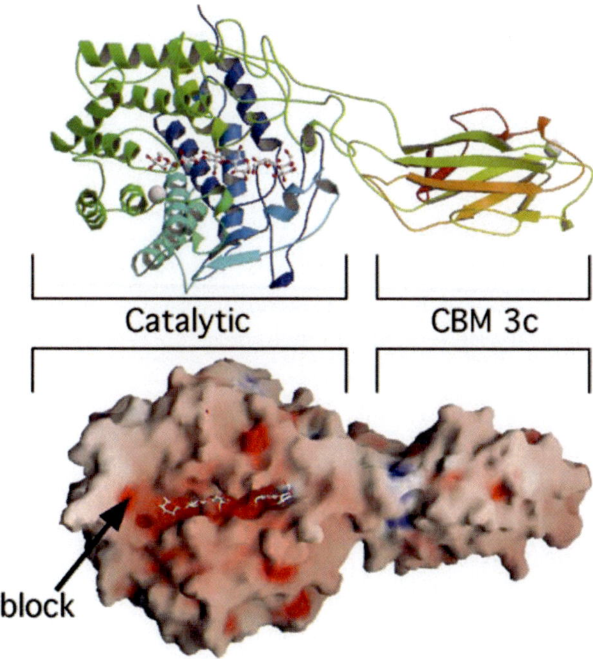

Figure 6. Ribbon and surface representation of family 9 endo/exocellulase E4 from *Thermomonospora fusca* (code 4tf4). Color coding as in Figs 1 and 5. The block preventing linear extension of substrate chain beyond site -4 is indicated (arrow). Four glucosyl residues occupying sites -1 to -4 (left) and two residues occupying sites +1 and +2 (right) are indicated in ball-and-stick mode.

In addition to the two *C. thermocellum* enzymes discussed above, a further type of GH-9 cellulase lacks the immunoglobulin-fold domain, but instead the $(\alpha/\alpha)_6$ barrel is tightly associated with a family 3c carbohydrate-binding module (CBM) linked to the C terminus of the catalytic domain. Examples found in cellulosomes are *C. thermocellum* Cel9B (formerly CelF), *C. cellulolyticum* Cel9G (formerly CelG), *C. josui* ORF2 [26], and *C. cellulovorans*

EngH [93]. Structural paradigms have been established for endoglucanase Cel9G from *C. cellulolyticum* [59] and endo/exocellulase E4 (now Cel9A) from *Thermomonospora fusca* [83] (Fig. 6). While most cellulases appear to be composed of catalytic and cellulose-binding domains that function independently, Cel9G and E4 are particular in that these two domains interact directly to form a flat platform onto which crystalline cellulose is suggested to bind. This particular arrangement is related to the unusual characteristics of these enzymes such as their high activity on bacterial microcrystalline cellulose or, in the case of E4, their double endo/exo-cellulase properties. The family 3c CBM consists of an antiparallel β-sandwich similar to the 3a CBM present in the scaffoldin (see below), but the aromatic platforms thought to interact with glucosyl residues are replaced by a strip of polar residues: this may account for the weaker affinity of the 3c CBM [27,83], which must allow sliding of the cellulose chain into the active site. Another noteworthy feature of the E4 structure is a block formed by residues from a protein loop (residues 245-255 at the non-reducing end of the substrate-binding site. The block leaves insufficient room for a hypothetical glucosyl residue to bind in a linear extension of the cellulose chain beyond site -4, a situation similar to that described for CbhA, where the block is located beyond site -2 (see above). This can explain at least in part the exoglucanase activity of E4 and the fact that it releases primarily cellotetraose [83].

Feruloyl-Esterases

Ferulic acid cross-links xylan chains in hemicellulose and attaches them to lignin. Feruloyl esterases (FAE) facilitate plant cell wall degradation by cleaving the ester bond formed between ferulic acid and the different carbohydrate moieties The presence of FAE domains has been recently recognized in two genes of *C. thermocellum*, Xyn10A and Xyn10B (formerly XynY), both of which also contain a family 10 xylanase domain [5]. The crystal structures of these two FAE domains [75,86] revealed a classical α/β-hydrolase fold with a central eight-stranded beta-sheet surrounded by alpha helices (Fig. 7). The overall structure and topology of these FAEs resemble those of other bacterial hydrolases, such as *M. tuberculosis* carbohydrate acetylase (PDB code 1DQY), the *S. exfoliatus* lipase (1JFR) and the *P. fluorescens* carboxyl esterase (1AUO).

The catalytic center of FAE contains the classical catalytic triad Ser-His-Asp first observed in serine proteases, suggesting that the hydrolytic reaction also takes place through a covalent acyl-enzyme intermediate. In various crystal structures of both the wild-type and inactive mutants of FAE with different feruloyl-sugar substrates, only the feruloyl moiety of the substrate was visible in the electron density [75,86]. Ferulate binds in a small surface depression, which consists mostly of hydrophobic residues and also provides recognition elements for both the methoxy and hydroxy ring substituents. Instead, the carboxylate group of ferulate (one of whose oxygen atoms is linked to the carbohydrate moiety in the natural substrate) points to the bulk solvent in the crystal structures (Fig. 7). This is in agreement with the requirements for substrate specificity in these enzymes, which must cleave the ester bond between ferulate and a number of distinct carbohydrate moieties. It is possible, as suggested by Prates et al [75], that FAEs have evolved to recognize specifically the feruloyl moiety while displaying a more permissive environment to accommodate the carbohydrate part of the substrate, thus making them capable of hydrolyzing the large ensemble of feruloyl xylans found in the plant cell wall.

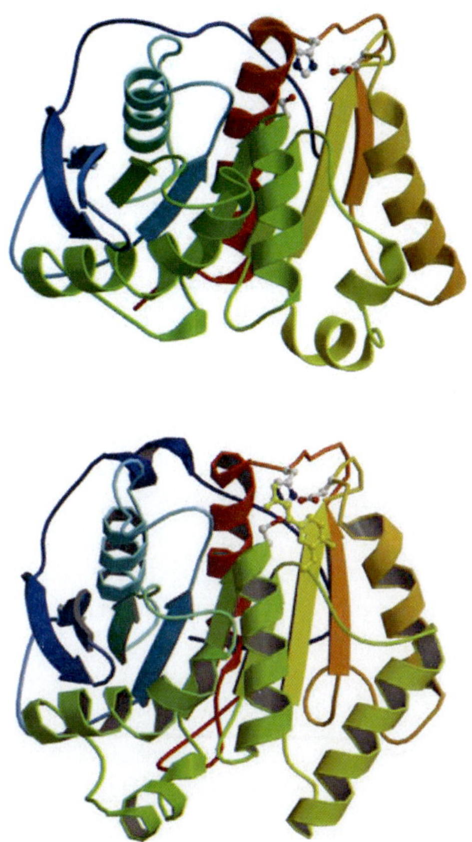

Figure 7. Backbone structures of the feruloyl esterase domains from xylanase 10A (code 1jff) (top) and xylanase 10B (code 1gkk) (bottom). Color code as in Fig. 1. The catalytic triad (Ser-His-Asp) and the ferulate ligand (yellow) are shown in ball-and-stick representation.

Carbohydrate-Binding Modules

The existence of independent carbohydrate-binding modules (CBMs) was first demonstrated for cellulases from *Trichoderma reesei* [98] and *Cellulomonas fimi* [48]. In cellulosome-producing organisms, scaffoldins usually comprise a CBM, which is thought to be responsible for anchoring the multienzyme complex to its substrate (although in the case of the ruminococcal cellulosome, no module has yet been positively identified as a CBM, so-called X domains of currently unknown function could turn out to fulfill this role).

A large body of evidence indicates that cellulose-binding modules enhance the degradation of crystalline cellulose significantly. Truncated enzymes deprived of their CBMs usually retain their activity on soluble or amorphous substrates, but loose part or all of their activity on crystalline cellulose [96] [30]. Conversely, providing a catalytic domain with a CBM can enhance its activity against crystalline cellulose [58]. The stimulation afforded by cellulose-binding domains may be ascribed to two phenomena. First, the presence of an independent binding module with a flexible linker should allow cellulases to perform several catalytic cycles while remaining attached to cellulose. Thus, the enzymes should remain in closer contact with their substrate, and limiting diffusion may help to focus enzyme action on spots whose structure has already been disrupted by previous attacks. Secondly, cellulose-binding modules seem to have a destabilizing effect on the macroscopic structure of cellulose. Exfoliation of the surface layers of cellulose fibers and release of small particles composed of bundles of microfibrils has been described with the isolated CBMs derived from *C. fimi* CenA [14,15] and *C. thermocellum* CbhA [42]. Thus, CBMs are expected to increase the surface area available for degradative enzymes. It must be stressed that such a disruption only concerns the macromolecular structure of the fibers: microfibrils whose association is relatively loose may be partly or completely dissociated from the macroscopic fiber by rupturing relatively few non-covalent bonds, but disrupting the actual structure of a cellulose crystallite would require far too much energy were it not for the concomitant hydrolysis of the cellulose chains.

Boraston *et al.* [6] recently reviewed the various structural types of CBMs and the features by which their affinity and specificity are modulated. Like catalytic domains, CBMs can be classified into different families based on sequence and structure similarities. Thirty-nine families have now been recognized { [6]; http://afmb.cnrs-mrs:fr/CAZY}, which can be further grouped into seven superfamilies sharing similar folds. In terms of function, three types of CBMs can be distinguished. Type A CBMs bind specifically to the flat surface of crystalline cellulose by means of aromatic residues lying in the same plane, while type B CBMs bind a single polysaccharide chain, which comes to lie within a groove. Such is the case of the CBMs binding to hemicellulose polysaccharides; in these CBMs, the geometry of the binding residues within the groove is adapted to the stereochemical constraints of the substrate. In contrast to type A, type B CBMs mostly display the same carbohydrate specificity as the catalytic modules to which they are linked. A third class, termed type C, binds soluble oligosaccharides. The CBMs found in cellulosome components belong to type A and type B.

Type A Carbohydrate-Binding Modules

Probably the most important module of the cellulosome is the type A cellulose-binding module borne by the scaffoldin (Fig. 8A), which is responsible for tethering the various glycoside hydrolases to crystalline cellulose. Sequence data are currently available for scaffoldins from several clostridia-type cellulosomes (for reviews see [2] and Bayer, this volume), and all of them comprise a type 3a CBM, which is located at the N terminus or in the middle of the sequence. Three-dimensional structures have been determined for the CBMs belonging to the scaffoldins of *C. thermocellum* and *C. cellulolyticum*, respectively [88,97]. Both modules are quite similar, consisting of a 9-stranded β-sandwich with a jelly roll topology (Fig. 8A). The structure comprises a Ca^{2+} ion, which does not participate in binding directly and most likely has a structural role. One face of the sandwich displays a shallow groove lined with conserved residues, whose role is still unclear, although it could possibly bind some as yet unidentified polysaccharide ligand. The other face is flat and displays a linear aromatic strip comprising a Trp, a Tyr and a His residue, and a salt bridge. Modelling shows that the residues and the salt bridge are in register to participate in stacking interactions with four glucosyl residues of one of the cellulose chains forming the edge (hydrophobic 110 face) of cellulose crystals. In addition, a set of polar residues would be adequately poised to establish hydrogen bonds with the second and third chain on one side of the aromatic strip. The model is supported by elegant electron microscopy studies showing that the CBM of *C. thermocellum* CipA does indeed bind to the 110 face of *Valonia* cellulose crystals [49].

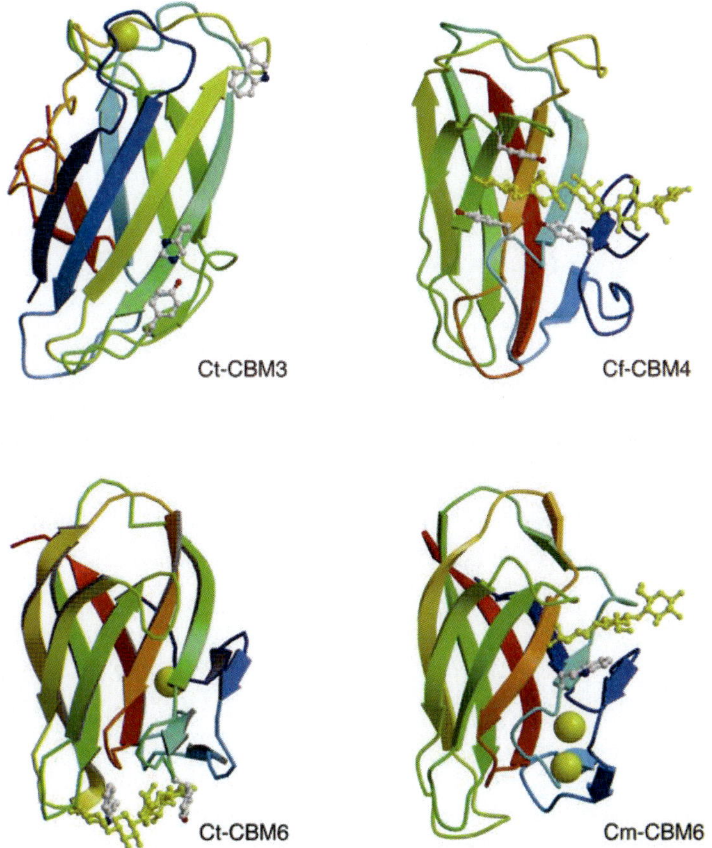

Figure 8. Backbone structures of cellulose-binding modules. Aromatic residues proposed or observed to bind cellulose are shown in ball-and-stick representation, and the bound carbohydrates are shown in yellow. Ct-CBM3, family 3 CBM of the scaffolding subunit from the *C. thermocellum* cellulosome (code 1nbc); Cf-CBM4, family 4 CBM of cellulase 9B from *Cellulomonas fimi* in complex with cellopentaose (code 1gu3); Ct-CBM6, family 6 CBM of xylanase 10A from *C. thermocellum* with xylopentaose bound to cleft A (code 1uxx), and Cm-CBM6, family 6 CBM of endoglucanase 5A from *Cellvibrio mixtus* with cellotriose bound to cleft B (code 1uyy).

The CBM component of the scaffoldins contributes quite significantly to the activity of the cellulosome on crystalline cellulose. CBMs derived from of *C. thermocellum* CipA and *C. cellulolyticum* CipC were incorporated into minicellulosomes in order to link them non-covalently to various cellulolytic components [8,22,23,41,69] see also Fierobe *et al.*, this volume. The presence of the scaffoldin CBM was shown to enhance cellulose degradation up to threefold as compared to homologous constructs devoid of CBM. Indeed, in a comparison between CBMs of different origins, the CBM of *C. thermocellum* CipA proved to be the most efficient at extending the percentage of the substrate accessible to *C. thermocellum* endoglucanase Cel9A [8].

In addition, numerous cellulosomal enzymes possess their own CBM. We discussed above the case of CBMs belonging to family 3c, which are actually part of the active site of some family 9 cellulases. True CBMs of type A, belonging to family 3a, can be found in *C. thermocellum* Cel5F (formerly CelO) [105], in a family 48 putative cellulosomes which is devoid of dockerin, and in several endoglucanases of family 9 (http://genome.jgi-psf.org/draft_microbes/cloth/cloth.home.html), particularly the cellobiohydrolase CbhA. In the latter enzyme, the CBM is flanked by a tandem of two fibronectin type 3 homology domains, which contribute to its destructuring activity [42].

Type B Carbohydrate-Binding Modules

Type B CBMs are also common, particularly among hemicellulases. Although members of families 4, 6, 11 and 22 have been grouped separately, they all belong to same vast superfamily as the type 3 CBMs discussed above, and share a β-sandwich framework with a jelly roll topology. The most commonly found type B CBMs belong to families 4 and 6 and are discussed below.

Family 4 CBMs are found associated mostly with family 9 glycoside hydrolases. The paradigm structure has been elucidated for the duplicated CBM present in *Cellulomonas fimi* Cel9B (formerly CenC) [7]. It consists of an 11-stranded β-sandwich with a jelly roll topology (Fig. 8b). One face of the sandwich displays a substrate-binding cleft, which accommodates cellooligosaccharides and soluble β-1,4-linked polymers of glucose such as hydroxyethylcellulose and β-1,3-1,4-glucans. In contrast to CBMs of type A, whose affinity is driven by entropy, binding of cellooligosaccharides CBMs of family 4 is driven by enthalpy, which is consistent with the formation of hydrogen bonds between the equatorial hydroxyl groups of the substrate and polar residues lining the binding cleft.

Family 6 CBMs are found associated mostly with xylanases and arabinofuranosidases. They are topologically (and probably phylogenically) related to CBMs of families 4, 11 and 22 [9,11], consisting likewise of a jelly roll β-sandwich. An interesting feature is that they comprise two carbohydrate-binding sites. The first site, termed cleft A, is formed by loops connecting the two β-sheets of the sandwich (Fig. 8C), while the second site, termed cleft B, lies on the concave side of the jelly roll (Fig. 8D). Cleft B corresponds to the binding cleft identified in CBMs of families 4, 11, and 22. In the case of the CBM6 belonging to *C thermocellum* Xyn11A (formerly XynU), cleft B is blocked by a proline residue belonging to a neighbouring loop, so that only cleft A is available for binding carbohydrates [11].

However, in the related CBM6 from *Cellvibrio mixtus* Cel5A, both sites have been shown to be accessible and to display distinct binding specificities [36,73]. For example, while cellooligosaccharides are able to bind to both clefts, xylooligosaccharides bind only to cleft A and β-1,3-β-1,4 mixed linked glucans to cleft B. Interestingly, both sites participate synergistically in binding to insoluble cellulose.

Conclusion

When the cellulosome concept was coined some twenty years ago, very little was known about the biochemistry of this mysterious and rather intractable complex: only two endoglucanases had been successfully purified and characterized by classical means from the culture medium of *C. thermocellum* [66,72]. The advent of recombinant DNA technology, together with convenient screening tests, rapidly made it possible to clone many genes encoding endoglucanases, as well as some hemicellulases. Although it was clear from the onset that the screening procedures were highly biased, there was some question for some time whether cellulosomes actually needed anything but endoglucanases to degrade cellulose. An attractive model was proposed suggesting that cellulosomes might dispense with cellobiohydrolases, which were known to be essential in fungal cellulase systems, by lining up endoglucanases so that they could perform simultaneous cutting events on cellulose chain [60]. It is now obvious that in fact, cellulosomes comprise the same types of glycoside hydrolases as the classical, unassociated cellulase systems: in addition to endoglucanases, they contain the two types of cellobiohydrolases hydrolyzing cellulose from the reducing or the non-reducing end of the chains, and the synergisms observed in the unassociated systems are now being reproduced with cellulosome components, both in the free form and upon being incorporated into (mini) cellulosomes ([22], see also Fierobe, this volume). In terms of structure and mechanism, the catalytic components of cellulosomes also differ very little from those found in unassociated systems: they belong to the same glycoside hydrolase families, and their activities and specificities are modulated by similar structural features. Nevertheless, cellulosome subunits raise several specific issues:

1. How did cellulosomes arise in evolution ? Most likely, catalytic modules predate the appearance of complexes based on cohesin-dockerin interactions, which have been "invented" twice independently and gave rise to the complexes typical of clostridia and anaerobic fungi. Was there a specific mechanism by which dockerins were fitted on the glycoside hydrolases ? Of course, gene duplications would account for the diversification of dockerin-bearing enzymes belonging to the same family. However, the variety of phylogenetically unrelated polypeptides harbouring these domains means that the fusion event probably occurred several times.
2. What is the significance of the very large number of *C. thermocellum* polypeptides bearing dockerins and thus potentially part of the cellulosome ? How many of them are redundant ? How many are really needed ? Other, mesophilic clostridia such as *C. cellulolyticum* or *C. cellulovorans* seem to make do with far fewer components (though their number will probably increase once the respective genomes are

sequenced). In this respect, *C. thermocellum* also differs from the mesophilic clostridia in that the latter contain large clusters of cellulase genes located downstream from the scaffoldin gene [40,70,93]. In contrast, the glycoside hydrolase genes of *C. thermocellum* are scattered throughout the genome, with rare clusters not exceeding four or five genes. This situation could perhaps be due to extensive duplications and rearrangements within the genome of *C. thermocellum*, which would erase any trace of the original organization of cellulosome genes. Indeed, the presence of over 100 transposase genes (http://genome.jgi-psf.org/draft_microbes/cloth/cloth.home.html) suggests that the *C. thermocellum* genome could be highly prone to such rearrangements.

3. Why do many of the cellulosome subunits possess their own CBMs ? It could be argued that these originated in organisms producing unassociated cellulases fitted with CBMs and that the genes were acquired as such by horizontal transfer. Whatever their origin, however, the CBMs appear to be functional, which argues that they are needed for full activity of the complex and that their function cannot be taken over completely by the CBM borne by the scaffoldin. Indeed, many of the CBMs are type B CBMs linked to hemicellulases and, unlike the type A CBM borne by the scaffoldin, their binding specificity matches that of the catalytic domain. Bayer *et al.* [2] suggested that upon binding to the substrate, the cellulosome may undergo a supramolecular rearrangement so that the subunits are redistributed to their target substrates. Thus, CBMs with different binding specificities would serve to direct the parent enzymes to the appropriate sites. It is also possible that having a CBM in the close vicinity of the catalytic site further enhances the synergism with the catalytic domain, as opposed to relying only on the somewhat distant CBM borne by the scaffoldin. The same proximity effect may also account for the fact that several polypeptides combine a xylanase and an esterase domain, which should act in concert for the degradation of substituted xylan.

REFERENCES

Alzari, P. M., H. Souchon, and R. Dominguez. 1996. The crystal structure of endoglucanase CelA, a family 8 glycosyl hydrolase from Clostridium thermocellum. *Structure 4*:265-275.

Bayer, E. A., J. P. Bélaïch, Y. Shoham, and R. Lamed. 2004. The cellulosomes: Multienzyme machines for the degradation of plant cell wall polysaccharides. *Annu. Rev. Microbiol. 58*:521-554.

Bayer, E. A., E. Morag, and R. Lamed. 1994. The cellulosome - a treasure-trove for biotechnology. *Trends in Biotechnol. 12*:379-386.

Birsan, C., P. Johnson, M. Joshi, A. MacLeod, L. McIntosh, V. Monem, M. Nitz, D. R. Rose, D. Tull, W. W. Wakarchuck, Q. Wang, R. A. Warren, A. White, and S. G. Withers. 1998. Mechanisms of cellulases and xylanases. *Biochem. Soc. Trans. 26*:156-160.

Blum, D. L., I. A. Kataeva, X. L. Li, and L. G. Ljungdahl. 2000. Feruloyl esterase activity of the Clostridium thermocellum cellulosome can be attributed to previously unknown domains of XynY and XynZ. *J. Bacteriol. 182*:1346-51.

Boraston, A. B., D. N. Bolam, H. J. Gilbert, and G. J. Davies. 2004. Carbohydrate-binding modules: fine tuning polysaccharide recognition. *Biochem. J. 382*:769-781.

Brun, E., P. E. Johnson, A. L. Creagh, P. Tomme, P. Webster, C. A. Haynes, and L. P. McIntosh. 2000. Structure and binding specificity of the second N-terminal cellulose-binding domain from Cellulomonas fimi endoglucanase C. *Biochemistry 39*:2445-2458.

Carrard, G., A. Koivula, H. Söderlund, and P. Béguin. 2000. Cellulose-binding domains promote hydrolysis of different sites on crystalline cellulose. *Proc. Natl. Acad. Sci USA 97*:10342-10347.

Carvalho, A. L., A. Goyal, J. A. Prates, D. N. Bolam, H. J. Gilbert, V. M. Pires, L. M. Ferreira, A. Planas, M. J. Romao, and C. M. Fontes. 2004. The family 11 carbohydrate-binding module of Clostridium thermocellum Lic26A-Cel5E accommodates beta -1,4 and beta -1,3-1,4-mixed linked glucans at a single binding site. *J. Biol. Chem.279*:34785-34793.

Chauvaux, S., P. Béguin, and J.-P. Aubert. 1992. Site-directed mutagenesis of essential carboxylic residues in Clostridium thermocellum endoglucanase CelD. *J. Biol. Chem. 267*:4472-4478.

Czjzek, M., D. N. Bolam, A. Mosbah, J. Allouch, C. M. Fontes, L. M. Ferreira, O. Bornet, V. Zamboni, H. Darbon, N. L. Smith, G. W. Black, B. Henrissat, and H. J. Gilbert. 2001. The location of the ligand-binding site of carbohydrate-binding modules that have evolved from a common sequence is not conserved. *J. Biol. Chem. 276*:48580-48587.

Davies, G., and B. Henrissat. 1995. Structures and mechanisms of glycosyl hydrolases. *Structure 3*:853-858.

Davies, G. J. 1998. Structural studies on cellulases. *Biochem. Soc. Trans. 26*:167-173.

Din, N., H. G. Damude, N. R. Gilkes, R. C. Miller, R. A. J. Warren, and D. G. Kilburn. 1994. C1-Cx revisited: Intramolecular synergism in a cellulase. *Proc. Natl. Acad. Sci. USA 91*:11383-11387.

Din, N., N. R. Gilkes, B. Tekant, R. C. Miller, Jr., R. A. J. Warren, and D. G. Kilburn. 1991. Non-hydrolytic disruption of cellulose fibres by the binding domain of a bacterial cellulase. *Bio/Technology 9*:1096-1099.

Ding, S. Y., M. T. Rincon, R. Lamed, J. C. Martin, S. I. McCrae, V. Aurilia, Y. Shoham, E. A. Bayer, and H. J. Flint. 2001. Cellulosomal scaffoldin-like proteins from Ruminococcus flavefaciens. *J. Bacteriol. 183*:1945-1953.

Divne, C., J. Stahlberg, T. T. Teeri, and T. A. Jones. 1998. High-resolution crystal structures reveal how a cellulose chain is bound in the 50 Ångström-long tunnel of cellobiohydrolase I from Trichoderma reesei. *J. Mol. Biol. 275*:309-325.

Dominguez, R., H. Souchon, M. B. Lascombe, and P. M. Alzari. 1996. The crystal structure of a family 5 endoglucanase mutant in complexed and uncomplexed form reveals an induced fit activation mechanism. *J. Mol. Biol. 257*:1042-1051.

Dominguez, R., H. Souchon, S. Spinelli, Z. Dauter, K. S. Wilson, S. Chauvaux, P. Béguin, and P. M. Alzari. 1995. The structures of endoglucanase CelC and xylanase XynZ from

Clostridium thermocellum reveal a common protein fold and similar active site in two distinct families of b-glycanases. *Nature Struct. Biol. 2*:569-576.

Egelseer, E. M., K. Leitner, M. Jarosch, C. Hotzy, S. Zayni, U. B. Sleytr, and M. Sára. 1998. The S-layer proteins of two Bacillus stearothermophilus wild type strains are bound via their N-terminal region to a secondary cell wall polymer of identical chemical composition. *J. Bacteriol. 180*:1488-1495.

Fanutti, C., T. Ponyi, G. W. Black, G. P. Hazlewood, and H. J. Gilbert. 1995. The conserved noncatalytic 40-residue sequence in cellulases and hemicellulases from anaerobic fungi functions as a protein docking domain. *J. Biol. Chem. 270*:29314-29322.

Fierobe, H. P., E. A. Bayer, C. Tardif, M. Czjzek, A. Mechaly, A. Bélaïch, R. Lamed, Y. Shoham, and J. P. Bélaïch. 2002. Degradation of cellulose substrates by cellulosome chimeras. Substrate targeting versus proximity of enzyme components. *J. Biol. Chem. 277*:49621-49630.

Fierobe, H. P., A. Mechaly, C. Tardif, A. Bélaïch, R. Lamed, Y. Shoham, J. P. Bélaïch, and E. A. Bayer. 2001. Design and production of active cellulosome chimeras. Selective incorporation of dockerin-containing enzymes into defined functional complexes. *J. Biol. Chem. 276*:21257-21261.

Fujino, T., P. Béguin, and J.-P. Aubert. 1992. Cloning of a Clostridium thermocellum DNA fragment encoding polypeptides that bind the catalytic components of the cellulosome. *FEMS Microbiol. Lett. 94*:165-170.

Fujino, T., P. Béguin, and J.-P. Aubert. 1993. Organization of a Clostridium thermocellum gene cluster encoding the cellulosomal scaffolding protein CipA and a protein possibly involved in the attachment of the cellulosome to the cell surface. *J. Bacteriol. 175*:1891-1899.

Fujino, T., S. Karita, and K. Ohmiya. 1993. Nucleotide sequences of the celB gene encoding endo-1,4-b-glucanase-2, ORF1 and ORF2 forming a putative cellulase gene cluster of Clostridium josui. *J. Ferm. Bioeng. 76*:243-250.

Gal, L., C. Gaudin, A. Bélaïch, S. Pagès, C. Tardif, and J.-P. Bélaïch. 1997. CelG from Clostridium cellulolyticum - A multidomain endoglucanase acting efficiently on crystalline cellulose. *J. Bacteriol. 179*:6595-6601.

Gebler, J., N. R. Gilkes, M. Claeyssens, D. B. Wilson, P. Béguin, W. W. Wakarchuk, D. G. Kilburn, R. C. Miller, R. A. J. Warren, and S. G. Withers. 1992. Stereoselective hydrolysis catalyzed by related b-1,4-glucanases and b-1,4-xylanases. *J. Biol. Chem. 267*:12559-12561.

Gerngross, U. T., M. P. M. Romaniec, T. Kobayashi, N. S. Huskisson, and A. L. Demain. 1993. Sequencing of a Clostridium thermocellum gene (cipA) encoding the cellulosomal SL-protein reveals an unusual degree of internal homology. *Mol. Microbiol. 8*:325-334.

Gilkes, N. R., R. A. J. Warren, R. C. Miller, Jr., and D. G. Kilburn. 1988. Precise excision of the cellulose binding domains from two Cellulomonas fimi cellulases by a homologous protease and the effect on catalysis. *J. Biol. Chem. 263*:10401-10407.

Guerin, D. M., M. B. Lascombe, M. Costabel, H. Souchon, V. Lamzin, P. Béguin, and P. M. Alzari. 2002. Atomic (0.94 A) resolution structure of an inverting glycosidase in complex with substrate. *J. Mol. Biol. 316*:1061-1069.

Guglielmi, G., and P. Béguin. 1998. Cellulase and hemicellulase genes of Clostridium thermocellum from five independent collections contain few overlaps and are widely scattered across the chromosome. *FEMS Microbiol. Lett. 161*:209-215.

Guimaraes, B. G., H. Souchon, B. L. Lytle, J. H. David Wu, and P. M. Alzari. 2002. The crystal structure and catalytic mechanism of cellobiohydrolase CelS, the major enzymatic component of the Clostridium thermocellum cellulosome. *J. Mol. Biol. 320*:587-596.

Henrissat, B., I. Callebaut, S. Fabrega, P. Lehn, J. P. Mornon, and G. Davies. 1995. Conserved catalytic machinery and the prediction of a common fold for several families of glycosyl hydrolases. *Proc. Natl. Acad. Sci. USA 92*:7090-7094.

Henrissat, B., M. Claeyssens, P. Tomme, L. Lemesle, and J.-P. Mornon. 1989. Cellulase families revealed by hydrophobic cluster analysis. *Gene 81*:83-95.

Henshaw, J. L., D. N. Bolam, V. M. Pires, M. Czjzek, B. Henrissat, L. M. Ferreira, C. M. Fontes, and H. J. Gilbert. 2004. The family 6 carbohydrate binding module CmCBM6-2 contains two ligand-binding sites with distinct specificities. *J. Biol. Chem. 279*:21552-21559.

Jenkins, J., L. Lo Leggio, G. Harris, and R. Pickersgill. 1995. β-Glucosidase, β-galactosidase, family A cellulases, family F xylanases and two barley glycanases form a superfamily of enzymes with 8-fold β/α architecture and with two conserved glutamate near the carboxy-terminal ends of b-strands four and seven. *FEBS-Lett. 362*:281-285.

Jindou, S., A. Soda, S. Karita, T. Kajino, P. Béguin, J. H. Wu, M. Inagaki, T. Kimura, K. Sakka, and K. Ohmiya. 2004. Cohesin-dockerin interactions within and between Clostridium josui and Clostridium thermocellum: binding selectivity between cognate dockerin and cohesin domains and species specificity. *J. Biol. Chem. 279*:9867-9874.

Juy, M., A. G. Amit, P. M. Alzari, R. J. Poljak, M. Claeyssens, P. Béguin, and J.-P. Aubert. 1992. Crystal structure of a thermostable bacterial cellulose-degrading enzyme. *Nature 357*:89-91.

Kakiuchi, M., A. Isui, K. Suzuki, T. Fujino, E. Fujino, T. Kimura, S. Karita, K. Sakka, and K. Ohmiya. 1998. Cloning and DNA sequencing of the genes encoding Clostridium josui scaffolding protein CipA and cellulase CelD and identification of their gene products as major components of the cellulosome. *J. Bacteriol. 180*:4303-4308.

Kataeva, I., G. Guglielmi, and P. Béguin. 1997. Interaction between Clostridium thermocellum endoglucanase CelD and polypeptides derived from the cellulosome-integrating protein CipA : stoichiometry and cellulolytic activity of the complexes. *Biochem. J. 326*:617-624.

Kataeva, I. A., R. D. Seidel, 3rd, A. Shah, L. T. West, X. L. Li, and L. G. Ljungdahl. 2002. The fibronectin type 3-like repeat from the Clostridium thermocellum cellobiohydrolase CbhA promotes hydrolysis of cellulose by modifying its surface. *Appl. Environ. Microbiol. 68*:4292-4300.

Koshland, D. E. J. 1953. Stereochemistry and the mechanism of enzymatic reactions. *Biol. Rev. 28*:416-438.

Kruus, K., A. Andreacchi, W. K. Wang, and J. H. D. Wu. 1995. Product inhibition of the recombinant CelS, an exoglucanase component of the Clostridium thermocellum cellulosome. *Appl. Microbiol. Biotechnol. 44*:399-404.

Kruus, K., W. K. Wang, and J. H. D. Wu. 1995. Exoglucanase activities of the recombinant Clostridium thermocellum CelS, a major cellulosome component. *J. Bacteriol. 177*:1641-1644.

Lamed, R., E. Setter, and E. A. Bayer. 1983. Characterization of a cellulose-binding, cellulase-containing complex in Clostridium thermocellum. *J. Bacteriol. 156*:828-836.

Lamed, R., E. Setter, R. Kenig, and E. A. Bayer. 1983. The cellulosome: a discrete cell surface organelle of Clostridium thermocellum which exhibits separate antigenic, cellulose-binding and various cellulolytic activities. *Biotechnol. Bioeng. Symp. 13*:163-181.

Langsford, M. L., N. R. Gilkes, B. Singh, B. Moser, R. C. Miller, Jr., R. A. J. Warren, and D. G. Kilburn. 1987. Glycosylation of bacterial cellulases prevents proteolytic cleavage between functional domains. *FEBS Lett. 225*:163-167.

Lehtio, J., J. Sugiyama, M. Gustavsson, L. Fransson, M. Linder, and T. T. Teeri. 2003. The binding specificity and affinity determinants of family 1 and family 3 cellulose binding modules. *Proc. Natl. Acad. Sci. USA 100*:484-489.

Leibovitz, E., and P. Béguin. 1996. A new type of cohesin domain that specifically binds the dockerin domain of the Clostridium thermocellum cellulosome-integrating protein *CipA. J. Bacteriol. 178*:3077-3084.

Leibovitz, E., H. Ohayon, P. Gounon, and P. Béguin. 1997. Characterization and subcellular localization of the Clostridium thermocellum scaffoldin dockerin binding protein SdbA. *J. Bacteriol. 179*:2519-2523.

Lemaire, M., I. Miras, P. Gounon, and P. Béguin. 1998. Identification of a region responsible for binding to the cell wall within the S-layer protein of Clostridium thermocellum. *Microbiol. 144*:211-217.

Lemaire, M., H. Ohayon, P. Gounon, T. Fujino, and P. Béguin. 1995. OlpB, a new outer layer protein of Clostridium thermocellum, and binding of its S-layer-like domain to components of the cell envelope. *J. Bacteriol. 177*:2451-2459.

Liu, C. C., and R. H. Doi. 1998. Properties of exgS, a gene for a major subunit of the Clostridium cellulovorans cellulosome. *Gene 211*:39-47.

Lupas, A., H. Engelhardt, J. Peters, U. Santarius, S. Volker, and W. Baumeister. 1994. Domain structure of the Acetogenium kivui surface layer revealed by electron crystallography and sequence analysis. *J. Bacteriol. 176*:1224-1233.

Lytle, B., C. Myers, K. Kruus, and J. H. D. Wu. 1996. Interactions of the CelS binding ligand with various receptor domains of the Clostridium thermocellum cellulosomal scaffolding protein, CipA. *J. Bacteriol. 178*:1200-1203.

Mader, C., C. Huber, D. Moll, U. B. Sleytr, and M. Sara. 2004. Interaction of the crystalline bacterial cell surface layer protein SbsB and the secondary cell wall polymer of Geobacillus stearothermophilus PV72 assessed by real-time surface plasmon resonance biosensor technology. *J. Bacteriol. 186*:1758-1768.

Maglione, G., O. Matsushita, J. B. Russel, and D. B. Wilson. 1992. Properties of a genetically reconstructed Prevotella ruminicola endoglucanase. *Appl. Environ. Microbiol. 58*:3593-3597.

Mandelman, D., A. Bélaïch, J. P. Bélaïch, N. Aghajari, H. Driguez, and R. Haser. 2003. X-Ray crystal structure of the multidomain endoglucanase Cel9G from Clostridium cellulolyticum complexed with natural and synthetic cello-oligosaccharides. *J. Bacteriol. 185*:4127-4135.

Mayer, F., M. P. Coughlan, Y. Mori, and L. G. Ljungdahl. 1987. Macromolecular organization of the cellulolytic enzyme complex of Clostridium thermocellum as revealed by electron microscopy. *Appl. Environ. Microbiol. 53*:2785-2792.

Mesnage, S., T. Fontaine, T. Mignot, M. Delepierre, M. Mock, and A. Fouet. 2000. Bacterial SLH domain proteins are non-covalently anchored to the cell surface via a conserved mechanism involving wall polysaccharide pyruvylation. *EMBO J. 19*:4473-84.

Mesnage, S., E. Tosi-Couture, and A. Fouet. 1999. Production and cell surface anchoring of functional fusions between the SLH motifs of the Bacillus anthracis S-layer proteins and the Bacillus subtilis levansucrase. *Mol. Microbiol. 31*:927-936.

Millet, J., D. Pétré, P. Béguin, O. Raynaud, and J.-P. Aubert. 1985. Cloning of ten distinct DNA fragments of Clostridium thermocellum coding for cellulases. *FEMS Microbiol. Lett. 29*:145-149.

Morag, E., E. A. Bayer, G. P. Hazlewood, H. J. Gilbert, and R. Lamed. 1993. Cellulase SS (CelS) is synonymous with the major cellobiohydrolase (subunit S8) from the cellulosome of Clostridium thermocellum. *Appl. Biochem. Biotechnol. 43*:147-150.

Navas, J., and P. Béguin. 1992. Site-directed mutagenesis of conserved residues in Clostridium thermocellum endoglucanase CelC. *Biochem. Biophys. Res. Commun. 189*:807-812.

Ng, T. K., and J. G. Zeikus. 1981. Purification and characterization of an endoglucanase (1,4-b-D-glucan glucanohydrolase) from Clostridium thermocellum. *Biochem. J. 199*:341-350.

Olabarría, G., J. L. Carrascosa, M. A. de Pedro, and J. Berenguer. 1996. A conserved motif in S-layer proteins is involved in peptidoglycan binding in Thermus thermophilus. *J. Bacteriol. 178*:4765-4772.

Pagès, S., A. Bélaïch, C. Tardif, C. Reverbel-Leroy, C. Gaudin, and J.-P. Bélaïch. 1996. Interaction betwen the endoglucanase CelA and the scaffolding protein CipC of the Clostridium cellulolyticum cellulosome. *J. Bacteriol. 178*:2279-2286.

Pagès, S., L. Gal, A. Bélaïch, C. Gaudin, C. Tardif, and J.-P. Bélaïch. 1997. Role of scaffolding protein CipC of Clostridium cellulolyticum in cellulose degradation. *J. Bacteriol. 179*:2810-2816.

Pagès, S., O. Valette, L. Abdou, A. Bélaïch, and J. P. Bélaïch. 2003. A rhamnogalacturonan lyase in the Clostridium cellulolyticum cellulosome. *J. Bacteriol. 185*:4727-4733.

Parsiegla, G., M. Juy, C. Reverbel-Leroy, C. Tardif, J.-P. Bélaïch, H. Driguez, and R. Haser. 1998. The crystal structure of the processive endocellulase CelF of Clostridium cellulolyticum in complex with a thiooligosaccharide inhibitor at 2.0 Å resolution. *EMBO J. 17*:5551-5562.

Pètre, J., R. Longin, and J. Millet. 1981. Purification and properties of an endo-β-1,4-glucanase from Clostridium thermocellum. *Biochimie 63*:629-639.

Pires, V. M., J. L. Henshaw, J. A. Prates, D. N. Bolam, L. M. Ferreira, C. M. Fontes, B. Henrissat, A. Planas, H. J. Gilbert, and M. Czjzek. 2004. The crystal structure of the family 6 carbohydrate binding module from Cellvibrio mixtus endoglucanase 5a in complex with oligosaccharides reveals two distinct binding sites with different ligand specificities. *J. Biol. Chem. 279*:21560-21568.

Piruzian, E. S., M. A. Mogutov, G. A. Velikodvorskaya, and V. K. Akimenko. 1985. Cloning and expression of structural genes of Clostridium thermocellum F7 cellulolytic complex endoglucanases in cells of Escherichia coli K12. *Doklady Acad. Nauk SSR 281*:963-965.

Prates, J. A., N. Tarbouriech, S. J. Charnock, C. M. Fontes, L. M. Ferreira, and G. J. Davies. 2001. The structure of the feruloyl esterase module of xylanase 10B from Clostridium thermocellum provides insights into substrate recognition. *Structure) 9:*1183-1190.

Reese, E. T., R. G. H. Siu, and H. S. Levinson. 1950. The biological degradation of soluble cellulose derivatives and its relationship to the mechanism of cellulose hydrolysis. *J. Bacteriol. 59*:485-497.

Reverbel-Leroy, C., S. Pagès, A. Bélaïch, J.-P. Bélaïch, and C. Tardif. 1997. The processive endocellulase CelF, a major component of the Clostridium cellulolyticum cellulosome - Purification and characterization of the recombinant form. *J.. Bacteriol.. 179*:46-52.

Ries, W., C. Hotzy, I. Schocher, U. B. Sleytr, and M. Sára. 1997. Evidence that the N-terminal part of the S-layer protein from Bacillus stearothermophilus PV72/P2 recognizes a secondary cell wall polymer. *J. Bacteriol. 179*:3892-3898.

Rincon, M. T., S. Y. Ding, S. I. McCrae, J. C. Martin, V. Aurilia, R. Lamed, Y. Shoham, E. A. Bayer, and H. J. Flint. 2003. Novel organization and divergent dockerin specificities in the cellulosome system of Ruminococcus flavefaciens. *J. Bacteriol. 185*:703-713.

Rincon, M. T., J. C. Martin, V. Aurilia, S. I. McCrae, G. J. Rucklidge, M. D. Reid, E. A. Bayer, R. Lamed, and H. J. Flint. 2004. ScaC, an adaptor protein carrying a novel cohesin that expands the dockerin-binding repertoire of the Ruminococcus flavefaciens 17 cellulosome. *J. Bacteriol. 186*:2576-2585.

Romaniec, M. P. M., N. G. Clarke, and G. P. Hazlewood. 1987. Molecular cloning of Clostridium thermocellum DNA and the expression of further novel endo-b-1,4-glucanase genes in Escherichia coli. *J. Gen. Microbiol. 133*:1297-1307.

Sakka, K., S. Furuse, and K. Shimada. 1989. Cloning and expression in Escherichia coli of the thermophilic Clostridium sp. F1 genes related to cellulose hydrolysis. *Agric. Biol. Chem. 53*:905-911.

Sakon, J., D. Irwin, D. B. Wilson, and P. A. Karplus. 1997. Structure and mechanism of endo/exocellulase E4 from Thermomonospora fusca. *Nature Struct. Biol. 4*:810-818.

Salamitou, S., M. Lemaire, T. Fujino, H. Ohayon, P. Gounon, P. Béguin, and J.-P. Aubert. 1994. Subcellular localization of Clostridium thermocellum ORF3p, a protein carrying a receptor for the docking sequence borne by the catalytic components of the cellulosome. *J. Bacteriol. 176*:2828-2834.

Salamitou, S., O. Raynaud, M. Lemaire, M. Coughlan, P. Béguin, and J.-P. Aubert. 1994. Recognition specificity of the duplicated segments present in Clostridium thermocellum endoglucanase CelD and in the cellulosome-integrating protein CipA. *J. Bacteriol. 176*:2822-2827.

Schubot, F. D., I. A. Kataeva, D. L. Blum, A. K. Shah, L. G. Ljungdahl, J. P. Rose, and B. C. Wang. 2001. Structural basis for the substrate specificity of the feruloyl esterase domain of the cellulosomal xylanase Z from Clostridium thermocellum. *Biochemistry 40*:12524-112532.

Schubot, F. D., I. A. Kataeva, J. Chang, A. K. Shah, L. G. Ljungdahl, J. P. Rose, and B. C. Wang. 2004. Structural basis for the exocellulase activity of the cellobiohydrolase CbhA from Clostridium thermocellum. *Biochemistry 43*:1163-1170.

Shimon, L. J., S. Pages, A. Bélaïch, J. P. Bélaïch, E. A. Bayer, R. Lamed, Y. Shoham, and F. Frolow. 2000. Structure of a family IIIa scaffoldin CBD from the cellulosome of Clostridium cellulolyticum at 2.2 A resolution. *Acta Crystallogr. D Biol. Crystallogr. 56*:1560-1568.

Shoseyov, O., M. Takagi, M. A. Goldstein, and R. H. Doi. 1992. Primary sequence analysis of Clostridium cellulovorans cellulose-binding protein A. *Proc. Natl. Acad. Sci. USA 89*:3483-3487.

Steenbakkers, P. J., A. Freelove, B. Van Cranenbroek, B. M. Sweegers, H. R. Harhangi, G. D. Vogels, G. P. Hazlewood, H. J. Gilbert, and H. J. Op den Camp. 2002. The major component of the cellulosomes of anaerobic fungi from the genus Piromyces is a family 48 glycoside hydrolase. *DNA Seq. 13*:313-320.

Steenbakkers, P. J., X. L. Li, E. A. Ximenes, J. G. Arts, H. Chen, L. G. Ljungdahl, and H. J. Op Den Camp. 2001. Noncatalytic docking domains of cellulosomes of anaerobic fungi. *J. Bacteriol.183:*5325-5333.

Takagi, M., S. Hashida, M. A. Goldstein, and R. H. Doi. 1993. The hydrophobic repeated domain of the Clostridium cellulovorans cellulose-binding protein (CbpA) has specific interactions with endoglucanases. *J. Bacteriol. 175*:7119-7122.

Tamaru, Y., S. Karita, A. Ibrahim, H. Chan, and R. H. Doi. 2000. A large gene cluster for the Clostridium cellulovorans cellulosome. *J. Bacteriol. 182*:5906-5910.

Tokatlidis, K., P. Dhurjati, and P. Béguin. 1993. Properties conferred on Clostridium thermocellum endoglucanase CelC by grafting the duplicated segment of endoglucanase CelD. *Protein Eng 6*:947-952.

Tokatlidis, K., S. Salamitou, P. Béguin, P. Dhurjati, and J.-P. Aubert. 1991. Interaction of the duplicated segment carried by Clostridium thermocellum cellulases with cellulosome components. *FEBS Lett. 291*:185-188.

Tomme, P., H. van Tilbeurgh, G. Pettersson, J. van Damme, J. Vandekerckhove, J. Knowles, T. Teeri, and M. Claeyssens. 1988. Studies of the cellulolytic system of Trichoderma reesei QM 9414. *Eur. J. Biochem. 170*:575-581.

Tormo, J., A. J. Chirino, E. Morag, E. A. Bayer, Y. Shoham, and T. A. Steitz. 1996. Crystal structure of a bacterial family III cellulose-binding domain - A general mechanism for attachment to cellulose. *EMBO J. 15*:5739-5751.

van Tilbeurgh, H., P. Tomme, M. Claeyssens, R. Bhikhabhai, and G. Pettersson. 1986. Limited proteolysis of the cellobiohydrolase I from Trichoderma reesei. *FEBS Lett. 204*:223-227.

Wang, Q., D. Tull, A. Meinke, N. R. Gilkes, R. A. J. Warren, R. Aebersold, and S. G. Withers. 1993. Glu280 is the nucleophile in the active site of Clostridium thermocellum CelC, a family A endo-β-1,4-glucanase. *J. Biol. Chem. 268*:14096-14102.

Wang, W. K., K. Kruus, and J. H. D. Wu. 1994. Cloning and expression of the Clostridium thermocellum celS gene in Escherichia coli. *Appl. Microbiol. Biotechnol. 42:*346-352.

Wu, J. H. D., and A. L. Demain. 1988. Proteins of the Clostridium thermocellum cellulase complex responsible for degradation of crystalline cellulose. In *FEMS Symposium No. 43*

Biochemistry and Genetics of Cellulose Degradation, Aubert, J.-P., Béguin, P., and Millet, J., eds. Academic Press, London & New York, pp. 117-131

Wu, J. H. D., W. H. Orme-Johnson, and A. L. Demain. 1988. Two components of an extracellular protein aggregate of Clostridium thermocellum together degrade crystalline cellulose. *Biochemistry 27*:1703-1709.

Yaron, S., E. Morag, E. A. Bayer, R. Lamed, and Y. Shoham. 1995. Expression, purification and subunit-binding properties of cohesins 2 and 3 of the Clostridium thermocellum cellulosome. *FEBS Lett. 360*:121-124.

Zou, J. Y., G. J. Kleywegt, J. Stahlberg, H. Driguez, W. Nerinckx, M. Claeyssens, A. Koivula, T. T. Teeri, and T. A. Jones. 1999. Crystallographic evidence for substrate ring distortion and protein conformational changes during catalysis in cellobiohydrolase Cel6A from Trichoderma reesei. *Structure 7*:1035-1045.

Zverlov, V. V., G. A. Velikodvorskaya, and W. H. Schwarz. 2002. A newly described cellulosomal cellobiohydrolase, CelO, from Clostridium thermocellum: investigation of the exo-mode of hydrolysis, and binding capacity to crystalline cellulose. *Microbiology 148*:247-55.

In: Cellulosome
Eds: V. Uversky and I. A. Kataeva, pp. 95-117
ISBN 1-59454-950-8

Chapter V

CELLULOSE HYDROLYSIS BY *CLOSTRIDIUM THERMOCELLUM*: A MICROBIAL PERSPECTIVE

Lee R. Lynd[1,2], *Paul J. Weimer*[3,4], *Gideon Wolfaardt*[5] *and Yi-Heng Percival Zhang*[1,6]

[1]Thayer School of Engineering,
[2]Department of Biological Sciences, Dartmouth College, Hanover, NH, USA;
[3]U.S. Dairy Forage Research Center, Agricultural Research Service, Madison, WI, USA;
[4]Department of Bacteriology, University of Wisconsin, Madison WI, USA;
[5]Department of Chemistry and Biology, Ryerson University, Toronto, ON, Canada;
[6]Current address: Department of Biological Systems Engineering, Virginia Tech, Blacksburg, VA, USA.

ABSTRACT

Cellulose hydrolysis must be considered as a microbial as well as enzymatic phenomenon in order to understand cellulose utilization in natural environments and realize the potential of biotechnological processes based on microbial conversion. Moreover, there are increasing indications that the role of microorganisms in cellulose hydrolysis extends far beyond serving merely as a source of cellulase enzymes. In this chapter, cellulose utilization is examined for the thermophilic anaerobe *Clostridium thermocellum*. Specific topics include the ecophysiology of this organism, the mechanism of cellulose hydrolysis, bioenergetics, control of cellulase synthesis, and mechanisms of adhesion. In addition, we suggest that it may be fruitful to apply tools and insights from the biofilm field to microbial cellulose utilization, which may be viewed as a reacting biofilm in which cellulose serves as both the solid substratum and the carbon source for heterotrophic growth. We close by summarizing recent progress addressing previously-posed questions, and by articulating some new questions.

Key words: Cellulosome, bioenergetics, biofilms, synergy

I. Introduction

Cellulose hydrolysis, of widely-recognized importance in both technological and ecological contexts, has been investigated primarily from the perspective of an enzymatically oriented intellectual paradigm. Key foci of this paradigm include cellulose hydrolysis-mediated by cellulose-enzyme complexes, fundamental studies emphasizing the structure and function of cellulase enzyme systems, and applied studies anticipating processes with a dedicated process step for cellulase production that is separate from the step(s) in which biomass is hydrolyzed and fermented to a desired product. An alternative microbially-oriented paradigm considers cellulose hydrolysis as being mediated in significant part by cellulose-enzyme-microbe complexes, and anticipates processes featuring cellulase production, cellulose hydrolysis, and fermentation in a single process step (Lynd et al., 2002).

Consideration of cellulose hydrolysis as a microbial, rather than purely enzymatic, phenomenon is of interest from several perspectives. If we are to understand cellulose utilization in natural environments, we need to understand not only the action of cellulase enzyme systems, but also the organisms that produce them. Microorganisms determine the amount of cellulase present in natural environments, and there are increasing indications, considered in this chapter, that the role of microorganisms in cellulose hydrolysis extends far beyond serving as sources of enzymes. From a biotechnological perspective, use of cellulose-fermenting microorganisms represents a potential breakthrough in low-cost processing of cellulosic biomass to commodity products (Lynd et al., 2002; Lynd et al., in press). Finally, elucidating the fundamentals of microbial cellulose utilization is fascinating from an intellectual standpoint and has considerable potential to shed light on the evolution of cellulolytic microorganisms and the enzyme systems they produce, as well as the general strategies involved in microbial utilization of insoluble substrates.

In this chapter, we consider microbial cellulose utilization by *Clostridium thermocellum*. We focus on this organism because it has among the highest growth rates on crystalline cellulose of described microorganisms (Lynd et al., 2002), because of its biotechnological potential, and because there is a significant body of recent work to consider. There are, of course, many other cellulolytic microorganisms that are also of interest, and which may have features of a fundamental and/or applied relevance that are quite different from *C. thermocellum*. Specific topics considered include the ecophysiology of *C. thermocellum*, the mechanism of cellulose hydrolysis, bioenergetics, control of cellulase synthesis, mechanisms of adhesion, and viewing microbial cellulose hydrolysis as a reacting biofilm. We close by offering a perspective on expanding frontiers of the microbial cellulose utilization field, including tabulation of recent progress addressing previously-posed questions as well as articulating some new questions.

II. ECOPHYSIOLOGY OF *CLOSTRIDIUM THEMOCELLUM*

C. thermocellum is a thermophilic, strictly anaerobic, spore-forming bacterium classified with other "low G+C" gram-positive bacteria in the phylum Firmicutes. The organism is an obligate carbohydrate-fermenting bacterium with a relatively narrow spectrum of growth substrates that include cellulose and its hydrolytic products (cellobiose and cellodextrins). Some strains have also been reported to grow on one or more additional carbohydrates, including glucose (Demain et al. 2005; Golovchenko et al., 1986; Strobel et al., 1995); laminaribiose (Demain et al. 2005); fructose (Patni and Alexander 1971b, Johnson et al. 1985, Sato et al. 1993); sorbitol (McBee 1950, Johnson et al. 1985); and/or esculin (Sato et al. 1993). Utilization of these substrates for growth is strain-specific and often depends on conditions under which inocula are grown (Demain et al. 2005), which may account for lack of agreement among researchers with regard to the utilization of a specific substrate by a specific strain. Typically, the growth on glucose and the non-cellulosic substrates requires a prolonged adaptation period (one or more days), after which the growth properties usually display some stability (see Demain et al. 2005 for a more complete discussion).

Cellobiose and cellodextrins arising from cellulose hydrolysis are metabolized to glucose-1-phosphate (G-1-P) via intracellular cellobiose- and cellodextrin phosphorylases in an ATP-independent reaction that allows substantial energy savings for the organism (Lynd et al. 2002; see section IV below). G-1-P is readily converted to Glucose-6-phosphate, which is also the initial product of glucose catabolism, and which is fermented by the Embden-Meyerhof fermentation pathway (Patni and Alexander 1971a). The resulting pyruvic acid intermediate is converted to a mixture of end products including ethanol, acetic acid, lactic acid, H_2, and CO_2 (McBee 1950, Lynd et al. 2002).

C. thermocellum has been isolated from a variety of natural and engineered environments, particularly soils, compost, animal feces, sewage digesters, and surface geothermal features (Bayer et al. 1983, Chuvil'skya et al. 1986, Han et al. 2002, McBee 1950, Ozkan et al. 2001). In these environments, *C. thermocellum* is thought to serve as a primary agent of plant cell wall hydrolysis, supplying both carbohydrate oligomers and fermentation end products to other members of the microbial community. *C. thermocellum* readily forms mutualistic associations with other anaerobic bacteria, e.g. *Methanothermobacter thermoautotrophicus* (formerly *Methanobacterium thermoautotrophicum*; Weimer and Zeikus 1977) and several thermophilic hemicellulolytic and/or saccharolytic species (Mori 1995, Weimer and Chou 1986). This trait has sometimes confounded laboratory "pure culture" studies from decades past (see McBee 1950 for summary) and up until quite recently (Erbeznik et al. 1997). These symbiotic interactions may have contributed to the development of multiple polysaccharidase functions within the cellulosome. In addition to its complement of cellulolytic enzymes, the cellulosome contains a variety of different hemicellulose-degrading enzymes, including not only xylanases, but also a mannanase, a lichenase, and a chitinase (Schwarz 2001). Most of the hydrolytic products of these hemicellulolytic enzymes are not utilizable by *C. thermocellum*, and benefit of these enzymes to this species may be due to the ability of these enzymes to "clear the way", by removing these hemicelluloses from the architecturally complex plant cell wall, enhancing the accessibility of the cellulosome to cellulose. However, the enhanced extent of degradation of plant cell wall materials by mixed cultures relative to monocultures attests to the importance of symbiotic interactions between *C. thermocellum* and its cohabitants, and suggests that the cellulosomal components may have co-evolved with enzymatic machinery of these symbionts.

III. Mechanism of Cellulose Hydrolysis

The cellulase enzyme systems of aerobic fungi, and *Trichoderma reesei* in particular, were the first to be studied in detail. As these systems are often taken as a frame of reference for other cellulase systems, the development of mechanistic understanding for these systems is briefly considered. Beginning with the pioneering work of Reese (1950), several schemes for the mechanism of cellulose hydrolysis by the *T. reesei* cellulase system have been proposed. All such schemes involve one group of enzymes that converts insoluble cellulose to a "reactive" (still insoluble) form and a second group of enzymes that convert reactive insoluble cellulose to one or more soluble products (Wood and McCrae, 1978). The mechanism widely accepted today features: 1) endoglucanases that cleave at interior portions of cellulose molecules resulting in a decreased degree of polymerization (DP), increased concentrations of chain-ends, and little solubilization; 2) cellobiohydrolases that cleave as the primarily agents of cellulose solubilization, cleaving cellobiose from chain ends accompanied by little change in substrate DP or the concentration of chain ends; 3) β-glucosidases that cleave cellobiose to glucose (Teeri, et al., 1998; Zhang and Lynd, 2004b). Consistent with this mechanism, in vitro studies involving the *T. reesei* cellulase system have consistently reported accumulation of cellobiose when cellobiohydrolase is present and β-glucosidase is not supplemented to levels higher than produced by naturally-occurring strains. Structural studies carried out in 1990s provide direct evidence that cellobiose is the primary product of hydrolysis mediated by CBH1 and CBH2 (the primary cellobiohydrolases in the *T. reesei* system; Rouvinen et al., 1990; Divne et al., 1994).

As detailed in other chapters of this volume, it is clear that the complexed cellulase enzyme systems of anaerobic microorganisms – whether prokaryotic or eukaryotic – have a different fundamental architecture relative to the non-complexed systems of aerobic microorganisms. Notwithstanding this deeply-rooted difference, cellulose hydrolysis by anaerobes has often been assumed to proceed via cellobiose-liberating exoglucanases as per the accepted mechanism for aerobes. This assumption is understandable in light of the well-established mechanism for aerobic cellulase systems together with the widespread observation that cellobiose accumulation is routinely observed in cell-free experiments involving anaerobic cellulase systems and under some conditions whole-cell experiments as well (Lynd et al,, 2002).

Recent studies provide considerable support for the proposition that solubilization by *C. thermocellum* in fact does not occur primarily via the action of cellobiohydrolases. One line of evidence comes from bioenergetic studies. Zhang and Lynd (2005b) developed an ATP balance equation representing ATP generation and consumption during fermentation of β-glucans by *C. thermocellum* (Lynd and Zhang, 2002; Lynd et al., 2002; Zhang and Lynd, 2004a). ATP generating processes considered include glycolysis, acetate kinase, and intracellular phosphorolytic cleavage of β-glucosidic bonds. ATP consuming processes include substrate transport, cell synthesis, cellulase synthesis, and maintenance. After determining values for several key parameters, the ATP balance equation was solved for G_{ATP}^{P-T}, corresponding to the net ATP from phosphorolytic cleavage of cellodextrins less the ATP required for transport, both in units of moles ATP/mole hexose. Experimental determination of G_{ATP}^{P-T} as a function of cellodextrin chain length based on data obtained using cellodextrin-fed chemostats agreed very well with the prediction G_{ATP}^{P-T} = (n - 2)/n, providing validation for the model and support for the proposition that cellodextrins up to cellohexaose (G_6) are taken up by *C. thermocellum* under the conditions examined without extracellular hydrolysis. A G_{ATP}^{P-T} value of 0.52 $\pm$ 0.06 was calculated for cellulose-grown measured for growth on cellulose. Equating this value to (n – 2)/n and solving for n gives a mean cellodextrin chain length of 4.2 $\pm$ 0.46 (Zhang and Lynd, 2005b).

A second, independent, line of evidence supporting uptake of hydrolysis products with higher molecular weight than cellobiose is provided by labeling studies. Substrate-exhausted cultures of *C. thermocellum* were exposed to ^{14}C-labeled cellulose and ^{14}C-labeled cellobiose, and the radioactivity of intracellular cellodextrins was monitored over 8 minutes. Very different cellodextrin profiles were observed for the two substrates, although similar patterns would have been expected if cellobiose (G2) were the primary hydrolysis product taken up by the cell. For labeled G2, the greatest amount of radioactivity was measured for G2, lesser amounts of labeled G1, G3 (cellotriose), and G4 (cellotetraose) were observed, and labeled G5 (cellopentaose) and G6 were not detected. For labeled cellulose, the greatest amount of radioactivity was measured for G5, with somewhat lesser amounts of G1, G2, and G6, and the lowest amounts of G3 and G4. The mean chain length of labeled cellodextrins was about 2 in the presence of labeled cellobiose, and about 4 for labeled cellulose (Zhang and Lynd, 2005b).

Based on these data, Zhang and Lynd (2005b) concluded that *C. thermocellum* assimilates cellodextrins with n ~ 4 during growth on cellulose, implying that the immediate products of cellulose solubilization have an average degree of polymerization of at least this value. Recent results obtained via rapid sampling of hydrolysis mediated by several purified components of the *C. thermocellum* cellulase system (Zverlov et al., 2003; W. Schwarz, personal communication) further support this interpretation. The distance between adjacent catalytic subunits in the *C. thermocellum* cellulosome has been estimated at 8 glucosidic bonds (Mayer et al., 1987). Events mediated by adjacent catalytic components would thus be expected to result in an insoluble G8 fragment, with any subsequent cleavage resulting in two soluble products with a mean chain length of 4.

Consumption of cellodextrins with average length ~ 4 can be reconciled with the observation of cellobiose accumulation in cell-free experiments involving *C. thermocellum* cellulase (above) by noting that enzymatic formation of cellodextrins from insoluble cellulose proceeds at a much slower rate than does the enzymatic hydrolysis of insoluble cellodextrins to cellobiose. Thus, in the absence of cells, cellodextrins react as soon as they are formed with the result that cellobiose accumulates and cellodextrins do not. Consuming cellodextrins with higher molecular weight than cellobiose has significant implications with respect to the bioenergetics and, we suspect, ecological strategies of *C. thermocellum*, as explored in subsequent sections.

IV. BIOENERGETICS

ATP-generating metabolic processes in *C. thermocellum* include intracellular phosphorolytic cleavage of β-glucosidic bonds by cellodextrin and cellobiose phosphorylase (Strobel et al., 1995; Zhang and Lynd, 2003a), glycolysis via the Embden-Meyerhoff pathway, and the action of acetate kinase. ATP-consuming metabolic processes include substrate transport via an adenosine-binding cassette system (Mitchell, 1998; Lynd et al., 2002; Zhang and Lynd, 2005b), cell synthesis, cellulase synthesis, and nonbiosynthetic maintenance functions.

Rates of phosphorolytic cleavage of β-glucan substrates for cell extracts of *C. thermocellum* ATCC 27405 have been determined and compared to the rates of hydrolytic cleavage. Phosphorolytic cleavage was found to exceed hydrolytic cleavage by ≥ 20-fold for both G2 and G5 over a 10-fold range of β-glucan concentrations (0.5 to 5 mM) and for G4 at a single concentration (2 mM) (Zhang and Lynd, 2004a). The ATP requirement for transport of cellodextrins was shown to be 1/n moles of ATP per mole hexose, where n is the mean cellodextrin chain length assimilated (Zhang and Lynd, 2005b). The added ATP available from phosphorolytic β-glucan cleavage and the savings in ATP expended per hexose for transport of cellodextrins with increasing chain length comprise bioenergetic benefits specific to growth on cellulose. The magnitude of these benefits is larger than the added ATP required for cellulase synthesis on cellulose as compared to cellobiose. Substrate transport using an adenosine binding cassette (ABC) mechanism, with very high affinity both in general and in the particular case of β-glucan transport by *C. thermocellum* (Strobel et al., 1995), puts the organism in a strong position to compete for products of cellulose hydrolysis. The observation that *C. thermocellum* cannot grow easily or well on glucose can readily be understood in light of the fact that substrate-level phosphorylation due to phosphorylase enzymes does not occur during growth on glucose and the cost of substrate transport per hexose, whether by an ABC system or some other active transport mechanism, is higher for glucose than for oligosaccharides. These features, which confer benefits that increase with increasing substrate chain length, provide support for the notion that *C. thermocellum* is a cellulose-utilizing "specialist".

Evaluation of the allocation of carbon and ATP to synthesis of cellulase as distinct from cells is complicated by the fact that a substantial fraction of cellulase is cell-associated under many conditions (Bayer et al., 1983&1985; Lamed et al., 1983&1986). Lynd and Zhang (2002) showed that cell and cellulase concentrations can be calculated with acceptable accuracy by measuring total protein in conjunction with a method specific to cellulase concentration, whereas calculations based on measurements of several other combinations of analytes resulted in unacceptably large errors. Subsequently, a method was developed for quantification of cellulosome concentration featuring an indirect ELISA based on antibodies raised against a 14 amino acid synthetic peptide with sequence taken from the cohesin domain of the scaffoldin protein. Using this methodology, cell and cellulase yields (g/g cellulose) at the end of batch culture were found to be 0.1 and 0.02, respectively, and about 20% of biosynthetic ATP (exclusive of maintenance and transport) was allocated to cellulase synthesis with the balance allocated to synthesis of cells (Zhang and Lynd, 2003).

V. Control of Cellulase Synthesis

Effect of Soluble Carbohydrates

The pioneering work of Johnson and Demain (Johnson et al. 1985) indicated that availability of readily assimilated monosaccharides repressed synthesis of extracellular cellulase in *C. thermocellum* ATCC 27405. Batch cultures grown on Avicel contained approximately 5- fold greater "true cellulase" (Avicelase) activity than did cellobiose-grown cultures. Transfer of cellobiose-grown cultures to glucose, fructose, or sorbitol resulted in long lag periods prior to re-initiation of growth. During this lag period cellulase activity increased approximately 5-fold, but declined precipitously once growth resumed, and intracellular concentrations of easily metabolized soluble sugars were presumably reduced. Moreover, no increase in cellulase activity was observed in G2-adapted cells upon addition of any of several potential inducers, including G3, G4, mixed cellodextrins, G-1-P, lactose, salicin, or thiocellobiose. This work established the general concept that cellulose degradation in anaerobic bacteria is affected by concentrations of soluble carbohydrates, but did not specifically indicate the relative effect of these carbohydrates on enzyme synthesis versus enzyme activity, nor did they examine cellulase activity in the surface-bound cellulosome organelle that was only later shown to harbor most of the bacterium's cellulolytic activity.

Subsequent characterization of the composition and structure of the cellulosome, and identification of genes encoding its components, has permitted a more detailed analysis of the regulation of cellulolysis. Mishra et al. (1991) demonstrated that batch cultures of *C. thermocellum* NCIB 10682 did not contain measurable levels of transcripts of the endoglucanase-encoding genes *celA*, *celD*, and *celF* until late in exponential growth phase, at which time cellobiose concentrations were presumed to have decreased below levels required for catabolite repression. Dror et al. (2003a, 2003b) used RNA protection assays to quantify transcripts of several cellulosome components and cell wall binding proteins in *C. thermocellum* YS. Batch cultures grown on cellobiose (0.32 to 0.35 h^{-1}) or cellulose (0.23 h^{-1}) displayed low levels of *celS* (the major exoglucanase), and *cipA* (scaffoldin), along with *olpB*, *orf2p*, and *sdbA*, which encode three cell wall proteins previously shown to bind the scaffoldin's type II dockerin domain. However, cellobiose-limited continuous cultures displayed increased levels of *cel S, cipA*, *olpB* and *orf2p* as growth rate was decreased from 0.21 h^{-1} to 0.04 h^{-1}. Increasing expression with decreasing growth rate was also observed when cultures were grown under nitrogen-limitation in chemostats fed G2 as an energy source. On the basis of these observations, the authors suggested that expression of these three genes was controlled by growth rate rather than by concentration of cellobiose (presumably elevated during nitrogen limitation), although concentrations of cellobiose or other soluble sugars were not reported in this study or the study of Mishra et al. (1991). The authors proposed that *C. thermocellum* responds to low growth rates both by increasing the concentrations of essential catalytic components of the cellulosome (e.g., CelS) and by increasing the levels of supporting structures (scaffoldin and cell wall-bound scaffoldin-anchoring proteins). Expression was not determined on cellulose at low growth rates, preventing a direct comparison of the effects of substrate type as a function of growth rate.

More recent real-time PCR studies (Stevenson and Weimer 2005) examined the differential effects of substrate type and growth rate on the expression of seventeen genes in cellobiose-limited and cellulose-limited continuous cultures of *C. themocellum* ATCC 27405 at dilution rates (D) of 0.02 to 0.16 h^{-1}. At low D (< 0.06 h^{-1}), *cipA* and *celS* were expressed to similar extents in cellulose- and cellobiose-limited cultures. With further increases in D, expression of these two genes (but not of the scafflodin-anchoring protein *sdbA*) increased slightly in cellulose-limited cultures, but declined substantially in cellobiose-limited cultures. The total concentrations and average degree of polymerization of soluble glucosyl units were similar under both cellulose- and cellobiose-limited cultures, suggesting that differential expression on the two substrates was not due to different concentrations of soluble substrates *per se*, but due instead to the presence of cellulose itself. Moreover, it appears that at low growth rates cells retain a constitutive level of transcription of essential cellulosomal components, regardless of substrate type.

Zhang and Lynd have examined cellulase synthesis under different growth conditions by simultaneously assaying cellulase enzyme activity and the mass amounts of cellulase complexes in *C. thermocellum* ATCC 27405, using the ELISA method described in section IV above) (Zhang and Lynd, 2003). In batch culture, cellulose-grown cells displayed 9-fold greater cellulase levels than did cellobiose-grown cells (Zhang and Lynd, 2005a). In continuous cultures fed 10g substrate/l, cellulose-limited cells displayed 1.3 to 2.4-fold greater levels of cellulase, depending on dilution rate. In continuous culture, the effect of steady-state G2 concentration ([G2], g/l) on cellulase yield ($Y_{E/X}$, mg cellulase/ g cell) was determined both by varying dilution rate at a single concentration (10 g/l) of G2 in the feed medium, or by varying G2 concentration (5.6 to 65 g/l) at a single dilution rate (0.056 h^{-1}). Linear regression of the data revealed a break point at 0.2 g cellobiose/l in the reactor; below this concentration, cellulase decreased dramatically with increasing cellobiose ($Y_{E/X} = -1443[G2] + 316$), while above this concentration cellulase decreased only slightly with increasing cellobiose concentration ($Y_{E/X} = -0.537[G2] + 26.8$, a change in slope of the line by a factor of ~2700).

Carbon Catabolite Repression (CCR)

C. thermocellum is the only organism known thus far to contain three different putative Hpr kinase/phosphatase genes, which are regarded as key components of CcpA-dependent CCR systems in gram-positive bacteria (Warner and Lolkema 2003). Thus, it would seem that this system should be a likely mechanism for regulation of cellulolysis by this organism. Despite this, Dror et al. (2003b), noted the lack of consensus sequences in genes encoding cellulolosomal components that would bind CcpA repressor proteins, and were unable to PCR-amplify putative *ccpA* homologs from *C. thermocellum* YS genomic DNA using degenerate primers. But, using the more degenerate CRE consensus sequence (WGWNANCGNTNNCW), many (> 100) putative CRE sequences, including two putative CREs inside the *cipA* structural gene (+ 953, +5231), are located in the *C. thermocellum* genomic sequence (Zhang and Lynd, 2005a). More recent work (Stevenson and Weimer 2005) using non-degenerate primer sets demonstrated that expression of the three putative *hpr* genes, as well as the gene for a putative serine kinase involved in Hpr phosphorylation (HprK), generally displayed slight increases with increasing growth rate on both cellulose and cellobiose. These expression levels were nevertheless rather modest (5- to 15-fold lower than that of *recA*). Moreover, two genes (*crpY* and *crpZ*) that encode putative cAMP-binding proteins (viz., catabolite gene activator and the regulatory subunit of cAMP-dependent protein kinases) displayed even weaker expression. Thus, cells grown on cellobiose or cellulose did not display obvious differences in expression of these regulatory genes that would suggest their involvement in controlling synthesis of cellulosomal components. On the other hand, it is quite possible that the phosphorylase system does differentially regulate gene expression, via the extent of phosphorylation of regulatory proteins that themselves may be present in similar amounts.

Genome Organization and Genetic Control

Most genes encoding polysaccharidases in *C. thermocellum* are distributed throughout the chromosome (Gugliemi and Beguin 1998). However, this species contains at least seven multigene clusters totaling 26 genes, and even larger gene clusters have been identified in the cellulolytic mesophiles *C. cellulovorans* and *C. cellulolyticum* (see Demain et al., 2005 for discussion). Transcriptional start sites and likely sigma factors have been identified for several cellulase genes. For the *cipA* operon, the transcriptional start sites are located at –81 and –50 bp upstream of the translational start site, and likely promoters include those with homology to the σ^A and σ^L promoters of *Bacillus subtilis* (Dror et al. 2003b). *CelS*, the gene encoding the major cellulosomal exoglucanase, has a transcriptional start site –140 and –145 bp upstream of the translational start site; likely promoters include those having homology to *B. subtilis* promoters σ^A and σ^A – perhaps the same promoters suggested to be involved in transcription of the endoglucanase genes *celA* (Beguin et al 1986) and *celD* (Mishra et al. 1991).

Mechanisms regulating the biosynthesis of the noncomplexed cellulases of the well-studied aerobic fungus *Trichoderma reesei* have been elucidated only with difficulty (Schmoll and Kubicek 2003), and thus it is no surprise that the regulatory mechanisms for synthesis of the *C. thermocellum* cellulosomal enzymes have also resisted characterization. There is tantalizing recent evidence (Demain et al. 2005) of negative regulation of at least some cellulosomal components. In particular, GlyR3, a homolog of the *E. coli* LacI family of negative regulators, has been identified in the *celC* gene cluster, and laminaribiose (3-*O*-β-glucosyl-glucopyranose) has been shown to inhibit binding of GlyR3 to the *celC* promoter region. Additional experimentation will reveal if negative regulation is more widely distributed in the synthesis of other cellulosomal components. The potential role of non-cellulose oligomers inducing cellulase synthesis finds analogy to cellulose degradation in the bovine rumen. In vitro exposure of mixed ruminal bacteria to pure cellulose results in a prolonged lag period (8 to 10 h) before cellulolysis begins (Schofield et al. 1994, Weimer et al., 1990), but exposure to partially purified plant cell walls results in a much slower lag (2 to 3 h) before cellulolysis begins (Gill et al. 1969, Van Soest 1973), suggesting that early hydrolysis of associated hemicellulose releases inducers of cellulase synthesis.

VI. MECHANISMS OF ADHESION

Occurrence

Adhesion of cells to cellulose appears to be an important strategy for cellulose utilization in anaerobic bacteria, and the cellulose-enzyme-microbe ternary complex is likely to be the major agent of hydrolysis (Lynd et al., 2002). In the case of *C. thermocellum*, mixing of cellulose (Avicel) and cellulose-free cultures results in visible clearing compared to control cultures with no cellulose added, indicative of cells rapidly binding to cellulose (Bayer et al. 1983). *C. thermocellum* cells are closely-associated with cellulose in early stages of batch culture (Lamed et al., 1983; Zhang and Lynd, 2003). In stationary-phase cultures, however, non-adherent cells are routinely observed and extracellular cellulase activity can readily be assayed (Bayer et al, 1983; Johnson et al., 1982&1985). In general, the tendency for cellulase to remain attached to cells, and also for cells (with cellulase activity) to attach to cellulose, appears to be somewhat less absolute for *C. thermocellum* as compared to ruminal cellulolytic bacteria (Weimer and Odt 1995). This difference remains largely unexplained in terms of both its mechanistic basis as well as its ecological implications. The state of non-adherent cells and the extent to which they are metabolically active, have access to substrate, and participate in cellulose hydrolysis (via secreted enzymes) are not clear at present for *C. thermocellum*, as considered subsequently in this chapter.

Cellulose Binding Domains

Prior to its identification as a catalytic complex responsible for cellulose hydrolysis, the cellulosome was originally described as a cell surface structure in *C. thermocellum* that bound cells to cellulose fibers (Bayer et al. 1983). Only a few of the polysaccharide-hydrolyzing enzymes of the cellulosome contain carbohydrate binding modules (CBMs). Instead, most of the adherence of the cellulosome to cellulose is mediated by a strongly binding CBM_{IIIa} contained within the scaffoldin protein itself (Schwarz, 2001). Crystallographic studies of the *C. thermocellum* CBM_{IIIa} at 1.75 Å resolution (Tormo et al., 1996) have revealed that one end contains a shallow groove of unknown function, while the other end has an exposed planar face thought to interact with crystalline cellulose. Interaction has been proposed to occur through a planar strip of aromatic amino acids that bind to a single cellulose chain, with additional anchoring to the adjacent cellulose chain on each side via polar amino acid residues.

Electron microscopic studies have confirmed the general features of cellulosome structure elucidated from biochemical, spectroscopic and genomic studies. At the fine structural level, Madkour and Mayer (2003) examined intact *C. thermocellum* cellulosomes by scanning and transmission electron microscopy, and were able to distinguish between condensed regions (presumably the catalytic proteins attached to the dockerin domains of the scaffoldin) and thin fibrillar regions (presumably the helical stretches of the scaffoldin between the dockerin domains). The CBM module itself was not visualized in this study.

Contact Corridors

On a somewhat larger scale, Lamed and Bayer (1986) have demonstrated that, in the presence of cellulose, polycellusomal structures on the *C. thermocellum* cell surface unfold into fibrillar structures that contact the cellulose at some distance (400 to 500 nm) from the bacterial cell surface. These fibrillar "contact corridors" were proposed to serve as a means of channeling hydrolytic products to the bacterial cell surface, ostensibly to provide a competitive advantage to the adherent cellulolytic bacterium (Lamed and Bayer, 1986). On a mechanistic level, it would seem that the elegantly organized array of catalytic proteins within the cellulosome must be brought into contact with the cellulose surface to allow effective cellulose hydrolysis. If contact corridors are established between the cells and the cellulose fibers, the scaffoldin and associated catalytic proteins would be expected to be localized at the distal end of the contact corridors, directly in contact with the cellulose chains and at some distance from the bacterial cell wall. The presence of an immunologically reactive cellulase component (the S1 protein of the cellulosome, now known as CipA, scaffoldin) at the distal ends of these fibrillar structures is consistent with such an arrangement, but additional research is necessary to establish these spatial relationships. The potential importance of these contact corridors in supplying fermentable hydrolytic products to the bacterial cell is difficult to evaluate at this time.

Involvement of Carbohydrates

The cellulosome contains 6 to 13% carbohydrate, primarily as *O*-linked galactopyranosyl residues attached to Thr residues in the hydrophilic "linker region" of the scaffoldin, between the catalytic domains and the dockerin domain involved in anchoring the scaffoldin to the cell membrane (Gerwig et al. 1993). Several of the cellulolytic enzymes attached to the scaffoldin are glycosylated as well. It has been proposed that glycosylation serves to provide resistance to proteolytic enzymes (Schwarz 2001). However, no definitive studies on the susceptibility of the cellulosome to proteolytic attack have been reported.

The widespread involvement of polysaccharides and sugars as agents of microbial adherence to solid surfaces suggests the intriguing possibility that carbohydrates, present either within the cellulosome or in ancillary cell surface structures contribute to the ability of the organelle to adhere to cellulose. Indeed, treatment of the solid residue of cellulose-grown cultures of *C. thermocellum* (as well as those of several cellulosome-producing

Ruminococcus species) with boiling sodium dodecyl sulfate solution removes adherent bacterial cells but leaves a residue containing only small amounts of galactosyl residues but substantial quantities of xylosyl and mannosyl residues, suggesting the presence of other polysaccharides that may contribute to adherence (Weimer et al. 2005). Pili, which appear to be of some importance in facilitating adherence of the cellulosome-producing *Ruminococcus* species to cellulose (Miron et al. 2001), have not been reported in *C. thermocellum*, although its unfinished genome contains numerous homologs of genes of pilin biosynthesis (D.M. Stevenson, personal communication).

VII. Reacting Biofilms

Overview of Current Understanding of Biofilms

The importance of the biofilm mode of microbial growth has been extensively described in medical, natural and engineered systems. Numerous advances have been made during the last two decades in response to the identified need for appropriate methodologies to study surface-associated cells. Many of these methods are based on microscopy in combination with image analysis and a range of reporter molecules, and it is now common practice to describe biofilms *in situ* in terms of species composition, cell viability, gene expression and function. Much progress has also been made in the development of techniques -- most notably micro-electrodes and fluorescent probes -- to define the physical and chemical structure of biofilms. As a result of these efforts, the general structure and function of biofilms are well described.

In a broad sense, biofilms are defined as assemblages of surface-associated microbial cells that, together with extracellular products and trapped abiotic particles, form a matrix that can substantially differ from the surrounding environment. These cells may have patterns of gene expression that differ notably from their planktonic counterparts (Sauer et al. 2001, 2002). With the exception of few instances in nature, such as extreme environments and cases where an infectious species succeeded in overcoming a host's immune system, and pure culture bioreactors in industry, biofilms typically consist of multiple species that are in associations representing most of the classes of interactions (mutualism, commensalism, competition, etc.) described in ecology. As a result of this heterogeneity, and the demonstrated ability of microbes to modify their biofilms in response to environmental fluctuations (Manz et al. 2003), biofilms with a variety of structures have been described and is there no single model for biofilm architecture. Even pure culture biofilms cultivated in the laboratory show huge variation in form and thickness (Heydorn et al. 2000).

The overwhelming majority of studies on biofilms have focused on cells associated with inert surfaces, where nutrients are obtained from the bulk liquid phase, which is a representative model for engineered systems such as bioreactors to treat industrial effluents, natural systems such as biofilms on a riverbed, and biofilms with medical relevance such as those that develop on medical implants. These biofilms typically have access to nutrients that show large variation in concentration and bioavailability; factors that contribute to variation in biofilm form and function, and usually have multiple layers of cells.

There is frequent mention in the biofilm literature of a number of benefits that microbial cells derive from the sessile mode of growth. Stoodley et al. (2002) provided a comprehensive review of the complex nature of biofilm communities. These authors pointed out that biofilm communities are the dominant form of prokaryotic growth in a diverse range of ecosystems, and that there is a high degree of structural adaptation and interaction that lead to integration and differentiation of metabolic activities within these communities that are made possible by the expression of gene sets remarkably different from those expressed by the individual members in the non-attached phase. Examples of such cooperation can be found in the areas of biodegradation of complex organic compounds (Rhee et al. 2004) and dental plaque (Aspiras et al. 2000). In addition to spatial organization in biofilms that favor cooperation and sequential degradation, another property that is often referred to is the establishment of a stable micro-environment. Extracellular polymeric substances (EPS) play a key role in the maintenance of the micro-environment (McSwain et al. 2005), which is characterized by steep gradients of gases, nutrients and metabolites. Interestingly, structural biofilm differentiation and EPS-mediated formation of a stable micro-environment are not limited to multi-species biofilm communities; in fact most work in this regard has been performed on pure-culture biofilms.

Consideration of Anaerobic Cellulose Utilization from a Biofilm Perspective

In contrast to the generalized biofilm model of cells associated with inert surfaces, where nutrients are obtained from the bulk liquid phase, cellulolytic bacteria derive their nutrients primarily from the attachment substrate (Figure 1). In fact, specific adhesion of bacteria to cellulose was described as an essential first step by cellulose-degrading species in the rumen (Costerton et al. 1987), while Walter et al. (1999) demonstrated that contact between *Streptomyces reticuli* and crystalline cellulose is a prerequisite for efficient hydrolysis of cellulose. It has been long known that *C. thermocellum* needs to strongly adhere to cellulose in order to utilize it efficiently (Shoham et al. 1999; Weimer et al. 2005). In this regard, the cellulosome has been the focus of numerous investigations because of its high cellulase activity and its role in adhesion to cellulose by *C. thermocellum* and other cellulolytic bacteria. Studies that recognized the importance of carbohydrate binding modules as part of the scaffolding protein are discussed in Section VI. Few if any studies have focused on *C. thermocellum* from a biofilm perspective, although this approach may yield new insights given the close association of this organism with the cellulose surface. Instead, the majority of studies have focused on cellulase-cellulose interactions with little consideration of the potential involvement of adherent cells as a third component of the association.

It remains to be demonstrated if, and to what degree the biofilm characteristics generally attributed to microbial survival in environments where nutrients are scarce and refractory also apply to cellulose degradation by *C. thermocellum*, since the need for cooperation and sequential degradation that is typically described for other environmental processes may not be relevant to cellulolytic organisms where specialized species apparently have evolved independently to overcome the challenge of growth on a refractory polymer. In contrast to the observed differentiation among multiple layers of cells in the generalized biofilm model, it has been demonstrated that adherence of ruminal bacteria to cellulose occur essentially as a monolayer of cells (Weimer et al. 1995).

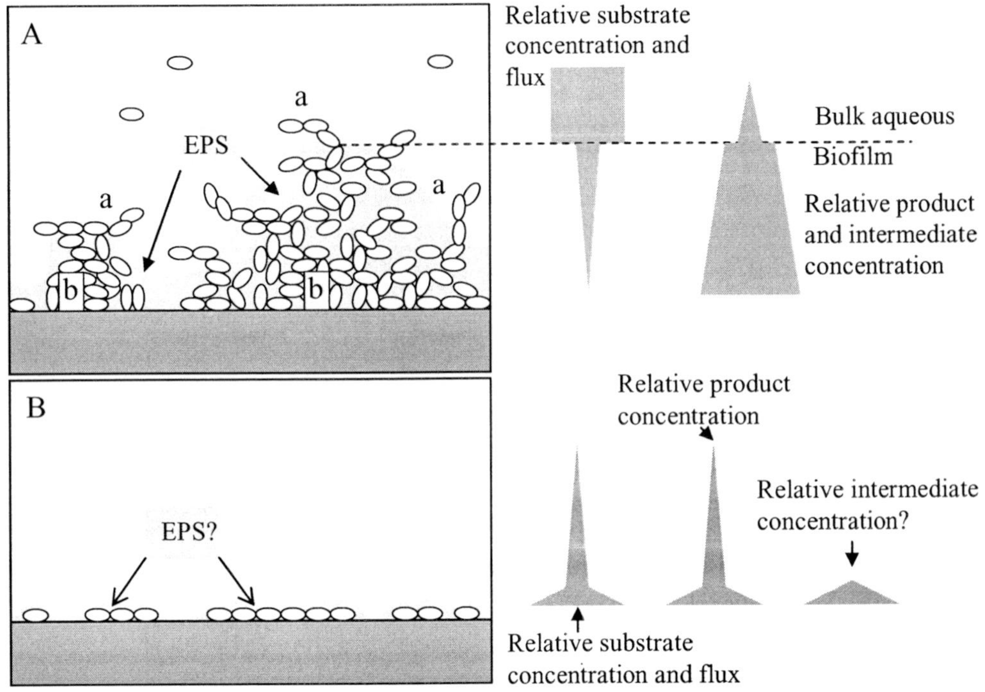

Figure 1. EPS have an important structural role to support spatial organization between cells and in the maintenance of a micro-environment that often differs substantially from the bulk surroundings. A generalized biofilm model is showed in (A) to illustrate the biofilm matrix consisting of cells and EPS. The cells nearest to the biofilm – aqueous interface (a) are optimally positioned to obtain nutrients from the bulk phase, are highly responsive to environmental fluctuations, and are responsible for most cell growth, while the cells closer to the inert attachment surface (b) are nutrient-limited. As a result, these cells are metabolically inactive and therefore more resistant to adverse conditions. In contrast, cellulolytic bacteria do not form biofilms with such a degree of differentiation, most likely because nutrient acquisition occurs from the attachment substrate (b). It remains to be demonstrated if EPS have an important role in cell-cellulose interaction, as do cells that colonize inert surfaces.

Despite the differences between the generalized biofilm model and adhesion of *C. thermocellum* to cellulose, it is probable that factors described for model biofilm strains also play a role in cellulose utilization. Indeed, it is possible that this area of research will benefit by adapting the methodologies and approaches that have been developed over the last two decades in other areas of science and engineering specifically to study biofilms. Costerton (2001) outlined the benefits of recognizing the biofilm phenotype to medical microbiology, and it is plausible that similar advances await the cellulose field. Gehin et al. (1995) described the importance of adhesion by *C. cellulolyticum* to cellulose to overcome carbon starvation. These authors demonstrated higher levels of adhesion by cells in the exponential growth phase than those in the stationary phase. Although they noted an increase in adhesion towards the late stationary phase, they suggested that adhesion by spores was responsible for this increase. This positive correlation between cell activity and adhesion to cellulose can be interpreted as evidence that biofilm formation facilitates cellulose degradation. More recently, Lynd et al. (2002) have hypothesized that "cellulase-microbe synergy", in which the effectiveness of a cellulase system is enhanced by acting as part of a cellulose-enzyme-microbe complex, may be operative.

Potential Role of Extracellular Polymers in Cellulose Utilization

An example where the interaction between bacterial cells and a relatively insoluble but metabolizable substrate has been investigated is the case of *Thiobacillus ferrooxidans*, an obligate acidophile that respires aerobically on pyrite. In this case, the cell creates a micro-environment that facilitates close interaction with the substrate. Extracellular polymeric substances (EPS) probably have an important function in these bacterial-substratum interactions by mediating contact between the bacterial cell and its sulfidic energy source, as suggested by Gehrke et al. (1998). These authors demonstrated the ability of cells to modify EPS composition in response to the properties of the substrate to form an extended extracellular reactive zone in which most of the activity occurs. It was further suggested (Sand and Gehrke, 1999) that the concentration of reactive compounds in the EPS layer between the outer membrane and the substrate interface is considerably higher than in the bulk environment. Others investigated the role of interaction between EPS and extracellular enzymes. For instance Dow et al. (2000) demonstrated the existence of a gene cluster that regulates the synthesis of extracellular protein and EPS in *Xanthomonas campestris*, which collectively contribute to the pathogenicity of this organism. Geesey and Van Ommen Kloeke (2005) made reference to other studies that described the role of EPS-associated extracellular enzymes in flocs that degrade toxic chemicals and transform complex molecules from the environment for subsequent uptake. These authors showed the importance of association between phosphatase enzymes with the EPS matrix of flocs in the activated sludge process.

Some work has been done to evaluate the role of EPS in cellulolytic microbes, for example Kenyon et al. (2005) provided evidence that linked EPS with cellulose utilization. These authors demonstrated the production of a curdlan-type (β-1,3-glucan) EPS by *Cellulomonas flavigena* when growing on microcrystalline cellulose, to form a matrix in which the cells and high numbers of cellulosome-like structures were embedded. It should be pointed out that although the curdlan EPS appeared to assist in adherence to cellulose, cells produced considerably more curdlan when grown on glucose or cellobiose than when grown on cellulose. Residues of cellulose fermentations from *Ruminococcus albus* and *R. flavefaciens* contain heteropolysaccharides that allow strong adherence of cells to cellulose (Weimer et al. 2003), and *C. thermocellum* appears to produce a similar material (Weimer et al. 2005). An interesting question is therefore whether *C. thermocellum* has the ability to maintain a micro-environment in which enzymes and intermediates are locally concentrated relative to their presence in the bulk phase. Such an arrangement will hold notable benefit to the cell, especially where the net gain of energy is limited. Addressing this question is a promising step towards efforts to delineate the ternary cell-enzyme-substrate interactions.

VIII. Expanding Frontiers

Writing in 2002, Lynd et al. posed a list of questions inherent to understanding microbial cellulose utilization that are most appropriately pursued by studies at a cellular level. Since then, studies involving *Clostridium thermocellum* have been completed that are relevant to many of these questions (Table 1). Of particular note, understanding has been extended significantly in the areas of hydrolysis product uptake, allocation of cellular resource (ATP, carbon) to cell and cellulase synthesis, and bioenergetics. A gene transfer system for *C. thermocellum* has also been developed (Tyurin et al., 2004), removing a significant barrier to advances in both fundamental and applied spheres. Other questions posed in the review of Lynd et al. are still outstanding, including the molecular basis for control of cellulase synthesis, the mechanistic basis of adhesion, the extent and mechanism of cellulase-microbe synergy, and elaboration of ecological strategies and their relation to cellular and subcellular features.

Table 1. Summary of insights relevant to microbial cellulose utilization gained through recent work with *Clostridium thermocellum.**

Question	New Information	Reference
Uptake of cellulose hydrolysis products		
What is the distribution of oligoglucan chain lengths?	A mean length of 4.2 has been calculated.	Zhang and Lynd, 2005b.
What is the bioenergetic requirement for substrate transport?	One ATP per transport event.	Zhang and Lynd, 2005b.
Metabolic control		
What is the allocation of substrate carbon to cellulase & cell synthesis?	The cell yield is 0.09~0.12 and the cellulose Yield is ~ 0.020 (both g/g cellulose).	Zhang and Lynd, 2005b.
Can this be reconciled at a molecular level?	Evidence for involvement of catabolite repression, much still to be to be elucidated.	Warner and Lolkema, 2002. Zhang and Lynd, 2005a.
Bioenergetics		
What is the allocation of ATP to cellulase and cell synthesis?	~9 % of available ATP to cellulase, 30~45 % of available ATP to cells.	Zhang and Lynd, 2005b.
What is the relative importance of phosphorolytic and hydrolytic cellodextrin cleavage?	The rate of phosphorolytic cleavage in higher by ~20-fold in cell extracts.	Zhang and Lynd, 2004.
Can the rate of ATP production be quantitively reconciled with rates of ATP consumption?	A comprehensive model was has been validated.	Zhang and Lynd, 2005b.

* Selected questions are taken from those posed by Lynd et al., 2002, presented here in abbreviated form.

Questions related to features and activity of adherent cellulolytic microorganisms represent a fruitful direction for expanding and elaborating the list of questions posed previously by Lynd et al. (2002). Natural selection rewards populations that arise from variant cells that are phenotypically different from the parent population and realize a benefit as a result of this difference. In the case of cellulolytic microbes, adherence and preferential access to hydrolysis products would appear to be necessary in order for a variant cell with a more active, more effective, or more plentiful cellulase system to have a selective advantage compared to non-variant cells (Fan et al., in press). While conceptually attractive, and suggested in the literature (Fan et al., in press; Weimer, 1996; Lynd et al., 2002, the hypothesis that capture of hydrolysis products by adherent microorganisms is a driving force for the evolution of cellulose-utilizing microorganisms has not in our view been experimentally proven. In addition to further work to characterize the occurrence and mechanism of microbial adhesion to cellulose, experimentally-approachable questions relevant to this "substrate capture hypothesis" include:

- To what extent do products of cellulose hydrolysis equilibrate, or not equilibrate, with the bulk solution away from cellulose surfaces?
- What fraction of hydrolysis products proceed from the cellulose surface directly to adherent cells as compared to proceeding from the cellulose surface to the bulk solution?
- What is the relative importance of cell-associated and cell-free cellulase enzymes in mediating hydrolysis?
- Why are cellulases tenaciously held on the cell surface by some cellulolytic microorganisms (e.g. rumen cellulolysis) while cell-free cellulase is commonly observed by others (Lynd et al., 2002)?
- What if any differences are there in adherent and non-adherent populations with respect to metabolic state, as manifested by indicators such as growth rate and gene expression?

If experiments motivated by questions such as those articulated above provide support for the substrate capture hypothesis, this would provide considerable impetus for an important new line of inquiry having significant conceptual and methodological overlap with the biofilm field. A particular focus of such inquiry would likely be the gap between the cellulose surface and adherent cellulolytic microbes. Properties of this cellulose-microbe gap presumably determine the extent to which substrate capture is achieved, and may well have been subject to selective pressure favoring increased capture. It is intriguing to speculate as to the mechanisms that might enable substrate capture. Modeling results from John Brady's group (Cornell, personal communication) indicate that water is highly structured at the surface of cellulose, and that this may constitute a barrier to diffusion, or "blanket", impeding diffusion of hydrolysis products away from such surfaces. The presence of an adherent cellulolytic microbe could well alter the structuring of water in the vicinity of the cellulose surface, and it seems logical that selection would favor alterations that enhance diffusion of hydrolysis products to the cell. Molecular structures in the cellulose-microbe gap, perhaps the "contact corridors" described by Lamed and Bayer (1986), might bind hydrolysis products as they are formed so that they progress to the cell without diffusing at all. At present, we are in a position to imagine how these and other potential mechanisms of substrate capture might be selectively advantageous. It remains, however, to show experimentally that such mechanisms are in fact operative.

Acknowledgements

This work was partially supported by grant DE-FG02ER15350 from the Department of Energy, grant 60NANB1D0064 from the National Institute of Standards and Technology, and agreement XCO-3-33033-01 from the Department of Energy. We are indebted to the anonymous reviewer of Zhang and Lynd (2005b) for insightful suggestions regarding differences between adherent and non-adherent cellulolytic populations, and in particular the possibility that such populations have significantly different growth rates.

References

Aspiras MB, Kazmerzak KM, Kolenbrander PE, McNab R, Hardegen N, Jenkinson HF. 2000. Expression of green fluorescent protein in *Streptococcus gordonii* DL1 and its use as a species-specific marker in co-adhesion with *Streptococcus oralis* 34 in saliva-conditioned biofilms *in vitro*. *Appl. Environ. Microbiol. 66*:4074-4083.

Bayer EA, Belaich J-P, Shoham Y, Lamed R. 2004. The cellulosomes: multienzyme machines for degradation of plant cell wall polysaccharides. *Annu. Rev. Microbiol. 58*:521-554.

Bayer EA, Kenig R, Lamed R. 1983. Adherence of *Clostridium thermocellum* to cellulose. *J. Bacteriol. 156*:818-827.

Bayer EA, Setter E, Lamed R. 1985. Organization and distribution of the cellulosome in *Clostridium thermocellum*. *J. Bacteriol. 163*:552-559.

Beguin P, Rocancourt M, Chebrou MC, Aubert JP. 1986. Mapping of mRNA encoding endoglucanase A from *Clostridium thermocellum*. *Mol. Gen. Genet. 202*:251-254.

Chuvil'skya NA, Golovchenko NP, Belokopytov BF, Akimenko VK. 1986. Isolation, identification and some physiological properties of *Clostridium thermocellum*. *Prikl. Biokhim. Mikrobiol. 22*:800-805.

Costerton JW, Cheng K.-J, Geesey GG, Ladd TI, Nickel JC. Dasgupta M, Marrie TJ. 1987. Bacterial biofilms in nature and disease. *Annu. Rev. Microbiol. 41*:435-464.

Costerton JW. 2001. Cystic fibrosis pathogenesis and the role of biofilms in persistent infection. *Trends Microbiol. 9*:50-52.

Demain AL, Newcomb M, Wu JHD. 2005. Cellulase, clostridia, and ethanol. *Microbiol. Mol. Biol. Rev. 69*:124-154.

Divne C, Stahlberg J, Reinikainen T, Ruohonen L, Pettersson G, Knowles JKC, Teeri TT, Jones TA. 1994. The three dimensional structure of cellobiohydrolase I from *Trichoderma reesei*. *Science. 265*:524-528.

Doi RH, Kosugi A. 2004. Cellulosomes: plant-cell-wall-degrading enzyme complexes. *Nat. Rev. Microbiol. 2*:541-551.

Dow JM, Feng J-X, Barber CE, Tang J-L, Daniels MJ. 2000. Novel genes involved in the regulation of pathogenicity factor production within the *rpf* gene cluster of *Xanthomonas campestris. Microbiology 146*:885-891.

Dror TW, Morag E, Rolider A, Bayer EA, Lamed R, Shoham Y. 2003a. Regulation of the cellulosomal *celS* (*cel 48A*) gene of *Clostridium thermocellum* is growth rate dependent. *J. Bacteriol. 185*:3042-3048.

Dror TW, Rolider A, Bayer EA, Lamed R, Shoham Y. 2003b. Regulation of expression of scaffoldin-related genes in *Clostridium thermocellum. J. Bacteriol. 185*:5109-5116.

Erbeznik M, Jones CR, Dawson KA, Strobel HJ. 1997. *Clostridium thermocellum* JW20 (ATCC 31549) is a coculture with *Thermoanaerobacter ethanolicus. Appl. Environ. Microbiol. 63*:2949-2951.

Fan Z, McBride JE, van Zyl WH, Lynd LR. 2005. Theoretical analysis of selection-based strain improvement for microorganisms with growth dependent upon extracytoplasmic enzymes. *Biotechnol. Bioeng.* in press.

Geesey GG, Van Ommen Kloeke F. 2005. Extracellular enzymes associated with microbial flocs from activated sludge of wastewater treatment systems. In: *Flocculation in Natural and Engineered Environmental Processes*, Droppo, IG, Leppard, GG, Milligan T., Liss SN (eds.), CRC Press, Boca Raton, FL, pp. 295-316.

Gehin AE, Gelhaye G, Raval, Pettitdemange H. 1995. *Clostridium cellulolyticum* viability and sporulation under cellobiose starvation conditions. *Appl. Environ. Microbiol. 61*:868-871.

Gehrke T, Telegdi J, Thierry D, Sand W. 1998. Importance of extracellular polymeric substances from *Thiobacillus ferrooxidans* for bioleaching. *Appl. Environ. Microbiol. 64*:2743–2747.

Gerwig GJ, Kamerling JP, Vliegenthart JFG, Morag E, Lamed R, Bayer EA. 1993. The nature of the carbohydrate-peptide linkage region in glycoproteins from the cellulosomes of *Clostridium thermocellum* and *Bacteroides cellulosolvens. J. Biol. Chem. 268*: 26956-26960.

Gill SS, Conrad HR, Hibb JW. 1969. Relative rate of in vitro cellulose disappearance as a possible estimator of digestible dry matter intake. *J. Dairy Sci, 52*:1687-1690.

Golovchenko NP, Chuvil'skaya NA, Akimenko VK. 1986. Regulation of the biosynthesis of cellulolytic enzymes and enzymes catalyzing the initial catabolism of glucose and cellobiose in *Clostridium thermocellum. Mikrobiologiya 55*:31-33.

Guglielmi G, Beguin P. 1998. Cellulase and hemicellulase genes of *Clostridium thermocellum* from five independent collections contain few overlaps and are widely scattered across the chromosome. *FEMS Microbiol Lett 161*:209-215.

Han R, Min H, Che M, Zhao Y. 2002. Isolation, identification and phylogenetic analysis of a thermophilic cellulolytic anaerobic bacterium. *Weishengwu Xuebao 42*:138-144 .

Heydorn A, Nielsen AT, Hentzer M, Sternberg C, Givskov M, Molin S. 2000. Quantification of biofilm structures by the novel computer program COMSTAT. *Microbiol. 146*:2325-2407.

Johnson EA, Bouchot F, Demain AL. 1985. Regulation of cellulase formation in *Clostridium thermocellum*. *J. Gen. Microbiol. 131*:2303-2308.

Kateyeva, IA, Golovchenko NP, Chuvil'skaya NA, Akimenko VK. 1992. *Clostridium thermocellum* beta-glucosidases A and B purification and properties, localization and regulation of biosynthesis. *Enzyme Microb. Technol. 14*:407-412.

Kenyon W, Esch S, Buller C.. 2005. The curdlan-type exopolysaccharide produced by *Cellulomonas flavigena* KU forms part of an extracellular glycocalyx involved in cellulose degradation. *Antonie van Leeuwenhoek 87*:143-148.

Lamed R, Bayer EA. 1986. Contact and cellulolysis in *Clostridium thermocellum* via extensile surface organelles. *Experientia 42*:72-73.

Lamed R, Setter E, Bayer EA. 1983. Characterization of a cellulose-binding, cellulase-containing complex in *Clostridium thermocellum*. *J Bacteriol. 156*:828-836.

Lelieveld HLM. 1982. The use of continuous cultures for selection and isolation of microorganisms producing extracellular enzymes adapted to extreme environments. *Biotechnol. Bioeng. 24*:1419-1425.

Lynd LR, van Zyl WH, McBride J, Laser M, Strobel H, Zhang Y-HP, Tyurin M, Lu Y, Shaw J, den Haan R. 2005. Consolidated bioprocessing: a 2005 update. *Curr Opin Biotechnol. 16*:577-583.

Lynd LR, Weimer PJ, van Zyl WH, Pretorius IS. 2002. Microbial cellulose utilization: fundamentals and biotechnology. *Microbiol Mol Biol Rev 66*:506-577.

Lynd LR, Zhang Y. 2002. Quantitative determination of cellulase concentration as distinct from cell concentration in studies of microbial cellulose utilization: analytical framework and methodological approach. *Biotechnol. Bioeng. 77*:467-475.

Madkour M, Mayer F. 2003. Structural organization of the intact bacterial cellulosome as revealed by electron microscopy. *Cell Biol. Intern. 27*:831-836.

Manz B, Volke F, Goll D, Horn H. 2003. Measuring local flow velocities and biofilm structure in biofilm systems with magnetic resonance imaging (MRI). *Biotechnol. Bioeng. 84*:424-432.

Mayer F, Coughlan MP, Mori Y, Ljungdahl LG. 1987. Macromolecular organization of the cellulolytic enzyme complex of *Clostridium thermocellum* as revealed by electron microscopy. *Appl. Environ. Microbiol. 53*:2785-2792.

McBee RH. 1950. The anaerobic thermophilic cellulolytic bacteria. *Bacteriol. Rev. 14*:51-63.

McSwain BS, Irvine RL, Hausner M,. Wilderer PA. 2005. Composition and distribution of extracellular polymeric substances in aerobic flocs and granular sludge. *Appl. Environ. Microbiol. 71*:1051–1057.

Miron J, Ben-Ghedalia D, Morrison M. 2001. Adhesion mechanisms of rumen cellulolytic bacteria. *J. Dairy Sci. 84*:1294-1309.

Mishra S, Béguin P, Aubert J. 1991. Transcription of *Clostridium thermocellum* endoglucanase genes *celF* and *celD*. *J. Bacteriol. 173*:80-85.

Mitchell, WJ. 1998. Physiology of carbohydrate to solvent conversion by clostridia. *Adv. Microbiol. Physiol. 39*:31-130.

Mori Y. 1995. Nutritional interdependence between Thermoanaerobacter thermohydrosulfuricus and Clostridium thermocellum. Arch. Microbiol. 164:152-154.

Ng TK, Weimer PJ, Zeikus JG. 1977. Cellulolytic and physiological properties of *Clostridium thermocellum*. *Arch. Microbiol. 114*:1-7.

Ozkan M, Desai SG, Zhang Y, Stevenson DM, Beane J, Guerinot ML, Lynd LR. 2001. Characterization of 13 newly isolated strains of anaerobic, cellulolytic, thermophilic bacteria. *J. Ind. Microbiol. Biotechnol. 26*:1-6.

Patni N.J, Alexander JK. 1971b. Catabolism of fructose and mannitol in *Clostridium thermocellum*: presence of phosphoenolpyruvate:fructose phosphotransferase, fructose-1-phosphate kinase, phosphoenolpyruvate:mannitol phosphotransferase, and mannitol-1-phosphate dehydrogenase in cell extracts. *J. Bacteriol. 105*:226-231.

Patni NJ, Alexander JK. 1971a. Utilization of glucose by *Clostridium thermocellum*: presence of glucokinase and other glycolytic enzymes in cell extracts. *J. Bacteriol. 105*:220-225.

Reese ET, Siu RGH, Levinson HS. 1950. Biological degradation of soluble cellulose deriatives. *J. Bacteriol. 59*:485-497.

Rhee SK, Liu X, Wu L, Chong SC, Wan X, Zhou J. 2004. Detection of genes involved in biodegradation and biotransformation in microbial communities by using 50-Mer oligonucleotide microarrays. *Appl. Environ. Microbiol. 70*:4303–4317.

Rouvinen J, Bergfors T, Reeri TT, Knowles J, Jones TA. 1990. Three-dimensional structure of cellobiohydrolase II from *Trichoderma reesei. Science 249*:380-386.

Sand W, Gehrke T. 1999. Analysis and function of the EPS from the strong acidophile *Thiobacillus ferroxxidans*. In: *Microbial Extracellular Polymeric Substances*. J. Wingender, T.R. Neu, and H.-C. Flemming (Eds). Springer International. Pp 127-141.

Sato K, Tomita M, Yonemura S, Goto S, Sekine K, Okuma E, Takagi Y, Hon-nami K, Saiki T. 1993. Characterization of ethanol hyper-production by *Clostridium thermocellum* I-1-B. Biosci. *Biotechnol. Biochem. 57:* 2116-2121.

Sauer K, Camper AK, Ehrlich GD, Costerton JW, Davies DG. 2002. *Pseudomonas aeruginosa* displays multiple phenotypes during development as a biofilm. *J. Bacteriol. 184*:1140–1154.

Sauer K, Camper AK. 2001. Characterization of phenotypic changes in *Pseudomonas putida* in response to surface-associated growth. *J. Bacteriol. 183*:6579–6589.

Schwarz WH. 2001. The cellulosome and cellulose degradation by anaerobic bacteria. *Appl. Microbiol. Biotechnol. 56*:634-649.

Shoham Y, Lamed R, Bayer EA. The cellulosome concept as an efficient microbial strategy for the degradation of insoluble polysaccharides. *Trends Microbiol. 7*:275-280.

Stevenson DM, Weimer PJ. 2005. Expression of 17 genes in *Clostridium thermocellum* ATCC 27405 during fermentation of cellulose or cellobiose in continuous culture. *Appl. Environ. Microbiol., 71*:4672-4678.

Stoodley PP, Sauer K, Davies DG, Costerton JW. 2002. Biofilms as complex differentiated communities. *Annu. Rev. Microbiol.56*:187-209.

Strobel HJ, Caldwell FC, Dawson KA. 1995. Carbohydrate transport by the anaerobic thermophilic *Clostridium thermocellum* LQRI. *Appl Environ Microbiol 61*:4012-4015.

Teeri TT, Koivula A, Linder M, Wolfaardt G, Divne C, Jones TA. 1998. *Trichoderma reesei* cellobiohydrolases: why so efficient on crystalline cellulose? *Biochem. Soc. Tranc. 26*:173-178.

Tormo J, Lamed R, Chirino AJ, Morag E, Bayer EA, Shoham Y, Steitz TA. 1996. Crystal structure of a bacterial family-III cellulose-binding domain: A general mechanism for attachment to cellulose. *EMBO J. 15*:5739-5751

Tyurin, MV, Desai, S, Lynd L. 2004. Electrotransformation of *Clostridium thermocellum. Appl. Environ. Microbiol. 70*:833-890.

Walter S, Rohde M, Machner M, Scrempf H. 1999. Electron microscopy studies of cell-wall-anchored cellulose (Avicel)-binding protein (AbpS) from *Streptomyces reticuli. Appl. Environ. Microbiol. 65*:886-892.

Johnson EA, Sakajoh M, Halliwell G, Madia A, Demain AL. 1982. Saccharification of complex cellulosic substrates by the cellulase system from *Clostridium thermocellum. Appl. Environ. Microbiol. 43*:1125-1132.

Warner JB, Lolkema JS. 2003. CcpA-dependent carbon catabolite repression in bacteria. *Microbiol. Mol. Biol. Rev. 67*:475-490.

Weimer, P. J. 1996. Why don't ruminal bacteria digest cellulose faster? *J. Dairy Sci. 79*: 1496-1502.

Weimer PJ, Chou YCT. 1986. Anaerobic fermentation of woody biomass pretreated with supercritical ammonia. *Appl. Environ. Microbiol. 52*:733-736.

Weimer PJ, Conner AH, Lorenz LF. 2003. Fermentation residues from *Ruminococcus* cellulose fermentations as novel components of wood adhesive formulations. *Appl. Microbiol. Biotechnol. 63*: 29-34.

Weimer PJ, Koegel RG, Lorenz LF, Frihart CR, Kenealy WR. 2005. Wood adhesives prepared from lucerne fiber fermentation residues of *Ruminococcus albus* and *Clostridium thermocellum*. *Appl. Microbiol. Biotechnol. 66:*635-640.

Weimer PJ, Lopez-Guisa JM, French AD. 1990. Effect of cellulose fine structure on kinetics of its digestion by mixed ruminal microorganisms in vitro. *Appl. Environ. Microbiol. 56*:2421-2429.

Weimer PJ, Odt CL. 1995. Cellulose degradation by ruminal microbes: physiological and hydrolytic diversity among ruminal cellulolytic bacteria. *ACS Symp. Ser. 618*:291-304.

Weimer PJ, Zeikus JG. 1977. Fermentation of cellulose and cellobiose by *Clostridium thermocellum* in the absence of *Methanobacterium thermoautotrophicum. Appl Environ Microbiol 33*:289-297.

Wiegel J, Mothershed CP, Puls J. 1985. Differences in xylan degradation by various noncellulolytic thermophilic anaerobes and *Clostridium thermocellum*. *Appl. Environ. Microbiol. 49*:656-659.

Schofield P, Pitt RE, Pell AN. 1994. Kinetics of fiber digestion from in vitro gas production. *J Anim Sci 72*:2980-2991.

Van Soest PJ. 1973. The uniformity and nutritive availability of cellulose. *Fed Proc 32*:1804-1808.

Wood TM, McCrae SI. 1978. The cellulase of *Trichoderma koningii. Biochem. J. 171*:61-72.

Zhang YH, Lynd LR. 2003. Quantification of cell and cellulase mass concentrations during anaerobic cellulose fermentation: development of an ELISA-based method with application to *Clostridium thermocellum* batch cultures. *Anal. Chem. 75*:219-227.

Zhang Y-HP, Lynd LR. 2004a. Kinetics and relative importance of phosphorolytic and hydrolytic cleavage of cellodextrins and cellobiose in cell extracts of *Clostridium thermocellum*. *Appl. Environ. Microbiol. 70*:1563-1569.

Zhang Y-HP, Lynd LR. 2004b. Toward an aggregated understanding of enzymatic hydrolysis of cellulose: non-complexed cellulase systems. *Biotechnol. Bioeng. 88*:797-824.

Zhang Y-HP, Lynd LR. 2005a. Regulation of cellulase synthesis in batch and continuous cultures of *Clostridium thermocellum*. *J. Bacteriol. 187*:99-106.

Zhang Y-HP, Lynd LR. 2005b. Cellulose utilization by *Clostridium thermocellum*: bioenergetics and hydrolysis product assimilation. *Proc. Natl. Acad. Sci. USA 102*:7321-7325.

Zverlov VV, Velikodvorskaya GA, Schwarz WH. 2003. Two new cellulosome components encoded downstream of cell in the genome of *Clostridium thermocellum*: the non-processive endoglucanase CelN and the possibly structural protein CseP. *Microbiology 149*:515-524.

In: Cellulosome
Eds: V. Uversky and I. A. Kataeva, pp. 119-151
ISBN 1-59454-950-8

Chapter VI

THE *C. THERMOCELLUM* CELLULOSOME: NOVEL COMPONENTS AND INSIGHTS FROM THE GENOMIC SEQUENCE

Vladimir V. Zverlov[1,2] *and Wolfgang H. Schwarz*[2*]
[1]Institute of Molecular Genetics, Russian Academy of Science, Moscow, Russia,
[2]Institute for Microbiology, Technische Universität München, Freising-Weihenstephan, Germany.

ABSTRACT

C. thermocellum is a very effective cellulose degrading bacterium which hydrolyzes a wide range of polysaccharides from lignocellulosic biomass. It is thus of increasing interest for the industrial utilization of lignocellulosic biomass in biotechnology. This bacterium ferments only the sugars derived from cellulose. Its extracellular, cell-bound multienzyme complex (cellulosome) consists of 13 major and an unknown number of minor catalytic components which are attached to structural proteins by protein-protein interaction. The major components of the cellulosome not only comprise the cellulases Cel48S, Cel8A, Cel9R, Cel9K, Cbh9A, Cel9N and Cel5G, but also the xylanases Xyn10C, Xyn10D and Xyn10Z as well as xyloglucanase Xgh74A, chitinase Chi18A. *C. thermocellum* has a potential to degrade a number of non-cellulosic polysaccharides, among them pectin by pectate lyases and polygalacturonases. Putative pectin- and xylan-esterases are connected to hydrolytic modules and support depolymerization as do debranching enzymes like the putative β-xylosidase, α-arabinofuranosidase and glucuronosidase activities for the hydrolysis of xylan. *C. thermocellum* may thus play a key role in crossfeeding of other bacteria in saprophytic ecosystems with breakdown products from hemicellulose and other non-cellulosic polysaccharides.

* Correspondence concerning this article should be addressed to Dr. Wolfgang H. Schwarz, Research Group Microbiology, Technische Universität München, Am Hochanger 4, D-85350 Freising-Weihenstephan, Germany, Fax: (+49)-8161-71-5475, E-mail: wschwarz@wzw.tum.de.

Newly detected enzyme components, putative genes and their structure are discussed on the basis of the genomic sequence and recent experimental findings. Besides the cellulosomal (cell-bound) enzyme system *C. thermocellum* seems to possess a non-cellulosomal (soluble) cellulase system. The potential utilization of the energetically more feasible larger cellodextrins in the context of the enzymatic activities of the major cellulosomal components is discussed.

Keywords: cellulose hydrolysis, enzyme complex, genome analysis, proteomics

Introduction

Clostridium thermocellum is a thermophilic, strictly anaerobic, chemo-organotrophic, spore forming bacterium with growth between 55 and 70 °C. It has relatively limited nutrient requirements and can be grown on synthetic media [24,42,81]. It degrades and ferments the recalcitrant substrate crystalline cellulose [65] and its soluble oligomers by a mixed acid fermentation. *C. thermocellum* is readily isolated worldwide from soil, compost, hot springs and other places where lignocellulosic biomass is abundant, although few strains were studied in detail. Such habitats often have anaerobic micro-environmental niches but are not necessarily warm (e.g. by self heating or geothermal activity): like many thermophilic bacteria *C. thermocellum* can be isolated from mesophilic environments where it seems to thrive well even at suboptimal temperatures ([58]; unpublished results). Isolates of *C. thermocellum* from various places at least in the Northern hemisphere appear to be physiologically and genetically quite homogeneous ([56,14,78]; unpublished results). *C. thermocellum* thus is an ubiquitously occurring biomass associated bacterium that gains its energy from degradation of lignocellulosic material, most probably exclusively by hydrolysis of different ß-glucans, especially crystalline cellulose and cellodextrins. It can be attributed an important role in the carbon cycle where it contributes considerably to mineralize the polysaccharides from plant biomass.

C. thermocellum has attracted high interest as a very effective cellulose decomposer at elevated temperatures. The consequence is that its exoenzymes have a high stability against a number of physical and chemical stress conditions like temperature, denaturing agents and proteases [25]. Under the artificial conditions of a laboratory it is growing slowly with tiny, whitish-opaque colonies in agar layers or on the surface of agar plates with cellulose as sole carbon source: the enzymes for the complete degradation of the insoluble cellulose particles in the agar are partially diffusing away from the cells and create a halo around the colonies (Fig. 1). The bacterium uses cellulose and cellodextrins readily as substrate and ferments them to the major products ethanol and acetate in varying ratios, H_2 and CO_2, and minor amounts of formate and lactate. Its ethanol production from the abundant and cheap substrate cellulose reaches up to 2 mole per mole glucose equivalent.

Mutants of *C. thermocellum* with enhanced resistance to the toxic solvent ethanol have been isolated which produced up to 30 g/l ethanol directly from cellulose [13]. This potential made the bacterium a highly interesting research target for the direct conversion of lignocellulosic plant biomass to fuel ethanol and other value added products, and thus for sustainable energy production [75,57]. The genomic DNA sequence of the type strain ATCC

27405 has been determined because of its importance for natural cellulose breakdown and its potential use in modern biotechnology for solvent or fuel production from lignocellulosic biomass (DOE Joint Genome Institute).

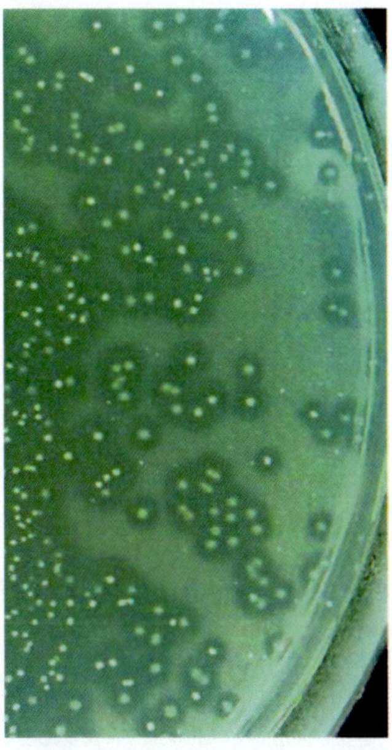

Fig. 1: Colonies of *C. thermocellum* on an 70 mm agar plate surface covered with a thin layer of crystalline cellulose (MN300). Halos indicate complete hydrolysis of cellulose crystals.

The accumulation of glucose and other reducing sugars on lignocellulose hydrolysis was reported by Park & Yu [67]. This makes *C. thermocellum* an ideal partner in co-fermentation or in pre-treatment of recalcitrant biomass for fermentation by other bacteria. *C. thermocellum* was used to deliver the necessary sugar and short-chain fatty acid components for the production of methane and solvents by other bacteria. Stable co-cultures were described with *Thermoanaerobacter thermohydrosulfuricus* for ethanol production [62], with *Methanosarcina* and *Methanobacterium* for methane gas production [76], or in sequential co-culture with *Klebsiella pneumoniae* for butanediol production [86].

C. thermocellum is highly specialized for the fermentation of cellulose and cellodextrins derived thereof. It was a matter of debate if it can utilize other substrates than cellodextrins: early reports showed that it is not well growing on glucose or fructose as sole carbon and energy source, where it also had a long lag phase [40,66,68,78,83]. Other monosaccharides, esp. pentoses, could not be fermented. We found it growing on ß-1,3(1,6)-glucan (laminarin) and ß-1,3-1,4-glucan (lichenan; [27]). But in general, due to the ease of contamination, reports of *C. thermocellum* growth on different substrates have to be interpreted with great care. However, cellobiose is widely used as a soluble and affordable substrate for culturing *C. thermocellum*. This substrate also solves another problem in working with *C. thermocellum*: the cells bind tightly to the crystalline cellulose through the extracellular cell bound cellulosomes [8,55] (Fig. 2). Isolation of protein and/or cells and their DNA is much easier when cells are freely suspended in the medium and insoluble substrates are absent.

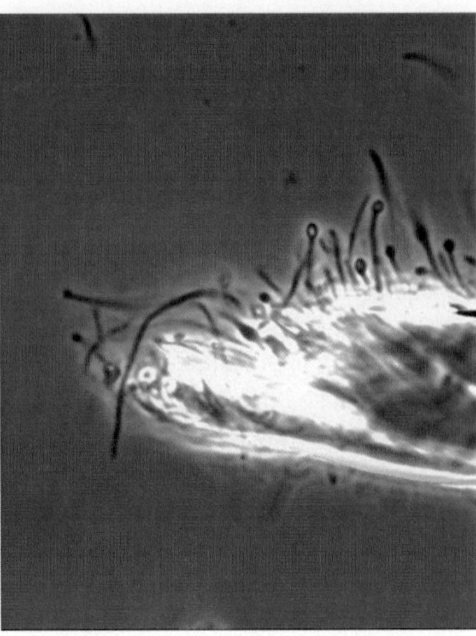

Fig. 2: Phase contrast microscopy of the end of a cellulose fibre (MN300) with attached *C. thermocellum* cells (partially sporulated).

CELLULOSE DEGRADATION BY *C. THERMOCELLUM*

C. thermocellum is the most effective cellulose degrading microorganism ever isolated in terms of enzymatic activity per mg of secreted protein (reviewed in [53]). The extraordinary efficiency of its hydrolytic machinery is due to the formation of a huge extracellular enzyme complex, the cellulosome, which has ~18 nm diameter and a mass in excess of $2x10^6$ Da [20,55]. Its composition and the mechanisms of cellulosome action were extensively reviewed [9,10,11,22,73]. In addition *C. thermocellum* seems to express a parallel soluble cellulase system (see below). The greatest advantage of an enzyme complex over a soluble enzyme system with diffusible proteins obviously is that the different enzymes in the complex are present in a high local concentration at the site of activity even if the overall concentration of protein is low. This adds to the economics of energy consumption for exoenzyme production in the anaerobic bacterium (a discussion of the energy consumption is to be found in [53,87,88]). The cellulosome components seem to have co-evolved for an optimal synergistic interplay of different catalytic subunits, of catalytic and substrate binding subunits and in positioning the different subunits in close neighbourhood on the surface of the topologically divers cellulose crystals. In addition, all components necessary for removing the network of the heterogeneous hemicellulose covering the cellulose are present in the complex (arabino- or glucuronoxylan, xyloglucan, galactomannan, pectin etc.). The lignin component of lignocellulose is obviously not attacked by the anaerobic bacteria.

The cellulase system of *C. thermocellum* can be prepared from culture filtrates of ageing cultures [43]. Purer preparations and preparations from actively growing cultures were obtained by washing the protein fraction adsorbed to cellulose fibres by the affinity-digestion technique which uses the adsorption to crystalline cellulose and its subsequent complete digestion [60]. The cellulose hydrolysis by isolated cellulosomes is optimal at 70 °C and a pH of 5.7-6.1. It is stimulated by addition of Ca^{++} and DTT, but activity is not abolished by oxygen [64,43]. Cellobiose is a potent end product inhibitor of the cellulase activity [51].

Since 1983, in a number of laboratories 23 hydrolytically active cellulosomal genes in *C. thermocellum* were identified solely by screening genomic DNA libraries in *E. coli* (reviewed in [34,73]). This screening was done with commercially available substrates, usually CMC, mixed linkage ß-1,3-1,4-glucan (lichenan), xylan or chromogenic aryl-ß-D-glucoside and -cellobioside. However, it is not surprising that only a fraction of the cellulosomal enzyme components present in the genome of *C. thermocellum* could be cloned this way. This is in part due to the limited number of substrates used for screening the genomic libraries. E.g. neither pectinases, nor xyloglucanases, esterases, nor glycosidases have been screened for, and except for accidentally found esterases such genes have not been isolated so far [16]. The accidental identification of unexpected genes in *C. thermocellum* was hampered by the obvious lack of cellulosomal gene clusters [11,73]. It was earlier concluded that, in stark contrast to the cellulosomal operons found in the mesophilic cellulolytic bacteria [22], the cellulosomal genes of *C. thermocellum* are scattered over the chromosome [34]. This is corroborated by the lack of contiguous reading frames with the newly detected ORFs in the genomic sequence (table 1: neighbouring genes would have consecutive frame numbers).

The glycosyl hydrolases were sorted by comparison of their primary sequences into 97 families many of which contain catalytic proteins of *C. thermocellum* [21]. To distinguish the glycanase genes easily, the number of the GH family is included in the protein designation: e.g. Cel8A for the gene product of *celA*, which is characterized by a GHF8 catalytic module [39]. Members of one family share an identical hydrolytic mechanism, either retaining or inverting the anomeric carbon. The substrate recognition range within a family is quite narrow in most cases so that by knowing the family (through sequence similarity analysis) the activity of the protein can be surmised, albeit not with 100 % certainty.

A characteristic feature of *C. thermocellum* is its binding to crystalline cellulose [8]. The same authors obtained a mutant of *C. thermocellum* by selecting for cells not adhering to cellulose fibres: they did not bind to cellulose. The non-adhesive mutants had greatly impaired cellulose hydrolysis ability [8]. The experiment was repeated in our lab (unpublished results) for investigating the mutated cellulosome complex with 2D-analysis of the proteins.

Binding of the enzyme to the crystalline substrate surface has an important function for cellulolysis: it may overcome part of the forces holding a cellulose molecule in the crystal; it should favour hydratization of the cellulose molecules which is a precondition for hydrolytic cleavage; binding by modules separate from the catalytic module adds flexibility to the enzyme substrate complex; it reduces diffusion of the substrate from the enzyme (or of the enzyme from the substrate) once a bond is hydrolysed so that the juxtaposed bond can be substrate for subsequent cleavage. At the same time the extracellular enzyme complex is bound to the cell surface by S-layer homologous modules (SLH) which anchor the protein within the cell wall (reviewed in [74]). The cell is thus near to the substrate surface which is much larger than the bacterium (Fig. 2). This enables the cell to take up with great efficiency the products of substrate hydrolysis immediately at the site of cellulose hydrolysis without loss by diffusion etc., i.e. the cellodextrins produced by endo- as well as exoglucanases (see below). As soon as the produced cellodextrins are small enough to be soluble, they may be taken up by the transport machinery of the cells, which "sees" a higher concentration before the products are diluted by diffusion or further degraded. This removes the reaction products from the enzymes and thus reduces end product inhibition as well as further degradation into small oligo- or even monosaccharides which are not nearly as good substrates for *C. thermocellum* [53].

Table 1: List of putative cellulosomal components in the whole draft genome sequence of *Clostridium thermocellum*: reading frames containing a dockerin module are ordered according to their putative function (derived from the glycosyl hydrolase family).

	Gene	Reading frame	M_w	Ref.	Properties and function	Module structure
		Structural component				
1.	CipA+	CtheDRAFT_1367 ZP_00510056	197	[28, 92]	scaffoldin, structural component	2(Coh1)-CBM3a-7(Coh1)-UN-Doc2
		Cellulase / ß-glucanase				
2.	CelO	CtheDRAFT_0386 ZP_00510834	75	[93]	cellobiohydrolase (from reducing end); ß-glucan, CMC, PASC, Avicel	CBM3b-GH5-Doc1
3.		CtheDRAFT_0957 ZP_00510548	103			GH5-CBM6-Fn3–Doc1
4.	CelB	CtheDRAFT_2770 ZP_00504345	64	[32, 19]	endoglucanase, CMC	GH5-Doc1
5.	CelG +	CtheDRAFT_1813 ZP_00505122	63	[52]	endoglucanase, CMC	GH5-Doc1
6.		CtheDRAFT_2160 ZP_00504816	58			GH5-Doc1
7.	CelA +	CtheDRAFT_3757 ZP_00503653	53	[12, 97, 19, 71]	endoglucanase, ß-glucan, CMC, PASC, Avicel	GH8-Doc1
8.	CbhA +	CtheDRAFT_2166 ZP_00504821	138	[91, 93, 80, 59]	cellobiohydrolase (from the non-reducing end)	CBM4-Ig-GH9-2(Fn3)-CBD3b-Doc1
9.	CelK +	CtheDRAFT_2165 ZP_00504820	101	[92, 46]	cellobiohydrolase	CBM4-Ig-GH9-Doc1
10.	CelD	CtheDRAFT_2210 ZP_00505048	72	[44, 45]	endoglucanase, CMC, PASC, pNPCb	Ig-GH9-Doc1
11.		CtheDRAFT_1914 ZP_00505158	110			GH9-CBM3c-CBM3b-Doc1
12.		CtheDRAFT_2704 ZP_00504466	105			GH9-CBM3c-CBM3b-Doc1
13.	CelN +	CtheDRAFT_3100 ZP_00504387	82	[97, 95]	endoglucanase non processive, barley ß-glucan, CMC, PASC, celloextrins > G4. Main product > G4	GH9-CBM3c-Doc1
14.	CelR +	CtheDRAFT_2821 ZP_00504387	85	[97]	processive endoglucanase, ß-glucan, CMC, PASC. Main product G4, no larger products.	GH9-CBM3c–Doc1
15.	CelQ +	CtheDRAFT_2570 ZP_00504610	80	[6]	endoglucanase. β-glucan, CMC	GH9-CBM3c-Doc1
16.	CelF	CtheDRAFT_2777 ZP_00504352	82	[63]	endoglucanase, CMC, β-glucan	GH9-CBM3c-Doc1

Table 1. Continued

	Gene	Reading frame	M_w	Ref.	Properties and function	Module structure
17.		CtheDRAFT_1915 ZP_00505159	80			GH9-CBM3c-Doc1
18.		CtheDRAFT_2181 ZP_00504836	89			GH9-CBM3c-Doc1
19.		CtheDRAFT_0785 ZP_00510325	82			GH9-CBM3c–Doc1
20.		CtheDRAFT_3633 ZP_00503648	63			GH9-Doc1
21.	CelT +	CtheDRAFT_2305 ZP_00504677	69	[49]	endoglucanase, CMC, ß-glucan, pNPCb, Avicel. Poor hydrolysis of G4.	GH9-Doc1
22.	CelS +	CtheDRAFT_2605 ZP_00504587	83	[82, 48]	exoglucanase (CMC viscosity): PASC, Avicel, CMC. No pNPCb! Avicel to G2+G3 (ratio 5:1); G4 to G2, G5 to G2+G3	GH48-Doc1
		Xylanases				
23.	XynD +	CtheDRAFT_3376 ZP_00503824	72	[96]	endo-xylanase: X3 and X4 to X+X2	CBM22-GH10–Doc1
24.	XynC +	CtheDRAFT_2387 ZP_00504483	70	[36]	endo-xlanase: X2 and X3 no cut, X4-8, and xylan to X1, X2, X3	CBM22-GH10-Doc1
25.	XynA, XynU +	CtheDRAFT_2454 ZP_00505196	74	[37]	endo-xylanase: X2 no activity, weak on X3, X4-8, to X1, X2, X3	GH11-CBM4-Doc1-CE4
26.	XynB, XynV +	Missing in genome seq. (connecting error)	50	[37]	endo-xylanase, high homology to XynA	GH11-CBM4-Doc1
		Other hemicellulases				
27.	LicB +	CtheDRAFT_3729 ZP_00503683	38	[89]	lichenase, endo-1,3-1,4-ß-glucanase: barley ß-glucan	GH16-Doc1
28.	ChiA +	CtheDRAFT_3758 ZP_00503652	55	[94]	endo-chitinase: pNP-ß-D-N,N'-diacetylchitobiose and -triose	GH18-Doc1
29.	ManA +	CtheDRAFT_2304 ZP_00504676	67	[35]	mannanase: mannan and manno-oligosaccharides (M3 and larger)	CBM-GH26-Doc1
30.		CtheDRAFT_3074 ZP_00503934	67			GH26-Doc1
31.		CtheDRAFT_1283 ZP_00510270	71			GH30-CBM6-Doc1
32.		CtheDRAFT_1461 ZP_00509877	47			GH53-Doc1
33.		CtheDRAFT_1622 ZP_00509828	86			GH81-Doc1
		Putative glycosidases				
34.		CtheDRAFT_0953 ZP_00510544	104			GH2-CBM6-Doc1
35.		CtheDRAFT_0550 ZP_00510823	88			GH39-2(CBM6)-Doc1
36.		CtheDRAFT_0954 ZP_00510545	59			GH43-CBM6-Doc1
37.		CtheDRAFT_1623 ZP_00509829	64			GH43-CBM13-Doc1
38.		CtheDRAFT_3790 ZP_00503587	75			GH43-2(CBM6)-Doc1

Table 1. Continued

	Gene	Reading frame	M_w	Ref.	Properties and function	Module structure
		Xyloglucanhydrolase				
39.	XghA +	CtheDRAFT_1463 ZP_00509879	92	[96]	endo-xyloglucan-hydrolase: tamarind xyloglucan to XXXG, XLXG and XLLG	GH74-CBM2-Doc1
		Putative carbohydrate esterases				
40.		CtheDRAFT_2282 ZP_00504911	91			Fn3-CE12-Doc1-CBM6-CE12
41.		CtheDRAFT_2234 ZP_00505069	55			CE3-CE3-Doc1
42.		CtheDRAFT_1808 ZP_00505120	55			Doc1-CE6
43.		CtheDRAFT_0956 ZP_00510547	54			CE1-CBM6-Doc1
		Putative pectinases				
44.		CtheDRAFT_2655 ZP_00504538	92			GH28-Doc1
45.		CtheDRAFT_1188 ZP_00510121	60			PL1-Doc1-CBM6
46.		CtheDRAFT_2861 ZP_00504275	98			PL1-Doc1-CBM6-PL9
47.		CtheDRAFT_1189 ZP_00510122	42			PL10-UN-Doc1
48.		CtheDRAFT_1006 ZP_00510295	89			Doc1-CBM6-PL11
		Multifunctional components				
49.	CelJ +	CtheDRAFT_2568 ZP_00504611	178	[2,3, 7]	Exo- or processive endo-glucanase: >G4, PASC or Avicel to G2, G3	CBM30-Ig-GH9-GH44-Doc1-UN
50.	CelH	CtheDRAFT_0662 ZP_00510594	102	[85]	endoglucanase, CMC, pNPCb, ß-glucan, larchwood xylan	GH26-GH5-CBM11-Doc1
51.		CtheDRAFT_0552 ZP_00510825	111			GH30-GH54-GH43-Doc1
52.		CtheDRAFT_1871 ZP_00505028 CtheDRAFT_1872 ZP_00505029 Frame shift?	79			GH54-Doc1-GH43
53.		CtheDRAFT_0551 ZP_00510824	66			GH54-GH43-Doc1
54.	XynZ +	CtheDRAFT_1505 ZP_00510011	92	[93, 97, 33, 16]	endo-xylanase: xylan to X2. CE1 activity yielding ferulic acid	CE1-CBM6-Doc1-GH10
55.	XynY	CtheDRAFT_1224 ZP_00509930	120	[26]	endo-xylanase: oat spelt, wheat and rye arabinoxylan, pNPCb	CBM22-GH10-CBM22-Doc1-CE1
56.	CelE +	CtheDRAFT_2235 ZP_00505070	90	[38]	endoglucanase: CMC, xylan	GH5-Doc1-CE2
		Putative protease and protease inhibitors				
57.		CtheDRAFT_2248 ZP_00504916	40			Subtilisin-like serine protease–Doc1-Doc1
58.		CtheDRAFT_3710 ZP_00503703	64			Fn3-Doc1-serpin

Table 1. Continued

	Gene	Reading frame	M_w	Ref.	Properties and function	Module structure
59.		CtheDRAFT_3709 ZP_00503704	68			Doc1-serpin
		Components with unknown function				
60.		CtheDRAFT_1002 ZP_00510288	117			2(UN)-UN-UN(CelP 550-870)-Doc1
61.		CtheDRAFT_0955 ZP_00510546	105			UN-CBM6-Doc1
62.	CseP +	CtheDRAFT_3069 ZP_00503930 (possibly structural component)	62	[95]		UN-Doc1
63.		CtheDRAFT_0895 ZP_00510083	58			Doc1-UN
64.		CtheDRAFT_3747 ZP_00503664	51			Doc1-U
65.		CtheDRAFT_2099 ZP_00504894	76			Doc1-UN
66.		CtheDRAFT_1199 ZP_00510210	65			2(UN)-Doc1
67.		CtheDRAFT_0399 ZP_00510814	135			UN-UN-UN-Doc1
68.		CtheDRAFT_2183 ZP_00504838	40			Doc1-UN
69.		CtheDRAFT_2277 ZP_00504920	47			UN-Doc1
70.		CtheDRAFT_3356 ZP_00503784	37			UN-Doc1
71.		CtheDRAFT_2419 ZP_00504513	236			2(UN)-UN-Doc2
72.		CtheDRAFT_1559 ZP_00509775	19			Doc1-Doc1

Abbreviations: PASC, phosphoric acid swollen cellulose; pNPCb, p-nitrophenyl-ß-cellobioside; G1, G2, glucose, cellobiose etc.; X2, xylobiose etc.; GH, glycosyl hydrolase family; Coh, cohesin module; Doc, dockerin module; CBM, carbohydrate binding module; UN, unknown module; Fn3, fibronectin homologous module; Ig, immunoglobulin like fold; CE, carbohydrate esterase module; PL, pectate lyase module; serpin, serine protease inhibitor.

The gene designation (if known), the molcular mass of the gene product (M_w), properties and function of the protein (if known), references for the gene and gene product, and the module structure are indicated. The locus_tag (CtheDRAFT...) and the protein ID (ZP...) are given (from URL: http://genome.ornl.gov/microbial/cthe/). A + sign behind a gene designation indicates experimental proof for cellulosomal localization. In the "properties and function" column the substrates on which an enzyme is active is indicated.

Cellulosomal Genes in the Genome of *C. Thermocellum*

The number of known cellulosomal genes was increasing with each screening of a genomic library [34]. Hitherto the genes were accidentally found by sequencing new interesting hydrolases and identifying a typical dockerin sequence which is an indicator for cellulosomal localization. With the increasing extent of genomic library screenings it became clear that the number of genes coding for cellulosomal components was by far greater than the 25 or so genes known to date (see URL: http://www.wzw.tum.de/mbiotec /celoscomp.htm). However, not all genes for "cellulosomal components" are necessarily expressed in *C. thermocellum* and incorporated into the extracellular complex (at least not under the growth conditions used in the lab). But most proteins could also be found to be present in the purified complex if an antibody search with purified cellulosomes was undertaken. However, quantification was difficult with this method and the sensitive antibody assay would detect even the trace amounts of a very rare protein, i.e. a protein present in only a small fraction of the cellulosomal particles. The cellulosome complex nevertheless appears to be incredibly heterogeneous, and in spite of the limited number of cohesin sites the distribution of an individual enzyme would come down to much less then one per particle for most of the components. This calculation is made on the basis of one scaffoldin CipA component per particle and does not take into account the formation of the so called polycellulosomes which must contain much more than one scaffoldin.

With the appearance of the "whole draft" genomic sequence of the type strain *C. thermocellum* ATCC 27405 by the Joint Genome Institute (URL: http://genome.jgi-psf.org/draft_microbes/cloth/cloth.home.html), it became possible to screen systematically for reading frames of potential cellulosomal components. Genes for the catalytically active cellulosomal components of *C. thermocellum* always contain a dockerin type I sequence as a unique signature sequence [54]. This consists of two repeated 21 amino acid motives which promote highly specific binding to the cohesin sequences of the scaffoldin CipA (reviewed in [11]). A distantly related type II dockerin binds the structural components like CipA to special cell wall components, but has no cross-reactivity with cohesins of type I.

Typical sequences of both types of dockerin sequences were included in a BLAST search against the draft genomic sequence. Apparent reading frames were checked manually for complete catalytic and non-catalytic modules known from other hydrolytic enzymes, for Shine-Dalgarno sequences, start and stop codons to avoid overannotation and to select potentially complete genes. The search identified 71 reading frames containing a dockerin module [96]. Their identity and structure is presented in table 1.

The number of 71 ORFs for cellulosomal components is surprisingly high considering the limited fermentation abilities of *C. thermocellum* which would only require degradation of the chemically simple polysaccharide cellulose. However, the structural complexity of cellulose crystals makes a great number of different enzymes with varying modes of activity and binding characteristics necessary [10,73]. In addition, less than one third of the putative cellulosomal genes in the genome (= 19) can be ascribed a direct role in cellulose hydrolysis: the endo- and exoglucanases of GHF 5, 8, 9, and 48 (table 2). One might assume that the other genes may be structural components, but the majority is presumably active on hemicellulose, pectin or chitin and oligosaccharides derived thereof, and thus in the unwrapping of the cellulose fibres. The putative subtilisin-like serine protease gene and the two potential protease inhibitor genes (serpin) in the cellulosome (pos. 57-59 in table 1) were surprising and their function remains to be elucidated.

Table 2: Number of ORFs and functional protein families of cellulosomal genes.

Number of ORFs	Potential enzymatic activity	Module families / comment / known genes
1	structural protein	CipA
21	cellulase, ß-glucanase	5 x GH5, 1 x GH8, 14 x GH9, 1 x GH48
4	xylanase	GH10, GH11
1	xyloglucanase	GH74
5	pectinase	GH28, PL1, PL9, PL10, PL11
2	carbohydrate-esterase	CE1, CE12
7	other hemicellulases	GH16, GH18, 2 x GH26, GH30, GH53, GH81
5	glycosidase	GH2, GH39, 3 x GH43
8	multifunctional enzyme	GH 9+GH44, GH5+GH26, GH5+CE2, 3 x GH43+GH54, 2 x GH10+CE1
10	unknown function	CseP

Reading frames (ORFs) in the whole draft genomic sequence of *Clostridium thermocellum* which are likely to express cellulosomal components. All ORFs contain a dockerin module, some have in addition non-catalytic modules which are ignored here. Short DNA fragments containing dockerin sequences are not listed.

Table 1 shows the putative genes identified in the genomic sequence of *C. thermocellum* in groups according to the suggested enzymatic activity of their catalytic modules (if present). Each single protein (gene) is listed including its designation, its enzymatic activity or function (if determined), and the number of the ORF in the data base. If known, the gene designation is indicated. Eight multifunctional components with more than one catalytic module and 13 components with modules of unknown function are listed at the end. Groups of putative function are summarized in table 2.

The cellulosomal genes previously cloned and investigated are all showing up in the genomic sequence. This is an indication for the relative completeness of its "whole draft" status, albeit a number of gaps and not connected contigs are still present. Altogether 34 hitherto unknown reading frames were newly detected and await characterization of their biochemical activity or function. These reading frames are potential genes for cellulosomal components, but expression and functionality has to be shown experimentally one by one.

Potential Cellulosomal Cellulase Genes in the Genome

In bacteria the enzymes containing catalytic modules of GHF5, GHF8, GHF9 and GHF48 were reported to contribute to cellulase activity [22,73]. All cellulosomal bacteria known so far contain one copy of a processive endoglucanase which is the most prominent protein in gel electrophoresis of denaturated cellulosomes: a GHF48 enzyme, in case of *C. thermocellum* named Cel48S. Only in *C. thermocellum* was a second GHF48 gene identified which, however, does not contain a dockerin module and thus is not a cellulosomal component (see below). Another singular cellulosomal enzyme was found: it contains a

GHF8 catalytic module, Cel8A, which is also found as a singular enzyme in the cellulosomes of *C. josui* and *C. cellulolyticum*, but not in the other cellulosomes known so far (reviewed in [22]). In contrast, potential ß-glucanase genes with GHF5 and GHF9 catalytic modules are abundant in genomes of cellulolytic bacteria: in *C. thermocellum* 5 genes with GHF5 and 12 with GHF9 modules (table 2). Many of them have in addition to the catalytic module one or more CBM3, CBM4 or CBM6 modules for substrate binding; two possess a fibronectin-type Fn3 module. The role of the different catalytic and non-catalytic modules in clostridial enzymes is reviewed in [74].

The module architecture of cellulosomal enzymes is variable but the modules are not statistically distributed (table 1): a pattern evolves. First of all, in most cases and in all ß-glucanases the dockerin module is located at the C-terminus of the enzyme. This seems to be a general rule for cellulosome components in all hitherto mapped cellulosomal genes in bacteria - with few exceptions esp. in combination with esterase modules (table 1; [22]). The GHF5 modules are located N-terminally with a C-terminal dockerin module; only in one still uninvestigated reading frame of this family is a CBM6 and a Fn3 module placed between the catalytic and the complex-mediating Doc module.

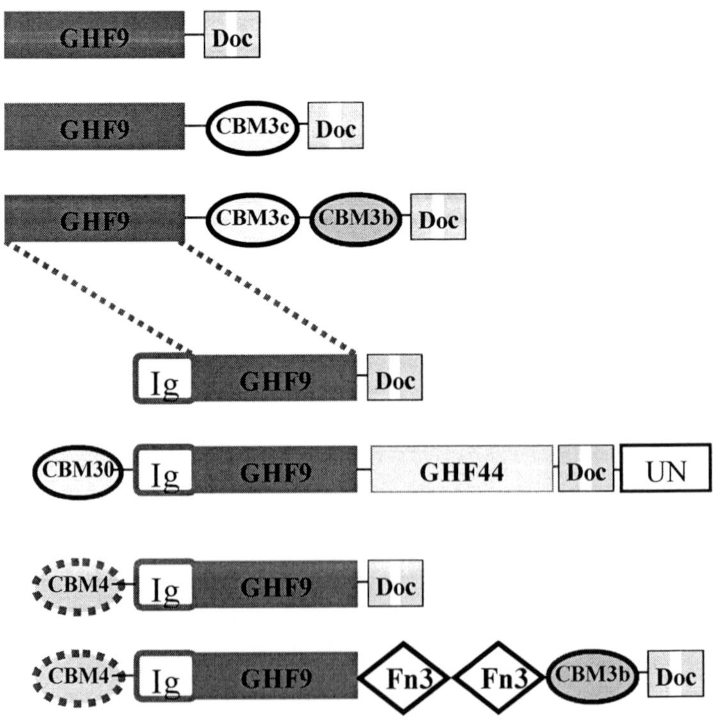

Fig. 3: Schematical drawing of the module architecture of all cellulosomal GHF9 components in the *C. thermocellum* genome.

The GHF9 genes represent a more complex pattern (Fig. 3). These were called themes A to D by Bayer et al. [10]. In some cases the combination with certain modules was shown to have an influence on the enzymatic mode. For the *Thermomonospora fusca* ß-glucanase E4 X-ray structure analysis and model building revealed the mechanism by which a cellulose molecule is held and fed into the active site pocket leading to a processive degradation of the polymer despite the active site pocket is open and could thus serve for a pure endoglucanase type of activity [41]. Cellulase E4 was characterized as a "processive endoglucanase" which produces a nick in the cellulose molecule followed by a processive cleavage of successive cellotetraose units from the non-reducing end.

The immunoglobulin-like fold (Ig module) in some GHF9 enzymes may function to modulate or stabilize the activity of the catalytic module. The Ig-connected catalytic modules form a distinct cluster in similarity trees [10]. It is interesting to note that the catalytic mode of GHF9 enzymes is quite divers: in *C. thermocellum* they appear as endo-glucanases (like in CelD, CelF, CelN, CelQ, CelT), processive endoglucanases (CelR, CelK) or as cellobiohydrolases (CbhA). One GHF9 module is coupled in a large multifunctional enzyme CelJ to a second catalytic module of GHF44 which also codes for an endo-ß-glucanase [2].

As shown in Fig. 3, two groups of GHF9-ß-glucanases evolve in the cellulosomal components: the first having an N-terminal catalytic module followed by the dockerin module and/or a CBM3c module. The binding module adjacent to the catalytic module was shown to have a stabilizing effect on the protein structure in a homologous thermostable enzyme from *Clostridium stercorarium* [70]. These enzymes form the majority of the GHF9 cellulases of *C. thermocellum* (11 out of 15) and most of them were characterized as endoglucanases, some as processive endoglucanases. At least some of them seem to be responsible for the production of cellotetraose (see below). The other group with the Ig-GHF9 module is the minority and does not seem to produce cellotetraose. The cellobiohydrolases are in this group.

Cellulosomal Hemicellulase Genes in the Genome

Six reading frames for potential xylanases of GHF10 and 11 were identified in the genomic sequence, three of them in combination with a carbohydrate esterase module of family CE1 or CE4 (table 1). Among them the three xylanases Xyn10C, Xyn10D and Xyn10Z were found as major components in the cellulosome [96]. It is presently not clear why only GHF10 enzymes are among the major components. The catalytic modules of the xylanases are combined in all ORFs with at least one of the carbohydrate binding modules CBM4, 6 or 22.

As is typical for the GHF10 xylanases, Xyn10C, Xyn10D and Xyn10Z have high activity on arabino- and glucurono-xylan as well as on mixed-linkage ß-glucan [33,36,96]. The xylanases including the newly characterized xylanase Xyn10D degrade xylan initially to higher xylo-oligosaccharides and later to mainly xylose and xylotriose. The enzymes thus are endo-xylanases. Short xylo-oligosaccharides can not be degraded further by those enzymes. For the fast and complete digestion of arabino- and glucuronoxylan the simultaneous presence of esterases and debranching glycosidases is a precondition (see below).

Only one potential cellulosomal xyloglucanase, Xgh74A, was identified in the genome. It was also found to be a major component in the cellulosome [96]. Similar to the ß-xylanases of GHF10 it has high activity on mixed-linkage ß-glucan. However, the relatively high molecular weight of the hydrolysis products from barley ß-glucan as well as from Tamarind xyloglucan indicates specificity for a limited number of target sites (96). The products from xylan were identified by capillary electrophoresis coupled to mass spectrometry as hexa- and heptamers of the xyloglucan, the typical products of GHF74 endo-xyloglucanases. They suggest cleavage on the reducing end of a non-substituted glucose residue.

Other reading frames identified in the genome comprise two mannanases, one of which was cloned and characterized as ManA (GHF26; [35]), a potential ß-1,6-glucanase of GHF30, an endo-ß-1,4-galactanase (GHF53) and a ß-1,3-glucanase or a family which hitherto was only found in eukaryotes (GHF81). It remains to be shown if it is active in bacteria.

An enzyme with extremely high activity on mixed linkage ß-1,3-1,4-glucan (barley ß-glucan or lichenan) is an obviously minor component in the cellulosome: Lic16B [89]. In combination with the ß-1,3-1,4-glucan active xylanases, the cellulosomal endoglucanases, the cellobiohydrolases and the non-cellulosomal Lic16A [27] there is a multitude of enzymes which act on the readily degradable ß-glucans of the mixed-linkage type. These glucans are especially prevalent in grasses, which possibly are a preferred substrate for *C. thermocellum* in nature. These glucans could be an easily accessible and fermentable substrate for the bacterium which was shown to grow on barley ß-glucan [27].

Glycosidase Genes in the Genome

Few glycosidase activities have been screened from genomic libraries of *C. thermocellum* and are biochemically characterized. The two β-glucosidase genes BglA and BglB have been cloned, but both turned out to be non-cellulosomal [31]. Only in one other cellulosomal bacterium, *C. cellulovorans*, glycosidases such as an α-arabinofuranosidase and a β-galactosidase (ArfA and BgaA resp.) were identified and shown to be non-cellulosomal glycosidases although many such genes must be present in saccharolytic bacteria [47]. It came thus as a surprise to find a number of cellulosomal components with putative glycosidase activity. These enzymes may, if they are expressed and integrated in the extracellular complex, have a role in the debranching of hemicellulose and may thus synergistically enhance the activity of the endo-acting polysaccharidases. The complete hydrolysis of the oligosaccharides produced by the depolymerases (e.g. cellulases, xylanases) may at best be a side effect e.g. for the delivery of rare monosaccharides for cell metabolism; the complete degradation of the polymeric energy substrate and carbon source cellulose is rather disadvantageous as for *C. thermocellum* the uptake of cellodextrins with a chain length as long as possible is much more advantageous [53]. The role of some of the two ß-glucosidases BglA and BglB may rather be in the detoxification of aryl-glucosides than in the degradation of cellodextrins which is preferably effected by cellodextrin- and cellobiose-phosphorylase [4,5,87].

An example for the putative debranching activity associated with the cellulosomal glycosidases is the occurrence of six GHF43 enzymes with potential activity for ß-xylosidic and α-arabinofuranosidic side groups. Three of the genes contain these modules in multifunctional combination with a GHF54 module which should have similar substrate specificity. The combination of these two modules in enzyme components and their multiplicity underline the crucial role of β-xylosidases and α-L-arabinofuranosidases in the degradation of xylan, and the importance of the presence of enzymes with slightly different enzymatic activity pattern or substrate recognition (Fig. 4; see also [1]). In one case there is even a third catalytic module attached at the N-terminus: a GHF30 module with putative ß-1,6-glucanase activity as in homology to already characterized enzymes of this protein family. The occurrence of three catalytic modules in one polypeptide is rather rare. It has to be experimentally shown that all three modules are hydrolytically functional.

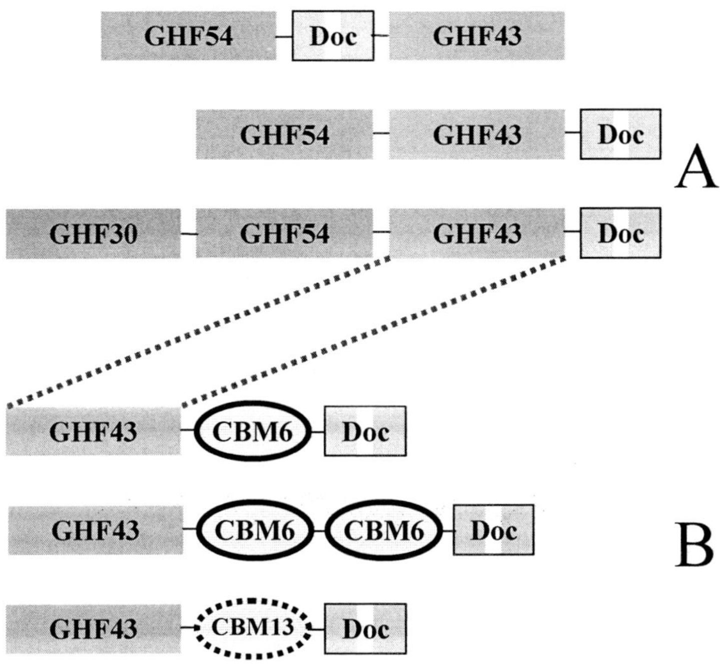

Fig. 4: Schematical drawing of the module architecture of all cellulosomal GHF43 (+GHF54) components in the *C. thermocellum* genome.

The combination of GHF43 with GHF54 modules marks a subgroup A within the GHF43 enzymes (Fig. 4): The GHF43 module is located C-terminally of the other functional modules; in two genes it is followed by a C-terminal dockerin module. In subgroup B all three genes have the GHF43 module at the N-terminus which is followed by one or two binding modules. It will be interesting to see, if the two isolated GHF43 modules from subgroup A and B have different substrate specificity. Another open question is, if the GHF53 modules are synergistically working together on the substrate(s), e.g. as the ß-xylosidase and the α-arabinofuranosidase part, respectively, whereby the alkyl- and not the aryl-glycosidase activity is relevant.

A conspicuous difference between subgroups A and B is the presence of carbohydrate binding modules. It can be surmised that the two subgroups have different substrate specificity and roles in hemicellulose degradation where, together with the xylan-esterases, they have a crucial function. The interplay of ß-xylosidases and α-arabinofuranosidases in the hydrolysis of arabinoxylan was recently demonstrated in the xylanase enzyme system of *C. stercorarium* [1]. A different ß-xylosidase with a GHF39 catalytic module is also present.

The presence of CBMs with the GHF43 and GHF39 enzymes underlines the importance of those enzymes in fibre hydrolysis.

The cellulosomal GHF2 enzyme (entry #34 in table 1) possibly is active as a ß-glucuronidase in analogy to some of the characterized members of that GH family. This enzyme activity is essential for the debranching of glucuronoxylan and was still missing in the catalogue of necessary depolymerization activities.

A Complete Pectinase System?

Pectin may be depolymerised in *C. thermocellum* cellulosomes by a polygalacturonase of GH28 and the pectin lyases of PL1, 9, 10 and 11. This spectrum of pectin, pectate and polygalacturonate hydrolyzing enzymes could constitute a complete pectin degradation system. *C. thermocellum* indeed has been reported to grow on pectin, although due to the long incubation times possible problems with contaminations have to be taken into account [76]. An active pectate lyase was identified in the cellulosome of the related cellulolytic bacterium *C. cellulovorans* [79].

Five genes for putative cellulosomal pectinase components were identified in the genomic sequence (entry #44-48 in table 1). Their existence in the cellulosome of *C. thermocellum* was hitherto unknown. Four of those genes have been cloned: entry #45-48 (table 1; unpublished results). The recombinant gene product of #45 was purified and characterized as a pectate lyase according to its hydrolytic activity on polygalacturonic acid as preferred substrate (unpublished results). The biochemical characterization of the pectinases is not yet completed, but all clones are enzymatically active and thus seem to constitute genes.

The preliminary results show that the *C. thermocellum* cellulosome contributes a complete hydrolysis system for the degradation of pectin and polygalacturonic acid. None of the proteins was identified among the 13 most prevalent cellulosomal components, but their expression may be enhanced by a yet unidentified inducer during growth on pectin as substrate.

The pectinases, polygalacturonases and pectate lyases are, as was mentioned above, supported by the carbohydrate esterases described below.

Other Hydrolase and Esterase Genes in the Genome

Downstream of the gene for the cellulosomal endo-ß-1,4-glucanase Cel8A, the cellulosomal gene *chiA* was identified by sequencing [94]. The purified heterologous protein was characterized as an active chitinase hydrolysing aryl-acetyl-chito-oligosaccharides such as *p*NP-chitobiose and *p*NP-chitotriose. The native molecular mass of the recombinant protein as well as of the native enzyme in the solubilized cellulosome was detected in denaturing SDS-PAGE after zymogram staining with methylumbelliferyl-ß-D-N,N′,N′′-triacetylchitotriose. Chi18A was shown as well with immunological methods as with MALDI-TOF as a cellulosomal component. With both methods the amount of Chi18A was calculated to be present in cellulose grown cellulosomes in approximately each 15. - 20. cellulosomal particle [96]. In addition it was shown by zymograms that Chi18A is the only chitinase in the cellulosome [94]. This was corroborated by the analysis of the whole-draft genomic sequence.

The role of a chitinase in a bacterial enzyme system designed for plant biomass degradation remains somehow obscure. However, the glucosamine from chitin catabolism may be an easily accessible ammonium source for the *Clostridium* which thrives in a generally nitrogen poor biomass. Moreover *C. thermocellum* is a successor e.g. in compost heaps were fungi and yeasts develop first. The fungal or yeast cell walls could be a substrate for the chitinase.

Eight different ORFs code for putative or proven esterases of 6 different protein families, carbohydrate esterases and acetyl-xylan esterases: two of them as multifunctional components with a GHF10, one with a GHF11 (xylanase) and one with a GHF5 module (ß-1,4-glucanase). In two genes esterase modules occur in conjunction with a CBM6 binding module. CE3 and CE12 modules appear also in duplicated form. Similar to debranching of hemicellulose by glycosidases, de-esterification is a precondition for effective de-polymerization of the polysaccharides, esp. of xylan, chitin and pectin. However, the only esterase family for which activity on pectin (and possibly rhamnogalacturonan) was reported is CE12 (present in entry #40). The other esterases seem to be active on xylan and different ß-glucans. Their activity and specificity has to be verified.

Genes for Non-Catalytic Components of the Cellulosome

Downstream of the *celN* gene a reading frame was identified, which contained a dockerin module and a module with unknown function, but no known catalytic module [95]. The purified recombinant protein was incubated with 20 different substrates containing various sugars, but no hydrolytic activity could be identified. However, the sequence was similar to that of a structural protein of *Bacillus subtilis* which is involved in spore morphogenesis. Thus a role in structure formation of the cellulosome was assumed. The presence of the protein in the cellulosome was proven with polyclonal antibodies and the protein was consequently named *cseP*: *cellulosomal structural element*. Nevertheless it was not found among the 13 major proteins identified in the cellulosome and thus is a minor component which may be present only during specific stages of cellulosome formation or under certain growth conditions.

Another reading frame contains a number of modules with unknown functions, one of them repeated and one with sequence similarity to CseP (entry #60). 10 more ORFs with unknown modules are present and there function could only be speculated upon.

It is interesting to note that a gene with similarity to a subtilisin-like serine protease and two genes for serine-protease inhibitors were recognized. Although it is not known if these are functional genes one might be tempted to speculate about a role in the decay of the cellulosomal particles which was reported for ageing cultures of *C. thermocellum* [51].

The Role of Carbohydrate Binding Modules

About half of the putative genes (33) contain a CBM. This high number was unprecedented and is esp. for the glycosidases difficult to explain. The CipA scaffoldin is contributing an almost irreversible binding of the whole complex to crystalline cellulose and has possibly a surface disrupting ability. Additional binding modules would principally not be necessary for keeping the enzyme complex near to the surface of the substrate. However, the function of the CBMs is more divers and seems also to be necessary to bind individual enzyme components to their specific substrates, especially at the molecular level of the single substrate molecule chains, like soluble ß-glucan or xylan strands.

In contrast the CBMs in the catalytic components belong to different CBM families and seem to attach the individual enzymes to their respective substrates or substrate sites, either by binding to amorphous regions in cellulose (as possible with CBM2, 3c, 6, 9), or to different types of hemicellulose (as with CBM2, 4, 6, 9, 13, 22) [11,17,21]. It is conspicuous that the binding module families connected to the esterases (CBM4, 6, 22) have binding characteristics with xylan and soluble ß-glucans, but not with crystalline cellulose. The pectinases are connected exclusively to CBM6.

The Module Architecture of Cellulosomal Proteins – Multifunctional Enzymes

The module architecture of the fungal cellulases and the soluble cellulase systems of aerobic bacteria is quite simple: the catalytic module is usually connected to a separately folding carbohydrate binding module (CBM) which fixes the enzyme to the crystalline or amorphous substrate. This type of module arrangement is found in fungal as well as in the soluble bacterial cellulases like *Thermobifida fusca* or *Clostridium stercorarium* [10,73]. In contrast, the extracellular soluble enzymes of the thermophilic anaerobic bacteria of the *Anaerocellum/Caldicellulosiruptor* group are extremely complex, consisting of e.g. two separate catalytic modules interspaced by linkers and connected to one or more CBMs in one polypeptide chain [30,90]. This architecture seems to enhance the enzyme synergism between two co-operating catalytic entities if they are tied together on the site of action [90]. This characteristic feature of the cellulase system of *Caldicellulosiruptor* or *Anaerocellum* was ascribed an intermediate role in the evolution between the soluble cellulase system of the aerobic bacteria (and the fungi) with enzymes of rather simple structure, and the cellulosomal protein complexes [10,73].

The cellulosome producing anaerobic bacteria found a different and much more elaborate, flexible and efficient way to combine various enzymes on a small substrate area: the cellulosome. In *C. thermocellum* the carbohydrate binding of the entire enzyme complex is exerted by a CBM3 module connected with the scaffoldin (or with one of the enzyme components in some mesophilic bacteria), but nevertheless most of the cellulosomal components have their own binding module, probably to bind to a different part of the large substrate particle. For binding to the cohesins on the scaffoldin they have a dockerin module.

In many cases additional spacer peptides or activity modulating modules are found. The function and structure of some of these modules is described in [74].

The mesophilic cellulosome producing bacteria have relatively simple enzymes with a catalytic and a dockerin module. Some have in addition a CBM [22]. But many of the ORFs and genes from *C. thermocellum* listed in table 1 have a highly complex module composition. Over 50 % of the putative cellulosomal proteins contain more than the minimum of two modules (the catalytic and the dockerin module) and even within the enzymes of the same GH family a great variation in module composition can be observed (see Fig. 3 and 4). This high complexity is a peculiarity of bacterial extracellular enzymes and seems to be connected to the enhanced economy and efficiency of protein production needed by anaerobic bacteria with their limited energy supply [53].

The potential cellulosomal genes of *C. thermocellum* contain about 45 structurally different modules, not included the unknown modules, for which no homologous sequences were detected in the data bases. They can be divided into 3 groups: the structural, binding and catalytic modules (table 3). The module arrangement is not random: in most cases the dockerin module is located at the C terminus, sometimes it separates two catalytic modules and only in two genes it is at the N-terminus. Examples of module arrangements are given in Fig. 3 and 4. The 15 ORFs containing GHF9 modules form variations by addition of further modules. The structures are visualized schematically in Fig. 3. Seemingly the position of additional modules at the C terminus of GHF9 allows for more flexibility, whereas the N terminus is substituted by only two types of CBMs in only 3 (of 15) genes.

Table 3: The modules occurring in cellulosomal components in three functional classes: Abbreviations see legend to table A. Unknown modules are not included.

Structural modules	Binding modules	Catalytic modules
Coh1	CBM2	GH2
Doc1, Doc2	CBM3a, 3b, 3c	GH5, 8, 9, 48
Fn3	CBM4	GH10, 11, 26
Ig	CBM6	GH16, 44, 74, 81
	CBM9	GH18
	CBM11	GH28
	CBM13	GH30
	CBM22	GH39, 43, 54
	CBM30	GH53
		CE1, 2, 3, 4, 6, 12
		PL1, 9, 10, 11
		Serpin

The modules are not combined randomly: with GHF9, an endo- or exo-ß-glucanase, only binding modules for soluble and insoluble cellulose strands are combined; the added GHF44 module in CelJ is also a ß-glucanase. Similarly, the GHF43 module is commonly combined with a GHF54 module (see above; Fig. 4); both represent ß-xylosidases or α-arabinosidases. These two modules are found in a great number of enzymes throughout the bacterial kingdom, although usually in separate polypeptides.

The complex module architecture leads to a considerable protein size which is rare even in bacteria. The two largest cellulases in the cellulosome, entries #8 (CbhA) and #49 (CelJ) in table 1, have a protein mass of 139 and 178 kDa and consist of 7 and 6 modules, respectively. Such huge proteins with more than thousand amino acid residues constitute a considerable burden for the bacterial energy balance and have to pay off the expense of their synthesis by a high activity on the substrate.

A problem with the modularity of the exo-enzymes in the cellulosomes is that they are prone to proteolytic degradation, especially in their non-structured proline-rich linker regions. This degradation is prevented by O-glycosylation of the secreted proteins in these peptide regions which are rich in the hydroxyl amino acids threonine and serine [29].

THE COMPONENTS ACTUALLY PRESENT IN CELLULOSE GROWN CELLULOSOMES

Since the discovery of the cellulosomes there protein composition was investigated and subject of keen interest. The first choice was the determination of their components by denaturing SDS-PAGE. The composition of the complex was analysed by 1D SDS-PAGE. However, in 1D gels the separation into only 14 not clearly separated bands (S1 to S14) was not unambiguous and some bands contained a mixture of cellulosomal components and their proteolytic fragments. Moreover, the band pattern changed with the preparation method and the age of the culture. It also was obvious that many more than 14 genes coded for cellulosomal components. 19 of the 28 cloned proteins were detected by different authors with immunological methods as cellulosomal components (table 1). However, due to the high sensitivity of the immunological assay it was not clear if these proteins were major components or just present in a few copies. But 1D gel electrophoresis gave a first hint that the scaffoldin CipA and the exoglucanase Cel48S were the most prevalent components in the cellulosome [28,92]. CelJ was also unequivocally identified as the largest band [2].

Attempts to separate the cellulosomes by native protein separation techniques or mild treatments failed [61]. Some disintegration was observed with calcium complexing agents like EDTA or with mild SDS denaturation, but disintegration of the components for *in vitro* reconstitution of the complete complex was not successful [15,18,61,84]. This might be due to the strong protein-protein interaction effected by the dockerin-cohesin binding were calcium ions play a role [10]. Were partial disintegration was successful, it often resulted in immediate proteolytic cleavage.

The abundance of numerous enzymes with very similar structure and activity pattern makes a functional analysis difficult, if the complex cannot be reconstituted from purified single components. Hence there is only the possibility to identify the constituents of actual cellulosomes and to characterize single recombinant components biochemically. It might be possible in the future to compose a cocktail of purified recombinant enzymes and reconstitute them *in vitro* for an active enzyme complex, albeit there could be problems with the unknown influence of lacking or wrong glycosylation in the recombinant host. First attempts to construct artificial "mini-cellulosomes" consisting of 2 enzyme components on a hybrid mini-scaffoldin are reviewed in [11].

To analyze the composition more thoroughly, the cellulosomal proteins were recently denaturated and separated by 2-D electrophoresis for a semi-quantitative protein determination [96]. The number of resolved spots in 2-DE was by far greater than the 14 bands visible in 1-D SDS-PAGE described earlier. Peptide mass fingerprints of single spots from stained 2-D gels were obtained with proteomic methods by MALDI-TOF and MALDI-TOF/TOF. Selected protein spots were in addition sequenced by MS/MS. Although not all visible spots gave clear data, all major spots could be identified and turned out to correspond to the already known cellulosomal components CipA, Cel48S, Xyn10C, Xyn10Z, Cel8A, Cel9K, Cbh9A, Cel9N, Cel5G and Chi18A (table 1). Most of the proteins were detected in more than one spot in horizontal or vertical distribution as a result of consecutive degradation.

In addition to the already known components three major spots gave unambiguous fingerprints and aligned perfectly with hitherto unknown reading frames from the draft genome sequence listed in table 1. According to the putative function of their catalytic modules the corresponding genes were called *celR*, *xghA* and *xynD* (entry #14, #39, #23 respectively in table 1) with catalytic modules of GHF9 (ß-1,4-glucanases), GHF74 (xyloglucanases) and GHF10 (endo-ß-1,4-xylanases), respectively [96]. Cellulase Cel9R is one of the most prominent proteins in the cellulosome. The identification of so many newly found ORFs (25 % of all major proteins!) indicated that even more new cellulosomal genes may be detected if the minor spots in the 2-D gels are identified.

It is of great interest for investigations of the cellulose hydrolysis mechanism to know which of the novel components are the most abundant in the mature cellulosomes. This gives a chance for a rational reconstitution of a synthetic cellulosome from recombinant proteins. The relative amount of the major components including their degraded forms was roughly estimated [96]. Taking into account the molecular mass of the respective proteins, it could be calculated how many molecules of a single component were present in an average cellulosomal particle (= per 1 scaffoldin CipA molecule) (table 1). The order of enzymes by molar abundance was Cel48S>Xyn10C>Cel8A>Cel9R>Xyn10Z>Cel9K>Xyn10D>Xgh74A >Cbh18A>Cel5G>Cel9N>Chi18A, with some uncertainty due to obvious protein degradation (smear of spots) and different staining behaviour, and with a possible underestimation of the major spots due to the staining method. Interestingly the earlier estimation of one Chi18A enzyme per 20 cellulosomal particles with activity determination was roughly in concordance with the number of 1 in 14 particles obtained with the quantification method in 2-D gels slabs [94]. It is also interesting to note that the three xylanases Xyn10C, Xyn10D and Xyn10Z together, all of them belonging to GHF10, seem to constitute about the same amount of protein in the cellulosome as all ß-glucanases together (without exoglucanase Cel48S). Xylanase Xyn10C is by far the most abundant enzyme after exoglucanase Cel48S. Cel48S and Xyn10C are besides CipA statistically present in every cellulosomal particle, followed by Cel8A in almost every particle and about 1 Cel9R molecule in every second particle.

The cellulases in the cellulosome are in the focus of the interest because of the efficiency of the cellulosomal activity on crystalline cellulose. The 7 major components active on ß-1,4-glucans have fundamentally different activities: Cel48S, Cel9K, Cel9R and Cbh9A are processive enzymes with Cel9K and Cbh9A producing cellobiose, Cel9R cellotetraose and

Cel48S a mixture of cellodextrins; the newly detected component Cel9R seems to be among the most important cellulases with Cel48S and Cel8A; Cel8A, Cel5G and Cel9N are endoglucanases with Cel8A being the most prominent. All types of hitherto detected cellulase activities are present and may interact synergistically.

Although most minor protein spots were not identified, the results are stimulating a number of questions about the synergistic function of the major cellulosomal components which have to be investigated further.

The Possible Role of Cel8A, an Unknown Player in the Cellulose Hydrolysis

The quantitative determination of the cellulosome components revealed that endoglucanase Cel8A was after exoglucanase Cel48S the apparently most frequent cellulolytic enzyme component in the cellulosome, albeit with some uncertainty of its exact quantity. It was characterized as an endoglucanase which fits well into the simple model of endo-exo-synergism favoured in the older literature. However, in most bacterial cellulase systems a GHF9 component was considered the most important endoglucanase component besides the GHF48 component [70,90].

Surprisingly in the *C. thermocellum* cellulosome it is a GHF8 catalytic module which obviously is more prevalent. At present the role of Cel8A can only be speculated upon and its role in synergistic interaction with the other cellulosomal components of GHF5, GHF9 and GHF48 is to be evaluated. Especially the interaction with the processive enzymes Cel48S, Cel9R, Cel9K and Cbh9A, according to the endo-exo-model, on one hand and the endoglucanases Cel5G and Cel9N on the other hand are of interest.

Cellulosome Composition Changes with Substrate

There is increasing evidence in the literature and from unpublished experiments that the cellulosome composition changes with the growth rate or the carbohydrate substrate used for growing the culture ([23,88]; unpublished observations). However, the data are solely based on the difference between cells grown on cellobiose and on cellulose. Other substrates were not tested yet and are difficult to find, because *C. thermocellum* is not growing readily or at all on other pure substrates. This is subject of further investigations. The quantification of mRNA (refereed in [11]) may not give a satisfying answer to the composition of the extracellular complex because too many factors may play a role in the determination how much of which protein is integrated in the cellulosome: expression of the gene, transcription control, secretion through the cell membrane and cell wall, anchoring on the cell surface and finally the incorporation into the complex by binding to one of the cohesin modules of the scaffoldin. Each of those steps could discriminate between proteins and prefer specific components.

The 2-D gel electrophoresis of cellulosome preparations is a good basis for the (semi)quantitative identification of the components changed in their prevalence in the complex and will be applied for such experiments on the protein level. However there are drawbacks for this approach too: the age of a culture or the preparation method may change the composition, and the staining method may discriminate certain proteins. But these are technical problems and can be overcome by scrutiny in the experimental design.

A Supplementary Non-Cellulosomal Cellulase System?

The cellulase gene *celI*, just upstream of *celN* in the genome, was found to code for a non-cellulosomal ß-glucanase with interesting traits: despite an identical module architecture and high sequence homology with the endoglucanase Cel9N (which suggests gene duplication) it is a processive ß-glucanase producing exclusively cellotetraose, which is only later and by far slower degraded further. The cellulosomal component Cel9R (see above) has an almost identical module structure and a similar hydrolysis mechanism which is seemingly not uncommon at least in *C. thermocellum* [95]. Cel9I possesses a CBM3c substrate binding module which usually is specific for crystalline cellulose. This is corroborating a role in hydrolysis of native cellulose.

Having found an active, soluble, not cellulosome-bound GHF9 cellulase, its similarity with Cs-Cel9Z of the cellulolytic bacterium *C. stercorarium* was obvious. This close relative of *C. thermocellum* does not possess cellulosomes but a completely soluble cellulase system consisting of two cellulases, Cs-Cel9Z and Cs-Cel48Y which sufficiently hydrolyzes crystalline cellulose [69]. The genomic sequence of *C. thermocellum* was therefore screened for a GHF48 component which is present in all true cellulase systems of bacteria as the crucial component besides the GHF9 enzyme. Such a gene was identified in the genome and called Cel48Y, consisting of a catalytic GHF48 module, connected through a PTS linker to a carbohydrate binding module, which is described in connection with other cellulases as a binding module for crystalline cellulose (table 4). The recombinant Cel48Y protein of *C. thermocellum* was purified and found to be weakly active, as typical for family GHF48 enzymes: it has activity on ß-1,3-1,4-glucan and lacks completely any activity on soluble carboxymethyl cellulose (CMC), whereas it is active on amorphic cellulose and weakly active on crystalline cellulose (unpublished data).

Table 4: Selected non-cellulosomal reading frames in the whole draft genomic sequence of *C. thermocellum* (this list is not complete). Abbreviations as in table A.

Locus tag	gene	protein ID	Mw	Module architecture
CtheDRAFT_3060	CelY	ZP_00503922	106	GH48-CBM3-CBM3
CtheDRAFT_3098	CelI	ZP_00503955	99	GH9-CBM3c-CBM3b
CtheDRAFT_2299		ZP_00504672	41	GH5
CtheDRAFT_2301	CelC	ZP_00504674	148	3(SLH)-GH16-
CtheDRAFT_0661	LicA	ZP_00510593	65	4(CBM4)-GH5

Cel9I and Cel48Y from *C. thermocellum* show a distinct synergism with an activity of 2 to 5 times higher in mixture than with the single components. The intensity of synergism is dependent on the type of cellulose used: the highest degree of synergism was found with filter paper as substrate (unpublished results). Both proteins have not yet been shown to be present in the culture supernatant of *C. thermocellum*. But the genes seem to be intact and the recombinant proteins are active and show the expected activities.

So far we can conclude that Cel9I and Cel48Y may constitute a parallel cellulolytic enzyme system besides the cellulosome. They could have a supplementary role in cellulose breakdown and support the cellulosome action, eventually by preparing the cellulose fibre surfaces for attachment of the large complex.

The degradation of fibre in biomass could be supported by another endo-ß-glucanase which is the major ß-1,3-glucanase in the culture supernatant: Lic16A [27]. This is a very large, highly modular enzyme which contains a GHF16 catalytic module with specificity for ß-1,3-linkages. It is at least partially located on the cell surface by virtue of an S-layer homologous module (SLH) [74]. Its gene is located in a region near the ß-glucanase gene *celC,* a *lacI*-type regulator gene, which eventually may be responsible for ß-glucanase expression, and a transposase gene. The presence of a transposase could possibly explain the evolution of the highly modular genes in this genomic region [27].

Another non-cellulosomal mixed-linkage ß-glucanase is Cel5C with activity on lichenan, barley mixed-linkage ß-1,3-1,4-gucan and cellodextrins [72]. Its gene is located upstream of *licA* [27]. Cel5C is an endo-ß-1,4-glucanase and degrades cellodextrins mainly to cellobiose. In combination with the non-cellulosomal (but mostly cell internal) ß-glucosidases BglA and BglB it degraded *in vitro* cellodextrins and mixed-linkage ß-1,3-1,4-dextrins to glucose.

The list of non-cellulosomal ß-glucanases shown in table 4 also includes a yet unknown GHF5 reading frame for which no activity was reported so far. Besides the soluble cellulase system Cel9I-Cel48Y the other genes could be responsible for the degradation of soluble ß-1,6-, ß-1,4-, ß-1,3-1,4- or ß-1,3-glucans which are part of the hemicellulose matrix enwrapping the cellulose fibres in secondary plant cell walls.

WHICH CELLODEXTRINS ARE THE PRODUCTS OF CELLULOSE HYDROLYSIS?

Early publications reported cellobiose as the major product of the cellulosome [50]. Consequently the search for cellobiohydrolases (CBH) was intensified and a number of cellobiohydrolases were identified. CBHs were the first processive cellulases characterized and the paradigm of processivity in cellulose hydrolysis stimulated further development of a mechanistic model for crystalline cellulose degradation. However later on, as a result of the experimental progress made with the CBHs, other processive enzymes were identified which degrade the cellulose molecule either from the non-reducing or the reducing end, and among them the processive endoglucanases: ß-glucanases which open new ends within a cellulose chain, stick to the site of activity and proceed processively. Some of those enzymes were producing cellotetraose *in vitro* as e.g. was shown for recombinant Cel9R, one of the major components of the cellulosome, or for the recombinant non-cellulosomal Cel9I. Even the purified major exoglucanase Cel48S is not producing exclusively cellobiose, but also higher cellodextrins, esp. at shorter times of incubation *in vitro* [48].

In addition, as can be seen from a kinetic *in-vitro* investigation of the hydrolysis products of endoglucanases, a mixture of cellodextrins is usually produced with an average degree of polymerization of 4 (cellotetraose) in most cases (Fig. 5). This can be demonstrated e.g. for the recombinant major cellulosomal endoglucanases Cel8A and Cel9N (Fig. 5). Cel8A produces G3 and G4 from phosphoric acid swollen cellulase (PASC) as end products, whereas the intermediately produced G4 and G5 from PASC by recombinant CelN were slowly further processed to G1, G2 and G3. A closer investigation of the other endoglucanases has to show which are the intermediate products of hydrolysis from amorphous (PASC) or crystalline cellulose (Avicel) as was e.g. done with the non-cellulosomal cellulase Cel9I [95].

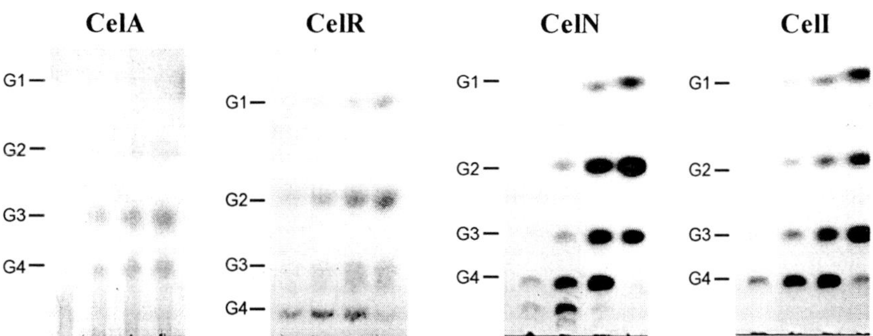

Fig. 5: Time course of the hydrolysis of phosphoric acid swollen cellulose (PASC) by purified recombinant preparations of Cel8A, Cel9R, Cel9N and Cel9I. Four points of equal time difference of incubation are shown for each enzyme. The markers are G1 to G4 (glucose to cellotetraose).

Zhang & Lynd [53] suggested that *C. thermocellum* could use cellodextrins (= larger than cellobiose) rather than glucose to yield additional energy for the energy demanding production of the huge extracellular enzyme complex for the degradation of natural cellulose fibres. The reasons for the higher energy yield could be:

1. the energy saving (per glucose equivalent) if cellodextrins are transported through the cell membrane and not each glucose residue separately: one ATP is used per molecule transported; and
2. the phosphorolytic cleavage of cellodextrins yielding glucose-1-phosphate instead of the phosphorylation of each glucose residue at the expense of one ATP [87].

This hypothesis is surprising in the light of cellobiose identified as the product of isolated cellulosomes, and it is presently not clear what the source for the longer cellodextrins could be if complete cellulosomes are applied. However, the conditions for the cellulose hydrolysis *in vivo* differ greatly from a reagent tube with its high concentration of purified recombinant enzymes and excessive incubation times. E.g. even with Cel9R, cellobiose can be found as the major end product in *in vitro* experiments with high enzyme concentration and long incubation [97]. If one designs the experiments carefully, one can see that with some cellulases upon incubation larger intermediates than cellobiose are quickly produced in a fast reaction which are then (probably with increasing product concentration) slowly degraded further (see Fig. 5). The bacterium would get a chance to take up the early (and larger) products because its surface is covering the spot of the hydrolysis by being hold to the substrate through the double binding of the cellulosome to substrate and cell. The close vicinity of the bacterial cell and the site of hydrolysis suggests that cellodextrins could be taken up as soon as they are released from the substrate, without a chance for further hydrolysis. This does not *a priori* imply that only larger cellodextrins than cellobiose are produced by the cellulosome. To reinvestigate this under the light of the new results from single components is a question of further kinetic experiments.

These theoretical considerations are corroborated by the finding of cellobiose- and cellodextrin-phosphorylases in *C. thermocellum* and the greater importance of the phosphorolysis than the hydrolysis of cellodextrins in *C. thermocellum* [4,5,87].

Whereas cellodextrins may be useful products from cellulose hydrolysis of *C. thermocellum* (but are used probably predominantly by *C. thermocellum* itself), other side products of cellulolysis or fermentation endproducts may be exploited by other microorganisms thriving together with *C. thermocellum*. Early investigators reported that *C. thermocellum* is secreting glucose when consuming cellodextrins and other sugars and oligosaccharides if grown on lignocellulosic biomass. If the bacterium is grown on cellobiose, the glucose release may be the result of phosphorolytic cleavage of cellobiose into glucose-1-phosphate and glucose, or the result of transglucosylation, i.e. by dispropotionisation of cellobiose into glucose and higher dextrins via phosphorolytic cleavage and chain elongation through glucose-1-phosphate. The remaining glucose is not phosphorylated by hexokinase but subsequently secreted out of the cell. The energy saving effect of the usage of larger cellodextrins seems to be profitable for the cell even if it has to expel some glucose. It was reported that the phosphorolytic cleavage of cellobiose and cellodextrins dominates intracellularily compared to the hydrolytic cleavage [87].

The phosphorolytic cleavage of cellodextrins and cellobiose may also be accompanied by transglucosylation which leads to larger cellodextrins as intermediates. In a search for the possible transglycosylation in *C. thermocellum*, cell extracts and concentrated culture supernatants were prepared and incubated with cellobiose. The products were analyzed by thin layer chromatography. Only with the cell extract, but not with the culture supernatants the products cellotriose, -tetraose and –pentaose were identified besides glucose (unpublished results). The intracellular enzymes responsible for this transglycosylation are not yet identified: all four ß-glucosidases BglA, BglB, BglC and BglD occurring in the genome of *C. thermocellum* were cloned and analyzed, but none of the recombinant ß-glucosidases had a detectable transglucosylation activity under the same conditions (to be published). The identification of the transglucosylating enzyme is subject of further investigations although the cellobiose- and the cellodextrin-phosphorylase are the most probable candidates [4,5].

In *C. thermocellum* cellodextrins larger than cellobiose may thus play a major role in the metabolic conversion of cellulose and some enzymes for the external production of those cellodextrins from cellulose were identified.

Conclusion

The genomic sequence of *C. thermocellum* allowed the analysis of the putative cellulosomal proteins. Altogether more than 71 genes code for cellulosomal components in the genome of *C. thermocellum*, almost all containing catalytic modules with some exceptions of proteins with structural or unknown function. Most of the genes can be assigned a putative function. The list includes cellulases (both endo- and exoglucanases), xylanases/xyloglucanases, mannanases, pectinases and pectate lyases, carbohydrate-esterases, glycosidases, a chitinase and a mixed-linkage ß-glucanase. A limited number of proteins with yet unknown function are present in the cellulosome. The cellulosomal components are modular proteins with catalytic and non-catalytic (e.g. carbohydrate binding) modules. Most of the reading frames have a surprisingly high complexity in module architecture and some of them are unusually large (up to 197 kDa).

A high number of endo-xylanases, a xyloglucanase, putative ß-xylosidases, α-arabinofuranosidases and a glucuronidase could be responsible for an effective degradation of the hemicellulose enwrapping the cellulose crystals. Pectin is degraded by functional pectin hydrolases and lyases. Depolymerization is supported by esterases and debranching glycosidases located in the cellulosome. The structure of the genes shows some regularity and almost half of the putative components bear a substrate binding module (CBM). Components with more than one catalytic module are rare but exist.

Cellulose crystals are hydrolyzed by the synergistic action of processive and non-processive ß-glucanases of GH families 5, 8, 9, and 48. Unexpectedly, Cel8A seems to play a key role in cellulose hydrolysis. In contrast to the aerobic cellulose degrading microorganisms, cellodextrins and not cellobiose may be the key products of cellulolysis. Some potentially cellodextrin producing enzymes have been identified among the major cellulosome components.

References

Adelsberger, H., Hertel, C., Glawischnig, E., Zverlov, V. V., & Schwarz, W. H. (2004). Enzyme system of *Clostridium stercorarium* for hydrolysis of arabinoxylan: reconstitution of the in vivo system from recombinant enzymes. *Microbiol. 150,* 2257-2266.

Ahsan, M. M., Kimura, T., Karita, S., Sakka, K., & Ohmiya, K. (1996). Cloning, DNA sequencing, and expression of the gene encoding *Clostridium thermocellum* cellulase CelJ, the largest catalytic component of the cellulosome. *J. Bacteriol. 178*, 5732.

Ahsan, M. M., Matsumoto, M., Karita, S., Kimura, T., Sakka, K., & Ohmiya, K. (1997). Purification and characterization of the family J catalytic domain derived from the *Clostridium thermocellum* endoglucanase CelJ. *Biosci. Biotech., Biochem. 61*, 427-431.

Alexander J. K. (1972a). Cellobiose phosphorylase from *Clostridium thermocellum. Method. Enzymol. 130*, 940-948.

Alexander J. K. (1972b). Cellodextrin phosphorylase from *Clostridium thermocellum. Method. Enzymol. 130*, 948-953.

Arai T., Ohara H., Karita S., Kimura T., Sakka K., & Ohmiya K. (2001). Sequence of *celQ* and properties of CelQ, a component of the *Clostridium thermocellum* cellulosome. *Appl Microbiol Biotechnol., 57*, 660-666.

Arai, T., Araki, R., Tanaka, A., Karita, S., Kimura, T., Sakka, K., & Ohmiya, K. (2003). Characterization of a cellulase containing a family 30 carbohydrate-binding module (CBM) derived from *Clostridium thermocellum* CelJ: importance of the CBM to cellulose hydrolysis. *J. Bacteriol. 185*, 504-512.

Bayer, E. A., Kenig, R., & Lamed, R. (1987). Adherence of *Clostridium thermocellum* to cellulose. *J. Bacteriol. 156*, 818-827.

Bayer, E. A., Shimon, L. J. W., Shoham, Y., & Lamed, R. (1998). Cellulosomes – structure and ultrastructure. *J. Struct. Biol. 124*, 221-234.

Bayer, E. A., Shoham, Y., and Lamed, R. (2000). Cellulose-decomposing prokaryotes and their enzyme systems. In M. Dworkin, S. Falkow, E. Rosenberg, K.-H. Schleifer, and E. Stackebrandt (eds.), *"The Prokaryotes: An Evolving Electronic Resource for the Microbiological Community", 3rd edition*. Springer-Verlag, New York.

Bayer, E. A., Belaich, J.-P., Shoham, Y., & Lamed, R. (2004). The cellulosomes: Multienzyme machines for degradation of plant cell wall polysaccharides. *Annu. Rev. Microbiol. 58*, 521-554.

Beguin, P., Cornet, P., & Aubert, J. P. (1985). Sequence of a cellulase gene of the thermophilic bacterium *Clostridium thermocellum J. Bacteriol. 162*, 102.

Béguin, P., Grépinet, O., Tailliez, P., Chauvaux, S., Girard, H., Millet, J., & Aubert, JP. (1989). Conversion of cellulose into ethanol by *Clostridium thermocellum*: genetic engineering of cellulases and improvement of ethanol production. *Appl. Environ. Microbiol. 55*, 207-211.

Bender, J., Vatcharapijarn, Y., & Jeffries, T. W. (1985). Characteristics and adaptability of some new isolates of *Clostridium thermocellum*. *Appl. Environ. Microbiol. 49*, 475-477.

Bhat, S., Goodenough, P. W., Bhat, M. K., & Owen, E. (1995). Isolation pf four major subuits from Clostridium thermocellum cellulosome and their synergism in the hydrolysis of crystalline cellulose. *Int. J. Biol. Macromol. 16*, 335-342.

Blum, D. L., Kataeva, I. A., Li, X. L., & Ljungdahl, L. G. (2000). Feruloyl esterase activity from *Clostridium thermocellum* cellulosome can be attributed to previously unknown domains of XynY and XynZ. *J. Bacteriol. 182*, 1346-1351.

Carrard, G., Koivula, A., Söderlund, H., & Béguin, P. (2000). Cellulose-binding domains promote hydrolysis of different sites in crystalline cellulose. *Proc. Natl. Acad. Sci. 97*, 10342.10347.

Choi, S. K., & Ljungdahl, L. G. (1996). Dissociation of the cellulosome of Clostridium thermocellum in the presence of ethylenediaminetetraacetic acid occurs with the formation of truncated polypeptides. *Biochem. 35*, 4897-4905.

Cornet, P., Millet, J., Béguin, P., & Aubert, J. P. (1983). Characterization of two *cel* (cellulose degradation) genes of *Clostridium thermocellum* coding for endoglucanases. *Bio-Techol. 1*, 589-594.

Coughlan, M. P., Hon-Nami, K., Hon-Nami, H., Ljungdahl, L. G., Paulin, J. J., & Rigsby, W. E. (1985). The cellulolytic enzyme complex of *Clostridium thermocellum* is very large. *Biochem. Biophys. Res. Commun. 130*, 904-909.

Coutinho, P.M. & Henrissat, B. (1999). Carbohydrate-Active Enzymes server at URL: http://afmb.cnrs-mrs.fr/CAZY/

Doi, R. H., & Kosugi, A. (2004). Cellulosomes: plant-cell-wall-degrading enzyme complexes. *Nature Rev. 2*, 541-551.

Dror, T. W., Morag, E., Rolider, A., Bayer, E. A., Lamed, R., & Shoham, Y. (2003). Regulation of the cellulosomal CelS (cel48A) gene of *Clostridium thermocellum* is growth rate dependent. *J. Bacteriol. 185*, 3042-3048.

Fleming, R. W., & Quinn, L. Y. (1971). Chemically defined medium for growth of *Clostridium thermocellum*, a cellulolytic thermophilic anaerobe. *Appl. Microbiol. 967-*968.

Fontes C. M., Hall J., Hirst B. H., Hazlewood G. P., Gilbert H. J. (1995a). The resistance of cellulases and xylanases to proteolytic inactivation. *Appl. Microbiol. Biotechnol. 43*, 52-57.

Fontes, C. M. G. A., Hazelwood, G. P., Morag, E., Hall, J., Hirst, B. H., & Gilbert, H. J. (1995b). Evidence for a general role for non-catalytic thermostabilizing domains in Xylanases from thermophilic bacteria. *Biochem. J. 307*,151.

Fuchs, K. P., Zverlov, V. V., Velikodvorskaya, G. A., Lottspeich, F., and Schwarz, W. H. (2003). Lic16A of *Clostridium thermocellum*, a non-cellulosomal, highly complex endo-ß-1,3-glucanase bound to the outer cell surface. *Microbiol. 149*: 1021-1031.

Fujino, T., Beguin, P., Aubert, J. P. (1992). Cloning of a *Clostridium thermocellum* DNA fragment encoding polypeptides that bind the catalytic components of the cellulosome. *FEMS Microbiol. Lett. 73*, 165-170.

Gerwig, G. J., Kamerling, J. P., Vliegenthart, J. F., Morag, E., Lamed, R., & Bayer, E. A. (1993). The nature of the carbohydrate-peptide linkage region in glycoproteins from the cellulosomes of *Clostridium thermocellum* and *Bacteroides cellulosolvens*. *J. Biol. Chem. 268*, 26956-26960.

Gibbs, M D., Reeves, R. A., Farrington, G. K., Anderson, P., Williams, D. P., & Bergquist, P. L. (2000). Multidomain and multifunctional glycosyl hydrolases from the extreme thermophile *Caldicellulosiruptor* isolate Tok7B.1. *Curr. Microbiol. 40*, 333-340.

Gräbnitz, F., and W. L. Staudenbauer (1988). Characterization of two ß-glucosidase genes from *Clostridium thermocellum*. *Biotechnol. Lett. 10*, 73-78.

Grepinet, O., & Beguin, P. (1986). Sequence of the cellulase gene of *Clostridium thermocellum* coding for endoglucanase B. *Nucleic Acids Res. 14*, 1791.

Grepinet, O., Chebrou, M.-C., & Beguin, P. (1988). Nucleotide sequence and deletion analysis of the xylanase gene (*xynZ*) of *Clostridium thermocellum*. *J. Bacteriol. 170*, 4582.

Guglielmi, G., Béguin, P. (1998). Cellulase and hemicellulase genes of *Clostridium thermocellum* from five independent collections contain few overlaps and are widely scattered across the chromosome. *FEMS Microbiol. Lett. 161*, 209-215.

Halstead, J. R., Vercoe, P. E., Gilbert, H. J., Davidson, K., & Hazlewood, G. P. (1999). A family 26 mannanase produced by *Clostridium thermocellum* as a component of the cellulosome contains a domain which is conserved in mannanases from anaerobic fungi. *Microbiol. 145*, 3101-3108.

Hayashi, H., Takagi, K., Fukumura, M., Kimura, T., Karita, S., Sakka, K., & Ohmiya, K. (1997). Sequence of *xynC* and properties of XynC, a major component of the *Clostridium thermocellum* cellulosome. *J. Bacteriol. 179*, 4246.

Hayashi, H., Takehara, M., Hattori, T., Kimura, T., Karita, S., Sakka, K., & Ohmiya, K. (1999). Nucleotide sequences of two contiguous and highly homologous xylanase genes *xynA* and *xynB* and characterization of XynA from *Clostridium thermocellum Appl. Microbiol. Biotechnol. 51*, 348.

Hazlewood, G. P., Davidson, K., Clarke, J. H., Durrant, A. J., Hall, J. Gilbert, H. J. (1990). Endoglucanase-E, produced at high-level in *Escherichia coli* as a *lacZ'* fusion protein, is part of the *Clostridium thermocellum* cellulosome. *Enz. Microb. Technol., 12*, 656-662.

Henrissat B., Bairoch A. (1996). Updating the sequence-based classification of glycosyl hydrolases. *Biochem. J. 316*, 695-696

Hernández, P. E. (1982). On the utilization of cellobiose and D-glucose by *Clostridium thermocellum*. *Rev. Espan. Fisiol. 38*, 473-474.

Irwin, D., Shin, D. H., Zhang, S., Barr, B. K., Sakon, J., Karplus, P. A., &Wilson D. B. (1998). Roles of the catalytic domain and two cellulose binding domains of *Thermomonospora fusca* E4 in cellulose hydrolysis. *J. Bacteriol. 180*, 1709-14.

Johnson, E. A., Madia, A., & Demain, A. L. (1981). Chemically defined minimal medium for growth of the anaerobic cellulolytic thermophile *Clostridium thermocellum*. *Appl. Environ. Microbiol. 41*, 1060-1062.

Johnson, E. A., Sakajoh, M., Halliwell, G., Madia, A., & Demain, A. L. (1982). Saccharification of complex cellulosic substrates by the cellulase system from *Clostridium thermocellum*. *Appl. Environ. Microbiol. 43*, 1125-1132.

Joliff, G., Beguin, P., & Aubert, J. P. (1986). Nucleotide sequence of the cellulase gene *celD* encoding endoglucanase D of *Clostridium thermocellum*. *Nucleic Acids Res. 14*, 8605.

Joliff, G., Beguin, P., Juy, M., Millet, J., Ryter, A., Poljak, R., & Aubert, J.-P. (1987). Isolation, crystallization and properties of a new cellulase of *Clostridium thermocellum* overproduced in *Escherichia coli*. *Bio/technol. 4*, 896- 900.

Kataeva, I., Li, X. L., Chen, H., Choi, S. K., & Ljungdahl, L.G. (1999). Cloning and sequence analysis of a new cellulase gene encoding CelK, a major cellulosome component of *Clostridium thermocellum*: evidence for gene duplication and recombination. *J. Bacteriol. 181*, 5288-5295.

Kosugi, A., Murashima, K., & Doi, R. H. (2002). Characterization of two noncellulsomal subunits, ArfA and BgaA, from *Clostridium cellulovorans* that cooperate with the cellulosome in plant cell wall degradation. *J. Bacteriol. 184*, 6859-6865.

Kruus, K., Wang, W. K., Ching, J., Wu, J. H. (1995). Exoglucanase activities of the recombinant *Clostridium thermocellum* CelS, a major cellulosome component. *J Bacteriol. 177*, 1641-1644.

Kurokawa, J., Hemjinda, E., Arai, T., Kimura, T., Sakka, K., Ohmiya, K. (2002). *Clostridium thermocellum* cellulase CelT, a family 9 endoglucanase without an Ig-like domain or family 3c carbohydrate-binding module. *Appl. Microbiol. Biotechnol. 59*, 455-461.

Lamed, R., Setter, E., & Bayer, E. A. (1983). Characterization of a cellulose-binding, cellulase-containing complex in *Clostridium thermocellum*. *J. Bacteriol. 156*, 828-836.

Lamed, R., Kenig,, R., Setter, E., & Bayer, E. A. (1985). Major characteristics of the cellulolytic system of *Clostridium thermocellum* coincide with those of the purified cellulosome. *Enz. Microb. Technol. 7*, 37-41.

Lemaire, M., & Beguin, P. (1993). Nucleotide sequence of the *celG* gene of *Clostridium thermocellum* and characterization of its product, endoglucanase CelG. *J. Bacteriol. 175*, 3353.

Lynd, L. R., Weimer, P. J., van Zyl, W. H., & Pretorius, I. S. (2002). Microbial cellulose utilization: fundamentals and biotechnology. *Microbiol. Molec. Biol. Rev. 66*, 506-577.

Lytle, B. L., Volkman, B. F., Westler, W. M., & Wu, J. H. D. (2000). Secondary structure and calcium-induced folding of the *Clostridium thermocellum* dockerin domain determined by NMR spectroscopy. *Arch. Biochem. Biophys. 379*, 237-244.

Mayer, F., Coughlan, M. P., Mori, Y., & Ljungdahl, L. G. (1987). Macromolecular organization of the cellulolytic enzyme complex of *Clostridium thermocellum* as revealed by electron microscopy. *Appl. Environ. Microbiol. 53*, 2785-2792.

McBee, R. H. (1954). The characteristics of *Clostridium thermocellum*. *J. Bacteriol. 67*, 505-506.

McMillan, J. D. (1997). Bioethanol production: status and prospects. *Renewable Energy 10*, 295-302.

McMullan, G., Christie, J. M., Rahman, T. J., Banat, I. M., Ternan, N. G., & Marchant, R. (2004). Habitat, applications and genomics of the aerobic, thermophilic genus Geobacillus. *Biochem. Soc. Transact. 32*, 214-217.

Mel'nik, M. S., Rabinovich, M. L., & Voznyi, I. V. (1991). Cellobiohydrolase from *Clostridium thermocellum*, synthesized by a recombinant *E. coli* strain. *Biokhimiia. 56*, 1787-1797.

Morag, E., Bayer, E. A., & Lamed, R. (1992). Affinity digestion for the near-total recovery of purified cellulosome from *Clostridium thermocellum*. *Enz. Microb. Technol. 14*, 289-292.

Morag, E., Yaron, S., Lamed, R., Kenig, R., Shoham, Y., & Bayer, E. A. (1996). Dissociation of the cellulosome of *Clostridium thermocellum* under nondenaturing conditions. *J. Biotechnol. 51*, 235-242.

Mori, Y. (1995). Nutritional interdependence between *Thermoanaerobacter thermohydrosulfuricus* and *Clostridium thermocellum*. *Arch. Microbiol. 164*, 152-154.

Navarro, A., Chebrou, M. C., Beguin, P., & Aubert, J. P. (1991). Nucleotide sequence of the cellulase gene *celF* of *Clostridium thermocellum*. *Res. Microbiol. 142*, 927.

Ng, T. K., Weimer P. J., & Zeikus, J. G. (1977). Cellulolytic and physiological properties of *Clostridium thermocellum*. *Arch. Microbiol. 114*, 1-7.

Nishiyama, Y., Kim, U. J., Kim, D. Y., Katsumata, K. S., May, R. P., & Langan, P. (2003). Periodic disorder along ramie cellulose microfibrils. *Biomacromolec. 4*, 1013-7.

Nochur, S. V., Roberts, M. F., & Demain, A. L. (1990). Mutation of *Clostridium thermocellum* in the presence of certain carbon sources. *FEMS Microbiol. Let. 71*, 199-204.

Park, W. S., & Yu, D. D. Y. (1983). Cellulolytic activities of *Clostridium thermocellum* and its carbohydrate metabolism. *J. Ferment. Technol. 61*, 563-571.

Patni, N. J., & Alexander, K. (1971). Utilization of glucose by *Clostridium thermocellum*: presence of glucokinase and other glycolytic enzymes in cell extracts. *J. Bacteriol. 105*, 220-225.

Riedel, K., Ritter, J., and Bronnenmeier, K. (1997). Synergistic interaction of the *Clostridium stercorarium* cellulases Avicelase I (CelZ) and Avicelase II (CelY) in the degradation of microcrystalline cellulose. *FEMS Microbiol. Lett. 147*: 239-243

Riedel, K., Ritter, J., Bauer, S., and Bronnenmeier, K. (1998). The modular cellulase CelZ of the thermophilic bacterium *Clostridium stercorarium* contains a thermostabilizing domain. *FEMS Microbiol. Let. 164*, 261-267.

Schwarz, W. H., F. Gräbnitz, & W. L. Staudenbauer (1986). Properties of a *Clostridium thermocellum* endoglucanase produced in *Escherichia coli*. *Appl. Environ. Microbiol. 51*, 1293-1299.

Schwarz, W. H., S. Schimming, K. P. Rücknagel, S. Burgschwaiger, G. Kreil, and W. L. Staudenbauer (1988). Nucleotide sequence of the celC gene encoding endoglucanase C of *Clostridium thermocellum*. *Gene 63*, 23-30.

Schwarz, W.H. (2001). The cellulosome and cellulose degradation by anaerobic bacteria. *Appl. Microbiol. Biotechnol. Appl. Microbiol. Biotechnol. 56*, 634-649.

Schwarz, W. H., Zverlov, V. V., & Bahl, H. (2004). Extracellular glycosyl hydrolases from clostridia. *Adv. Appl. Microbiol. 56*, 215-261.

Shin, H. S., Zeikus, J. G., & Jain, M. K. (2002). Electrically enhanced ethanol fermentation by *Clostridium thermocellum* and *Saccharomyces cerevisiae*. *Appl. Microbiol. Biotechnol. 58*, 476-481.

Smiti, N., Ollivier, B., & Garcia, J.L. (1986). Thermophilic degradation of cellulose by a triculture of *Clostridium thermocellum*, *Methanobacterium* sp. and *Methanosarcina* MP. *FEMS Microbiol. Let. 35*, 93-97.

Spinnler H. E., Lavigne B., Blachere H. (1986). Pectinolytic activity of *Clostridium thermocellum*: Its use for anaerobic fermentation of sugar beet pulp. *Appl Microbiol Biotechnol 23,* 434 – 437

Stainthorpe, A. C., & Williams, R. A. D. (1988). Isolation and properties of *Clostridium thermocellum* from Icelandic hot springs. *Int. J. Syst. Bacteriol. 38*, 119-121.

Tamaru, Y., & Doi, R. H. (2001). Pectate lyase A, an enzymatic subunit of the *Clostridium cellulovorans* cellulosome. *PNAS, 98*, 4125-4129.

Tuka, K., Zverlov, V. V., Bumazkin, B. K., Velikodvorskaya, G. A., Strongin, A.Y. (1990). Cloning and expression of *Clostridium thermocellum* genes coding for thermostable exoglucanases (cellobiohydrolases) in *Escherichia coli* cells. *Biochem. Biophys. Res. Commun. 169*, 1055-1060

Vuillet, S., Spinnler, H. E., & Blachere, H. (1986). Analysis of amino acid requirements of *Clostridium thermocellum*. *Appl. Microbiol. Biotechnol. 23*, 496-498.

Wang, W. K., Kruus, K., & Wu, J. H. D. (1993). Cloning and the DNA sequence of the gene coding for the *Clostridium thermocellum* cellulase Ss (CelS), a major cellulosome component. *J. Bacteriol. 175*, 1293.

Weimer, P. J., & Zeikus, J. G. (1977). Fermentation of cellulose and cellobiose by *Clostridium thermocellum* in the absence of *Methanobacterium thermoautotrophicum*. *Appl. Environ. Microbiol. 33*, 289-297.

Wu, J. H. D., Orne-Johnson, W. H., & Demain, A. L. (1988). Two components of an extracellular protein aggregate of *Clostridium thermocellum* together degrade crystalline cellulose. *Biochem. 27*, 1703-1709.

Yaguee, E., Beguin, P., Aubert, J.-P. P. (1990). Nucleotide sequence and deletion analysis of the cellulase- encoding gene *celH* of *Clostridium thermocellum. Gene 89*, 61.

Yu, E. K. C., Chan, M. K. H., & Saddler, J. N. (1985). Butandiol production from lignocellulosic substrates by *Klebsiella pneumoniae* grown in sequential co-culture with *Clostridium thermocellum. Appl. Microbiol. Biotechnol. 22*, 399-404.

Zhang, Y. H., & Lynd, L. R. (2004). Kinetics and relative importance of phosphorolytic and hydrolytic cleavage of cellodextrins and cellobiose in cell extracts of *Clostridium thermocellum. Appl. Environ. Microbiol. 70*, 1563-1569.

Zhang, Y. H., & Lynd L. R. (2005). Regulation of cellulase synthesis in batch and continuous cultures of *Clostridium thermocellum. J. Bacteriol. 187*, 99-106.

Zverlov, V. V., Fuchs, K. P., Schwarz, W. H. and Velikodvorskaya, G. A. (1994). Purification and cellulosomal localization of *Clostridium thermocellum* mixed linkage ß-glucanase LicB (1,3-1,4-ß-D-glucanase). *Biotechnol. Let. 16*, 29-34.

Zverlov, V. V., Mahr, S., Riedel, K., and Bronnenmeier, K. (1998a). Properties and gene structure of a bifunctional cellulolytic enzyme (CelA) from the extreme thermophile "*Anaerocellum thermophilum*" with separate glycosyl hydrolase family 9 and 48 catalytic domains. *Microbiol. 144*, 457-465.

Zverlov, V. V., Velikodvorskaya, G. V., Schwarz, W. H., Bronnenmeier, K., Kellermann, J., & Staudenbauer, W. L. (1998b). Multidomain structure and cellulosomal localization of the *Clostridium thermocellum* cellobiohydrolase CbhA. *J. Bacteriol. 180*, 3091.

Zverlov, V. V., Velikodvorskaya, G. A., Schwarz, W. H., Kellermann, J., and Staudenbauer, W. L. (1999). Duplicated *Clostridium thermocellum* cellobiohydrolase gene encoding cellulosomal subunits S3 and S5. *Appl. Microbiol. Biotechnol. 51*, 852-859.

Zverlov, V. V., Velikodvorskaya, G. A., & Schwarz, W. H. (2002a). A newly described cellulosomal cellobiohydrolase, CelO, from *Clostridium thermocellum*: investigation of the exo-mode of hydrolysis, and binding capacity to crystalline cellulose. *Microbiol. 148*, 247- 255.

Zverlov, V. V., Fuchs, K.-P., and Schwarz, W. H. (2002b). Chi18A, the endochitinase in the cellulosome of the thermophilic, cellulolytic bacterium *Clostridium thermocellum*. Appl. Environ. *Microbiol. 68*, 3176-3179

Zverlov, V. V., Velikodvorskaya, G. A., Schwarz, W. H. (2003). Two new cellulosome components encoded downstream of *celI* in the genome of *Clostridium thermocellum*: the non-processive endoglucanase CelN and the possibly structural protein CseP. *Microbiol. 149*, 515-524.

Zverlov, V. V., Kellermann, J., & Schwarz, W. H. (2005). Functional subgenomics of *Clostridium thermocellum* cellulosomal genes: Identification of the major catalytic components in the extracellular complex and detection of three new enzymes. Proteomics **5**, 3646-3653.

Zverlov, V. V., Schantz, N., & Schwarz, W. H. (2005). Cel9R, a new major component in the cellulosome of *Clostridium thermocellum*: a processive endo-ß-1,4-glucanase producing cellotetraose. FEMS Microbiol. Let. 249: 353-358.

In: Cellulosome
Eds: V. Uversky and I. A. Kataeva, pp. 153-168

ISBN 1-59454-950-8

Chapter VII

THE CLOSTRIDIUM CELLULOVORANS CELLULOSOME

*Roy H. Doi**
Section of Molecular & Cellular Biology The University of California, Davis, CA, USA.

ABSTRACT

The *Clostridium cellulovorans* cellulosome is an extracellular enzyme complex that can degrade plant cell wall material very efficiently. It is comprised of a non-enzymatic scaffolding protein called CbpA and a number of cellulosomal glycosyl hydrolase enzymes capable of degrading cellulose, hemicellulose, and pectin. The CbpA has a mass of 189,000 kDa and a cellulose binding domain (CBD) at its N-terminus, four conserved hydrophilic domains (HLDs), two near the N-terminus and two near the C-terminus, and nine homologous cohesin (Coh) domains. The CBD preferentially binds to crystalline cellulose and the cohesins bind the cellulosomal enzymes through their dockerin domains. Current studies indicate that the HLDs bind to the substrate as well as to the cell surface. The cellulosomal enzymes work in synergy with CbpA as well as with each other and with non-cellulosomal glycosyl hydrolase enzymes to degrade the various components of plant cell walls. The cellulosomal enzyme EngE binds not only to the CbpA, but also to the cell surface through its surface layer homology (SLH) domains and tightly binds the cellulosome to the cell surface and allows an efficient uptake of hydrolytic products. One large cluster of cellulosomal genes occurs, and also there are a number of cellulosomal genes unlinked to the cluster. The complex is useful for basic studies on plant cell wall degradation, for construction and expression of designer cellulosomes, for converting non-cellulose degraders to cellulose utilizers, and for various aspects of biotechnology involving the conversion of biomass to useful products.

* Correspondence concerning this article should be addressed to Roy H. Doi, Section of Molecular & Cellular Biology The University of California, Davis, CA 95616, USA. Tel: (530) 752-3191; Fax: (530) 752-3085; E-mail: rhdoi@ucdavis.edu.

INTRODUCTION

Clostridium cellulovorans was isolated by Mah and his colleagues from a wood chip pile [51]. It properties are summarized in Table 1. This organism was chosen for our studies based on the recommendation of Robert Mah, who described it as a producer of copious amounts of cellulase. As Table 1 indicates, *C. cellulovorans* is capable of growth on cellulose, xylan, pectin, cellobiose and a number of other sugars.

Table 1. Properties of Clostridium cellulovorans.

Growth	anaerobic, catalase negative, spore former,
Growth optima	37°C, pH 7
Growth substrates	cellulose, xylan, pectin, cellobiose, glucose, fructose, maltose, galactose, sucrose, lactose, mannose
Fermentation products	H_2, CO_2, acetate, buytyrate, formate, lactate during fermentation of cellulose or cellobiose
Substrates not used for growth	gum Arabic, rhamnose, melzitose, sorbitol, trehalose, xylose, arabitol, arabinose, glycerol, erythritol, lactate, pyruvate
Substrates not hydrolyzed	gelatin, starch, casein
DNA composition	26-27 mol% G+C

From Sleat et al. [51]

Scanning electron microscopic studies [4] have shown the presence of protuberances on the surface of the cell and it is believed that they are comprised of cellulosomes and polycellulosomes [2,3].

Our studies on *C. cellulovorans* cellulosomes have been directed towards analyzing their subunit structure, characterizing the genes encoding their subunits, investigating the synergy of the enzymatic subunits, determining the factors that regulate the expression of the cellulosomal genes, improving the function of cellulosomal enzymes, studying the interaction of cellulosomes and non-cellulosomal glycosyl hydrolases, and converting microorganisms into cellulose utilizers [6-10,55]. We have been interested in expressing the cellulosomal genes in other bacterial species to determine whether cellulosomes or minicellulosomes could be synthesized by heterologous systems. If in fact, this can be done, it may be possible to convert many non-cellulolytic organisms that make valuable products into utilizers of cellulosic biomass as an inexpensive substrate.

COMPOSITION AND STRUCTURE OF THE *C. CELLULOVORANS* CELLULOSOME

The cellulosome is comprised of a scaffoldin, CbpA, with a mass of 189 kDa and a number of enzyme subunits [45,47]. Figure 1 illustrates the domain structure of CbpA. The

CbpA consists of a cellulose binding domain (CBD) at its N-terminus, four hydrophilic domains (HLDs) and nine conserved hydrophobic cohesin (Coh) domains. Each of the domains has a mass of about 15-20 kDa. The CBD binds preferably to crystalline cellulose portions of the substrate [15,39] and binds the cell to the substrate via the cellulosome. The association of the cellulosomal enzymes through their dockerin domains to the cohesins of CbpA constitutes the main assembly process for formation of the cellulosome. The CBD is able to bind to chitin as well as cellulose, although no chitinase activity has been detected in our studies [15]. Mutations in various locations of CBD render it incapable of binding to cellulosome [14].

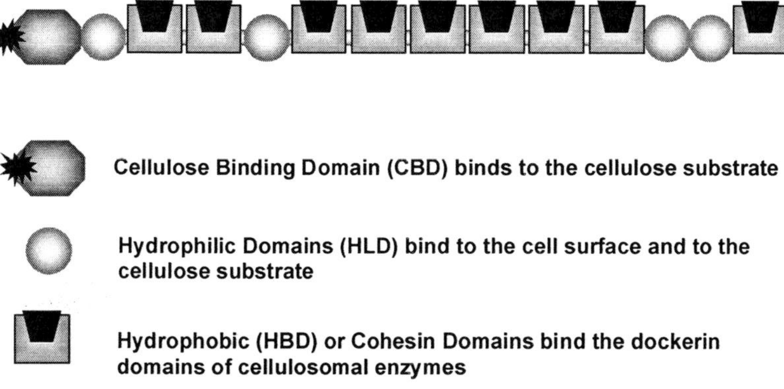

Fig. 1. A model of the scaffolding protein CbpA from C. cellulovorans. CbpA contains a single cellulose binding domain (CBD), four hydrophilic domains (HLD) and nine cohesin domains.

Much effort to determine the function of the HLD domains has led to a tentative conclusion that this domain has a dual function [24]. Binding studies have shown that it has an affinity to the cell surface as well as to cellulose. This suggests that HLD along with EngE binds the cellulosome to the cell surface and that HLD along with CBD binds the cellulosome to the substrate.

The cohesins bind the cellulosomal enzymes to the CbpA through their interactions with the dockerins present in all cellulosomal enzymes. The cellulosomal enzymes that have been identified to this date are listed in Table 2. They include cellulases and hemicellulases that are able to attack cellulose, xylan, mannan, and pectin. These functions make the cellulosome an effective enzymatic complex for the degradation of plant cell walls. Several more cellulosomal enzymes have been found associated with cellulosomes that are not represented in Table 2 and have not been characterized as yet except for their N-terminal amino acid sequences [34].

A predominant form of cellulosomal enzymes is the family 9 glycosyl hydrolases. So far five family 9 members have been identified in *C. cellulovorans* [54]. When their substrate specificity and products were analyzed, it was found that the pattern of products formed was different [1]. This suggests that each family 9 member may attack a complex substrate in a slightly different manner and that the presence of several family 9 enzymes facilitates the degradation of cellulose to cellobiose and glucose.

Table 2. Cellulosomal and Non-Cellulosomal Enzymes and Proteins.

Cellulosomal Enzymes	Modular Structure	No. of Residues	MW (kDa)
CbpA	CBM-HLD-(Coh)2-HLD-(Coh)6-HLD2-Coh	1,848	189,149
HbpA	HLD-Coh	240	24,940
EngB	GH5-DS	440	48,823
EngE	(HLD)3-GH5-X-DS	1,030	111,777
EngH	GH9-CBMIII-DS	715	79,321
EngK	CBMIV-Ig-GH9-DS	892	97,024
EngL	GH9-DS	522	57,629
EngM	CBMIV-Ig-GH9-DS	876	96,373
EngY	GH9-DS	738	80,290
ExgS	GH48-DS	727	80,485
ManA	GH5-DS	425	47,156
PelA	X-CBMIII-PL4-DS	914	94,458
XynA	GH11-DS-NodB	520	57,037
XynB	CBMIV-IX-GH10-DS	607	65,976
Non-cellulosomal Enzymes			
ArfA	GH51	492	55,730
BgaA	GH42	659	76,462
EngD	GH5-CBMII	515	55,974
EngF	GH5-CBMXVII	557	60,130

CBM: carbohydrate binding module; HLD: hydrophilic domain; Coh: cohesin domain;
GH: glycosyl family; PL: pectate lyase family; DS: dockerin domain; Ig: immuno-globulin-like domain; X: unknown domain.

The EngE is a particularly interesting component of the cellulosome, since it has three surface layer homology (SLH) domains at its N-terminus and a dockerin domain [52]. The dockerin domain interacts with the cohesins in CbpA and the SLH domains bind to the cell surface [26]. Thus EngE is not only an enzymatic component of the cellulosome, but also functions to bind the cellulosome to the cell surface. A current model of the *C. cellulovorans* cellulosome is presented in Fig. 2.

In addition to cellulosomal enzymes, there are non-cellulosomal enzymes that are involved in cellulose and hemi-cellulose degradation. These are listed in Table 2. These enzymes are believed to work in concert with the cellulosome to degrade plant cell wall materials. For instance arabinofuranosidase and β-galatosidase are believed to remove substituent groups from xylan to facilitate the degradation of xylan by cellulosomal xylanase [27,28].

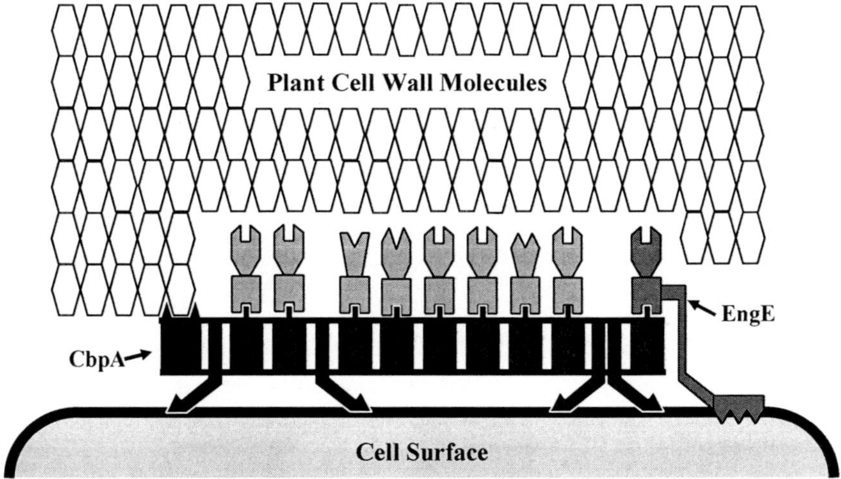

Fig. 2. A model of the *C. cellulovorans* cellulosome. The cellulosome consists of the scaffolding protein CbpA, and nine enzyme units attached to cohesins through their dockerin domains. The cellulosome is bound to the cell surface by EngE and its SLH domains and by the four hydrophilic domains which also bind to the substrate.

Although most of the cellulosomal enzymes, such as EngB [13], when expressed in *E. coli* form an insoluble inclusion body, the non-cellulosomal enzyme EngD [16] was found to be expressed as a soluble protein in *E. coli*. EngB and EngD have highly homologous N-terminal catalytic domain, but differ primarily in their C-termini. EngB has a C-terminal dockerin domain while EngD has a C-terminal CBD. To test the hypothesis that the EngD CBD conferred solubility to the enzyme, it was found that removal of the CBD from EngD resulted in an insoluble catalytic domain of EngD and a soluble CBD peptide. Furthermore when the EngB C-terminal dockerin was replaced by the EngD CBD, the EngB-CBD_{EngD} was now soluble in *E. coli* [38]. These results indicated that the C-terminal CBD of EngD played an important role in the soluble expression of EngD in *E. coli*. It would be interesting to see whether the fusion of CBD_{EngD} to other ordinarily insoluble proteins can "solubilize" these proteins.

ORGANIZATION OF CELLULOSOMAL GENES

The genes for cellulosomal proteins and enzymes are found in one large gene cluster [54] and in a number of genes unlinked to this cluster. The linked genes are shown in Fig. 3 and they include the genes for CbpA, ExgS, EngH, EngK, HbpA, EngL, ManA, EngM and EngN. The function for HbpA is unknown, and it appears to be similar to the ORFXp of *C. cellulolyticum* which has been postulated to play a role in the assembly of the cellulosome [40].

The cellulosomal genes unlinked to the gene cluster include EngB, EngE, EngY, PelA [53], XynA [25,27] and XynB [20]. In addition there are several non-cellulosomal genes that have been identified and include three glucanases EngD [15], EngF [23,44], and EngO [22] and two enzymes that are involved in degrading xylans, ArfA, and BgaA [28]. These enzymes are not components of the cellulosome, but are extracellular enzymes that work in conjunction with the cellulosome to degrade plant cell wall polymers.

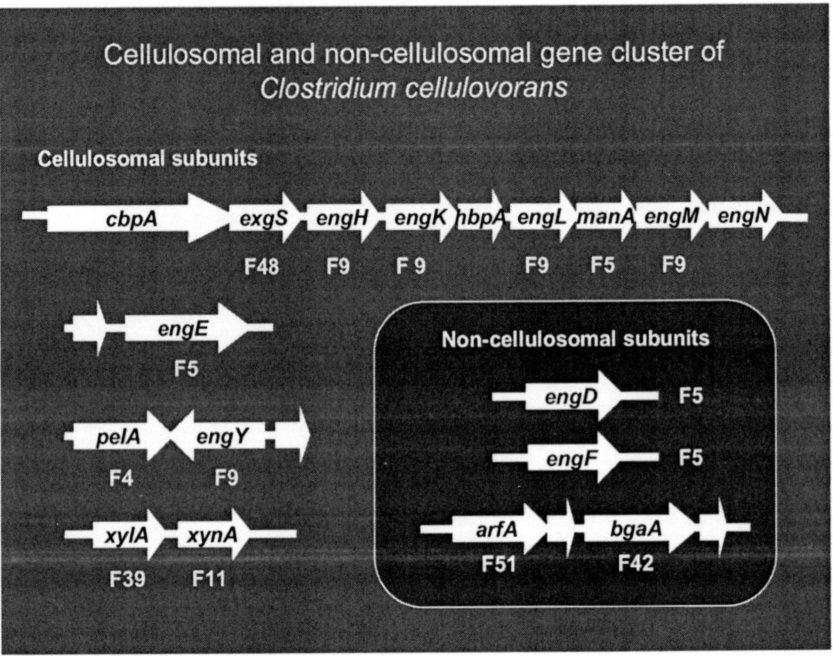

Fig. 3. The *C.cellulovorans* cellulosomal gene cluster. Only one large gene cluster exists for the scaffolding protein and several cellulosomal enzymes. Many of the cellulosomal enzymes also are unlinked to the large gene cluster. There are also several non-cellulosomal enzymes that work in concert with the cellulosome to degrade plant cell wall materials.

Cellulosomes are Heterogenous in Subunit Composition

The subunit composition of *C. cellulovorans* cellulosomes can vary depending on several factors. These factors include the limited number of cohesin sites on the scaffolding protein CbpA, the large number of different cellulosomal enzymes, and the different types of cellulosomal enzymes produced on different growth substrates.

Since the scaffolding protein CbpA contains nine cohesin sites, the maximum number of enzymes bound to it are nine. The nine bound enzymes could be one set of nine different enzymes, or another set of nine different enzyme, or the CbpA in theory could bind nine identical enzyme subunits and be composed of just two kinds of proteins. Therefore, it is possible that a large number of different cellulosome could exist within the cell. When cellulosomes have been purified and fractionated, it appears that a limited number of cellulosomes with different enzyme compositions are present. For instance in *C. papyrosolvens*, there appeared to be about 7 different sub-populations of cellulosomes [42,43]. In the case of *C. cellulovorans*, fractionation of the cellulosomes has also yielded about 6 or 7 subpopulations of cellulosomes [21].

Another major factor affecting subunit composition of cellulosomes is the carbon source on which the cells are grown. This is shown both by subunit composition and by enzymatic activity present in the cellulosome [17]. Two dimensional gel fractionation of cellulosome subpopulations indicates that the subunit population is quite different when the growth substrate is either cellulose, xylan, pectin or a mixture of these three carbon substrates.

The rather limited number of subpopulations observed suggest that the assembly of cellulosomes is not a random process. If the process were completely random, one would expect a larger number of subpopulations. The assembly process occurs outside of the cell, since all cellulosomal subunits contain a signal peptide. Therefore the assemble of the cellulosome includes secretion of the CbpA and enzymatic subunits, refolding of the secreted proteins into their mature forms, and a non-random association of the enzyme subunits with the CbpA. This mechanism of assembly is still unknown and presents a great challenge to future investigators.

Transcription and Regulation of Expression of Cellulase Genes

How are the genes in the *cbpA* gene cluster expressed? Are they expressed in one giant operon? The sequence studies showed the presence of putative transcription termination sites within the gene cluster [54]. Transcription studies indicated the presence of several operons within the cluster [18,19]. Fig. 4 shows the transcripts that were produced from the gene cluster. These included transcripts of the following genes: *cbpA-exgS; cbpA-exgS-engH-engK; engK-hbpA-engL; hbpA; manA;* and *engM.*

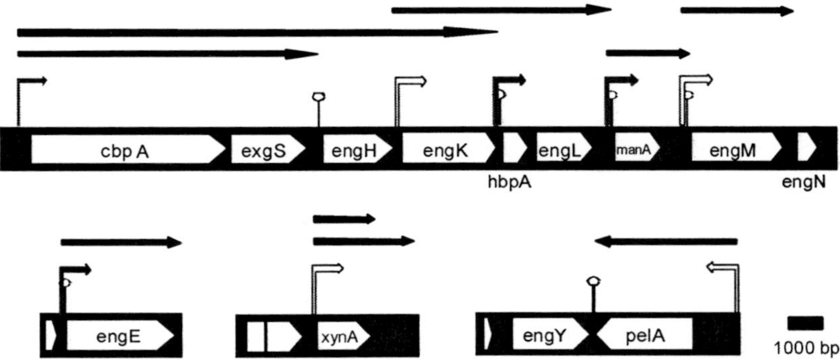

Fig. 4. Transcription map of the *C. cellulovorans* gene cluster. The mRNA transcripts are designated by black arrows. The bent black arrows indicate promoter sites. The white arrows are secondary transcripts. The lollypops indicate transcription termination sites.

Furthermore it was found that monocistronic transcripts were found for *engE, xynA*, and *pelA*. The *engY* gene may be part of an operon, although this has not been confirmed. The promoter regions for *cbpA* , *engE,, manA* and *hbpA* contained a consensus promoter sequence TAGAAA at –35 and TATAAT at –10 which is similar to that found for σ^A promoters of gram positive organisms [56].

The transcription of cellulosomal genes is determined in part by the growth conditions [17]. The cellulase genes *cbpA* and *engE* and the hemicellulase gene *xynA* were coordinately expressed when cells were grown in medium containing cellulose or cellobiose. When cells were grown on a medium containing cellulose, xylan and pectin, most genes (*cbpA-exgS, engH,bhpA, manA, engM, engE, xynA* and *pelA*) were expressed actively. Expression of *xynA* and *pelA* were specifically induced when cells were grown on xylan and pectin, respectively. Pectin not only induces *pelA* expression, but also stimulates the expression of *cbpA-exgS* and *engE*, thus demonstrating that pectin can induce not only *pelA* but also the cellulase genes. Although there is very low constitutive synthesis of cellulase and hemicellulase genes, the presence of glucose reduces the expression of these genes.

Designer Cellulosomes

Designer cellulosomes are cellulosomes that are constructed from CbpA or parts of CbpA (i.e. mini-CbpA) and from cellulosomal enzymes. The mini-CbpA consists usually of a CBD and one or more cohesins (Coh)(Fig. 5). The mini-cellulosome is assembled by adding cellulosomal enzymes whose dockerins can bind specifically to the cohesins present in mini-CbpA. Since there is species specificity in the dockerin-cohesin interaction [11,12], one can construct mini-scaffolding proteins with cohesins from different species to ensure that only enzymes with dockerins specific for particular cohesins will bind.

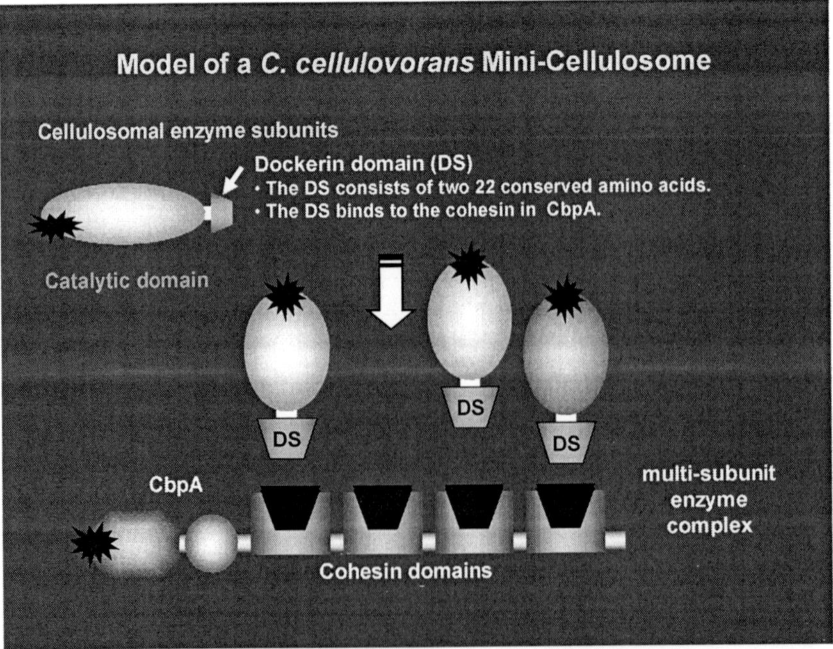

Fig. 5. A model mini-cellulosome. A mini-CbpA with four cohesin sites is able to bind four enzymes and form a mini-cellulosome. Mini-CbpA with cohesins that bind with specific dockerins can be constructed to form mini-cellulosomes with defined enzymatic composition.

Designer cellulosomes for *C. cellulovorans* were first utilized to study the function of hydrophilic domains (HLDs) and hydrophobic domains (HBD or Cohesin) of CbpA [6,7]. Mini-cellulosomes were constructed between a mini-CbpA containing CBD-HLD_1-Coh_1 and EngE and ExgS that could hydrolyze pebbled milled cellulose. This initial study has led to a number of investigations in which designer cellulosomes have been constructed to study a number of different functions of the cellulosome [11]. The use of mini-CbpA also overcomes the difficulty of obtaining the whole length CbpA, which is degraded when expressed in *E. coli.* There is less degradation of mini-CbpAs expressed in *E. coli*. Furthermore mini-CbpAs containing species-specific cohesins can be obtained [12].

Basically, the constructs of designer cellulosomes have consisted of CBD, HLD, and cohesins in different combinations with cellulosomal enzymes to study synergy, specificity of cohesins, and cellulosome structure in *C. cellulovorans*. These studies have been facilitated by the fact that the CBD, HLD and Coh domains can function quite independently after being expressed in *E. coli* [15,36,37,41].

Synergy Studies with Mini-Cellulosomes

Since cellulosomes consists of a cluster of enzymes associated with a scaffolding protein, do the enzymes act synergistically? One type of study utilizing mini-cellulo-somes showed that synergy existed between cellulosomal cellulases EngE (family 5), EngH (family 9) and ExgS (family 48) [36] when the substrate was insoluble cellulose. Moreover, the activity of the enzymes was enhanced when they were associated with the mini-CbpA that contained a CBD, HLD_1 and Coh_1 and Coh_2. In sequential addition reactions, it was shown that the addition of EngH and EngE prior to the addition of ExgS resulted in synergy indicating that the initial reactions of EngH and EngE promoted cellulose degradation by ExgS.

Since cellulosomes contain not only cellulases, but also hemi-cellulases such as xylanase [27], mini-cellulosomes containing cellulase and mini-cellulosomes containing xylanase were tested for activity on corn cell wall and on purified xylan and crystalline cellulose [29.37]. The mixture of minicellulosomes containing cellulase and xylanase were able to degrade corn cell wall synergistically, but not xylan or cellulose. The synergistic effect was observed during simultaneous addition of the mini-cellulosomes, but not during sequential addition of the mini-cellulosomes. This is in contrast to the synergy observed with the cellulases in which there was sequential action. The structure of corn cell wall, which includes a network of xylan and cellulose, may present a structure in which synergy can occur when both cellulases and xylanases act simultaneously and not sequentially.

DNA Shuffling to Improve Cellulosomal Enzyme Properties

In order to improve the heat stability property of cellulases, DNA shuffling experiments were carried out by using the genes for cellulosomal EngB and the non-cellulosomal EngD

[35]. EngB and EngD are highly homologous in their N-terminal catalytic domains with over 80% identity. However, EngB has a C-terminal dockerin domain and EngD has a C-terminal CBM. When the DNA shuffling experiments were carried out initially with the two wild type genes, recombinant enzymes could not be obtained. However, when the CBM of EngD replaced the dockerin domain of engB so that both genes contained an identical C-terminal domain, recombinant genes were obtained [35]. A two-selection method was used [28], since it was previously observed that increasing the heat stability of an enzyme often reduced its catalytic activity. In the first selection, recombinants were selected that showed the same activity as the parental enzyme EngB. Of the 10,000 clones selected by the first selection, in the second selection 800 clones were selected that showed some heat stable enzyme activity at 55°C. From these 800 clones, three clones were finally obtained that produced enzymes that were significantly more stable at 55°C than the parental enzymes that were stable at 45°C. Thus it is possible to obtain cellulases with novel properties by directed evolution. By use of high technology robotic systems that can screen large number of clones, it should be possible to select cellulases that are more heat stable, more active at high or low pH, and more active under various conditions, such as high ionic concentrations and the presence of inhibitors.

HETEROLOGOUS EXPRESSION OF CELLULOSOMAL GENES

There are several reasons for desiring the expression of cellulase genes in heterologous microorganisms. For instance, if one can express cellulosomes in aerobic microorganisms, the faster growing microorganisms could use cellulose or cellulosic biomass as inexpensive carbon sources for the production of valuable products. A *C. cellulovorans* cellulosome was produced *in vitro* from cellulosomal EngB made by *Bacillus subt*ilis and miniCbpA made by *Escherichia coli* [32]. The miniCbpA contained a CBD, a HLD, and an Coh. When the miniCbpA and EngB were mixed, an active mini-cellulosome resulted.

In a further extension of this strategy, both mini-CbpA and EngB were produced by *B. subtilis* and active minicellulosomes were produced *in vivo* [5]. In these experiments the expression construct consisted of the P43 promoter, a ribosome binding sequence, and a signal peptide all derived from *B. subtilis*. The *mini-cbpA* gene and the *engB* genes were each fused to this type of *B. subtilis* sequence. Furthermore a *B. subtilis* WB800 strain containing deletions of all 8 extracellular protease genes was used as the host. The active cellulosome was able to degrade cellulose.

The success with *B. subtilis* illustrates the potential for converting any organism into a cellulose utilizing organism that could utilize low cost cellulosic biomass as a substrate to produce valuable products such as ethanol, amino acids, solvents, organic acids and perhaps even hydrogen gas.

Use of CBD for Bioapplications

Since CBD binds efficiently to cellulose, CBD fusion proteins have been used for a number of applications [30,46]. These CBD fusion proteins have been used as protein purification systems [48] and as bioreactors [49], since cellulose is a relatively inexpensive, inert, matrix that can be utilized in large scale operations. CBD has also been used as a plant growth modulator in transgenic plants [50] and thus can be used to modify the chemical and physical properties of plants both *in vivo* and *in vitro.*

Conclusions and Comments

Assembly Process

One of the most intriguing questions about cellulosomes is "How are cellulosomes assembled?" There appear to be at least two aspects to the assembly process. First, since all the components are secreted, they must fold into a mature form extracellularly, assemble into the enzyme complex through cohesin-dockerin interactions, and then finally associate with the cell surface during growth. Second, there is convincing evidence for the presence of discrete subpopulations of cellulosomes and their compositions do not appear to be random, i.e. they appear to have fixed compositions. This suggests that after proper folding of the enzyme subunits that they bind to a scaffolding protein in a non-random fashion. How is the composition of these subpopulations determined?

Genetics of Cellulosomes

The genetics of cellulosomes is in its infancy. It is now apparent that transformation techniques are available for studying many aspects of cellulosome synthesis, regulation, and modification. Preliminary results indicate that *C. cellulovorans* will be amenable to such investigations in the near future.

Synergy

It is apparent that the cellulosomal enzyme subunits act synergistically during degradation of a complex substrate such as plant cell walls. The cellulases and hemicellulases act synergistically. Cellulosomes and non-cellulosomal enzymes also act in concert. When all the enzymes that degrade plant cell walls have been identified, it may give a clue as to which combinations are most effective and which combinations can be assembled to further improve cellulosome efficiency.

Designer Cellulosomes for More Efficient Activity

Both *in vitro* and *in vivo* construction of designer cellulosomes should lead to more efficient degradation of various substrates. Furthermore the modification of enzyme properties by DNA shuffling could lead to designer cellulosomes with different types of activities. The specificity of cohesin-dockerin interaction may lead to assemblies of enzymes for many purposes including bioreactors.

Cellulosome Production by Heterologous Microorganisms

The successful production of *C. cellulovorans* minicellulosomes by *B. subtilis* indicates that many bacteria can be converted to use cellulose or cellulosic biomass as their carbon source. What is needed is the proper combination of a mini-CbpA and a mixture of enzymes that are able to attack crystalline cellulose. By use of current molecular biology techniques this should be a feasible result.

Acknowledgements

The research reported from our laboratory was partially supported by grant DE-DDF03-92ER20069 from the U.S. Department of Energy and a grant from the Research Institute of Innovative Technology for the Earth (RITE), Kyoto, Japan. The author thanks Akihiko Kosuge and Sung Ok Han for providing the illustrations and his graduate students, postdoctoral fellows, and visiting scientists who contributed to results reported from his laboratory.

References

Arai, T., A. Kosugi, H. Chan, R. Koukiekolo, H. Yukawa, M. Inui, M. and R.H. Doi, 2006. Properties of cellulosomal family 9 cellulases from Clostridium cellulovorans. *Appl. Microbiol Biotechnol.* 70: in press.

Bayer, E.A. and R. Lamed. 1986. Ultrastructure of the cell surface cellulosome of Clostridium thermocellum and its interaction with cellulose. *J. Bacteriol. 167*:828-836.

Bayer, E.A., E. Morag, and R. Lamed. 1994. The cellulosome – a treasure trove for biotechnology. *Trends Biotechnol. 12*:378-386.

Blair, B.G. and K.L. Anderson 1998. Comparison of staining techniques for scanning electron microscopic detection of ultrastructural protuberances on cellulolytic bacteria. *Biotech. Histochem 73*: 107-113.

Cho, H.-Y., Yukawa, H., Inui, M., Doi, R.H. and Wong, S.-L. 2004. Production of minicellulosomes from Clostridium cellulovorans in Bacillus subtilis WB800. *Appl. Environ. Microbiol. 70*: 5704-5707.

Murashima, K., A. Kosugi, and R.H. Doi. 2002. Thermostabilization of cellulosomal endoglucanase EngB from Clostridium cellulovorans by in vitro DNA recombination with non-cellulosomal endoglucanase EngD. *Molec. Microbiol. 45*:617-626.

Murashima, K., A. Kosugi, and R.H. Doi. 2002. Synergistic effects on crystalline cellulose degradation between cellulosomal cellulases from Clostridium cellulovorans. *J. Bacteriol. 184*: 5088-5095.

Murashima, K., A. Kosugi, and R.H. Doi. 2003. Synergistic effects of cellulosomal xylanase and cellulases from Clostridium cellulovorans on plant cell wall degradation. *J. Bacteriol. 185*: 1518-1524.

Murashima, K., A. Kosugi, and R.H. Doi. 2003. Solubilization of cellulosomal cellulases by fusion with cellulose binding domain of non-cellulosomal cellulase EngD from Clostridium cellulovorans. *Protein: Structure, Function, Genetics 50*:620-628.

Murashima, K., A. Kosugi, and R.H. Doi. 2005. Site-directed mutagenesis and expression of the soluble form of the family IIIa cellulose binding domain from the cellulosomal scaffolding protein of Clostridium cellulovorans. *J. Bacteriol.* 187: 7146-7149.

Pages, S., A. Belaich, H.-P. Fierobe, C. Tardif, C. Gaudin, and J.-P. Belaich. 1999. Sequence analysis of scaffolding protein CipC and ORFXp, a new cohesin-containing protein in Clostridium cellulolyticum: comparison of various cohesin domains and subcellular localization of ORFXp. *J. Bacteriol. 181*:1801-1810.

Park, J.-S., Y. Matano, and R.H. Doi, 2001. Cohesin-dockerin interactions of cellulosomal subunits of Clostridium cellulovorans. *J. Bacteriol. 183*:5431-5435.

Pohlschroder, M., E. Canale-Parola, S. B. Leschine. 1995. Ultrastructural diversity of the cellulase complexes of Clostridium papyrosolvens C7. *J. Bacteriol. 177*:6625-6629.

Pohlschroder, M.D., S.B. Leschine, and E. Canale-Parola. 1994. Multicomplex cellulase-xylanase system of Clostridium papyrosolvens C7. *J. Bacteriol. 176*:70-76.

Sheweita, S.A., A. Ichi-ishi, J.-S. Park, C.-C. Liu, L.M. Malburg, Jr. and R.H. Doi, 1996. Characterization of engF, a gene for a non-cellulosoma Clostridium cellulovorans endoglucanase. *Gene 182*:163-167.

Shoseyov, O. and R.H. Doi. 1990. Essential 170 kDa subunit for degradation of crystalline cellulose by Clostridium cellulovorans cellulase. *Proc. Natl. Acad. Sci. USA 87*:2192-2195.

Shoseyov, O., I. Levy, Z. Shani, and S.D. Mansfield, 2003. Modulation of wood fibers and paper by cellulose binding domains. In *Applications of Enzymes to Lignocellulosics.* pp. 116-131. S.D.Mansfield and J.N. Saddler, eds., American Chemical Society, Washington, DC.

Shoseyov, O., M. Takagi, M. Goldstein, and R.H. Doi. 1992. Primary sequence analysis of Clostridium cellulovorans cellulose binding protein A (CbpA). *Proc. Natl. Acad. Sci. USA 89*:3483-3487.

Shpigel, E., D. Elias, I.R. Cohen,and O. Shoseyov. 1998. Production and purification of a recombinant human hsp60 eptiope using the cellulose-binding domain in Escherichia coli. *Protein Expr. Purific. 14*:185-191.

Shpigel, E., A. Goldlust, G. Efroni, A. Avraham, A. Eschel, M. Dekel and O. Shoseyov, 1999. Immobilization of recombinant heparinase I fused to cellulose binding domain. *Biotechnol. Bioengin. 65*:17-23.

Shpigel, E., I. Roiz, R. Goren, and O. Shoseyov. 1998. Bacterial cellulose binding domain modulates in vitro elongation of different plant cells. *Plant Physiol.117*:1185-1194.

Sleat, R., R.A. Mah, and R. Robinson. 1984.Isolation and characterization of an anaerobic, cellulolytic bacterium, Clostridium cellulovorans sp. nov. *Appl. Environ. Microbiol. 48*:88-91.

Tamaru, Y. and R.H. Doi. 1999 Three surface layer homology domains at the N-terminus of the Clostridium cellulovorans major cellulosomal subunit, EngE. *J. Bacteriol. 181*: 3270-3276.

Tamaru, Y. and R.H. Doi, 2001. Pectate lyase A, an enzymatic subunit of the Clostridium cellulovorans cellulosome. *Proc. Natl.Acad. Sci. USA 98*: 4125-4129.

Tamaru, Y., S. Karita, A. Ibrahim, H. Chan, and R.H. Doi, 2000. A large gene cluster for the Clostridium cellulovorans cellulosome. *J. Bacteriol. 182*:5906-5910.

Tamaru, Y., A. Kosugi, K. Murashima, and R.H. Doi. 2003. The Clostridium cellulovorans cellulosome and its application to plant biomass degradation. *Recent Res. Devel. Microbiol. 7*:1-18.

Young, M., N. P. Minton, and W. L. Staudenbauer. 1989. Recent advances in the genetics of the clostridia. *FEMS Microbiol. Rev. 5*:301-325.

In: Cellulosome
Eds: V. Uversky and I. A. Kataeva, pp. 169-190
ISBN 1-59454-950-8

Chapter VIII

THE CELLULOSOMES OF CLOSTRIDIUM JOSUI AND RUMINOCOCCUS ALBUS

Shuichi Karita[1,*], *Sadanari Jindou*[1], *Akio Shimamura*[2], *Tsutomu Kajino*[3], *Tetsuya Kimura*[1], *Kunio Ohmiya*[1] and *Kazuo Sakka*[1]

[1]Faculty of Bioresources, Mie University, Tsu, Japan
[2]Technological Development Group, Katakura Industries Co. Ltd., Sayama, Japan
[3]Special Research Laboratory II, Toyota Central R&D Labs. Inc., Aichi, Japan

ABSTRACT

Catalytic components of the *Clostridium josui* cellulosome are quite similar to those of the well-characterized *Clostridium cellulolyticum* cellulosome although the size of the scaffolding protein of the former is smaller than that of the latter, resulting in the formation of smaller cellulosome. *C. josui* is closely related to *C. cellulolyticum* judging from their 16S rDNA sequences, but the optimal growth temperature of the former (45 °C) is higher than that of the latter (33-35 °C), suggesting that properties of their cellulosomes are different from each other. Species specificity was observed in the cohesin-dockerin interaction between *C. josui* and *Clostridium thermocellum*. We constructed a dockerin-cohesin affinity purification system for recombinant proteins expressed in the baculovirus expression system.

SDS-PAGE and Zymogram analyses of the cellulolytic complex isolated from the culture supernatant of *Ruminococcus albus* F-40 revealed fifteen proteins including eight endoglucanase-active and nine xylanase-active proteins. Immunological analysis done with the antiserum raised against Cel5D-dockerin confirmed that assembly of the *R. albus* cellulosome was made by the cohesin-dockerin interactions.

---

* Correspondence concerning this article should be addressed to Shuichi Karita, Department of Sustainable Resource Science, Faculty of Bioresources, Mie University, 1577 Kurimamachiyacho, Tsu 514-8507, Japan

Keywords: Clostridium josui, Ruminococcus albus, cohesin-dockerin interaction, affinity tag, cohesin-dockerin affinity chromatography

Multienzyme complexes having high activity against crystalline cellulose, known as a cellulosome, were identified and characterized in some cellulolytic clostridia such as *C. cellulolyticum*, *C. cellulovorans*, and *C. thermocellum*, and anaerobic cellulolytic fungi such as *Neocallimastix patriciarum* and *Piromyces* sp.. In this chapter, we describe the cellulosome systems of two mesophilic anaerobic bacteria, *Clostridium josui* and *Ruminococcus albus.*

1. The Cellulosome of *Clostridium Josui*

1-1. Taxonomic Position of *C. Josui*

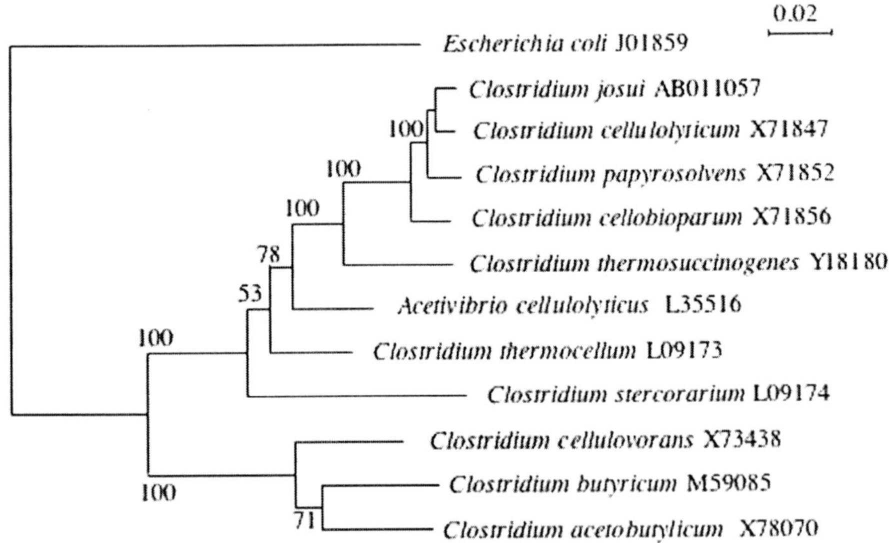

Fig. 1. Phylogenetic tree of 16S rDNA sequences of *C. josui* and some related clostridial species constructed by the ClustalW program.

C. cellulolyticum and *C. cellulovorans* are well known as mesophilic cellulosome-forming clostridia and their cellulosomes are well characterized as described elsewhere. Since *C. josui* is also a mesophilic cellulosome-forming strain with an optimal growth temperature of 45°C, the taxonomic position of *C. josui* should be specified among these mesophilic and cellulolytic clostridia. The phylogenetic tree was constructed from several clostridial 16S rDNA sequences by the ClustalW program. As shown in Fig. 1, *C. josui* is closely related to *C. cellulolyticum* and *C. papyrosolvens* on the basis of the comparison of 16S rDNA sequences although the optimal growth temperatures of them are different noticeably from each other, i.e., the optimal growth temperature of *C. josui* is 45°C [1] and *C. cellulolyticum* and *C. papyrosolvens* optimally grow at 33 – 35°C and 25 – 30°C, respectively [2]. Difference in their optimal growth temperatures suggests difference in their enzyme properties, at least in optimal temperatures of their cellulolytic activities, probably leading to difference in their specific activities. Therefore, it is worth studying the cellulolytic system of *C. josui* even though a scaffolding protein and some catalytic components of the *C. josui* cellulosome closely resemble those of *C. cellulolyticum* as described below. On the other hand, *C. cellulovorans* is taxonomically less related to *C. josui*, *C. cellulolyticum* and *C. papyrosolvens*.

1-2. A Large Gene Cluster Responsible for a Scaffolding Protein (CipA) and Several Catalytic Components

A common feature of the clostridial cellulosomes is that they consist of a large number of catalytic components arranged around noncatalytic scaffolding proteins (or cellulosome-integrating protein, Cip). Many cellulase genes are known to be scattered on the *C. thermocellum* chromosomal DNA but with some exceptions [3]. By contrast, a large gene cluster was found to consist of genes encoding a scaffolding protein and several catalytic components in *C. cellulolyticum* (Fig. 2) [4]. We also sequenced the DNA region of about 21 kb of *C. josui* including the *cipA* gene and found that this region contained a large gene cluster responsible for several catalytic components in addition to the scaffolding protein CipA, i.e., six cellulase genes (*cel48A*, *cel8A*, *cel9E*, *cel9F*, *cel9G*, and *cel9H*), a mannanase gene (*man5A*), and an open reading frame of unknown function (*orfX*) (Fig. 2) [5, 6; unpublished results]. The gene cluster of *C. josui* is quite similar to that of *C. cellulolyticum*, i.e., the order of genes in the *C. josui* cluster is identical to that in the *C. cellulolyticum* cluster and each gene product of *C. josui* is highly homologous with a corresponding gene product of *C. cellulolyticum* (sequence identities: 80 to 95%). However, it is noteworthy that although *C. josui* CipA is also quite similar to *C. cellulolyticum* CipC (sequence identity: 80%), the former is smaller than the latter: *C. josui* CipA consists of a carbohydrate-binding module (CBM) and six cohesin modules but *C. cellulolyticum* CipC contains eight cohesin modules in addition to a CBM (Fig. 3A).

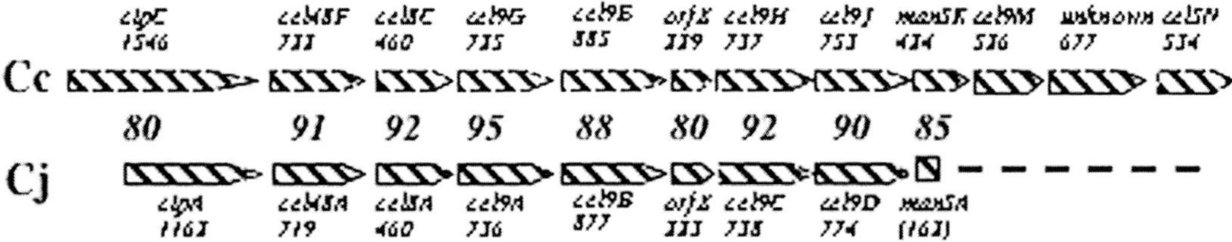

Fig. 2. Comparison of gene clusters including scaffolding protein-encoding genes of *C. josui* and *C. cellulolyticum*. Percentage figures shown between *C. josui* and *C. cellulolyticum* gene clusters are identities between amino acid sequences of gene products deduced from nucleotide sequences. Numbers beneath respective genes are numbers of amino acids of deduced gene products.

All genes for catalytic components are not necessarily exist in a gene cluster [7]: the *cel5A* and *aga27A* genes that encoded a noncellulosomal cellulase and a cellulosomal α-galactosidase, respectively, were found in a DNA fragment different from the fragment containing the large gene cluster.

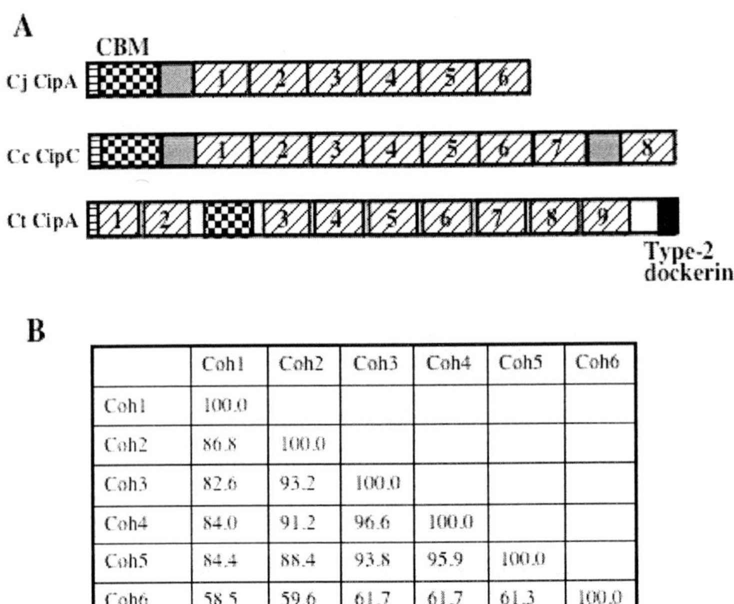

	Coh1	Coh2	Coh3	Coh4	Coh5	Coh6
Coh1	100.0					
Coh2	86.8	100.0				
Coh3	82.6	93.2	100.0			
Coh4	84.0	91.2	96.6	100.0		
Coh5	84.4	88.4	93.8	95.9	100.0	
Coh6	58.5	59.6	61.7	61.7	61.3	100.0

Fig. 3. Schematic diagram of scaffolding proteins (A) of *C. josui* (Cj), *C. cellulolyticum* (Cc) and *C. thermocellum* (Ct), and amino acid sequence identities among cohesin modules of *C. josui* CipA (B).

1-3. Catalytic Components of the *C. josui* Cellulosome

Molecular architecture of catalytic components deduced from the nucleotide sequences is summarized in Table 1. The cellulosome fraction could be isolated as high molecular mass cellulolytic complexes from the supernatant of the *C. josui* culture grown on ball-milled cellulose by a HiPrep Sephacryl S-300HR column chromatography [6]. When a fraction containing protein complexes of ca. 700 kDa was subjected to SDS-PAGE and zymogram (activity staining with Congo red dye after SDS-PAGE) analyses, more than 15 proteins were detected in the gel by Coomassie brilliant blue staining and more than 10 proteins showed endoglucanase activity in the gel containing carboxymethyl-cellulose (CMC) or xylan by zymogram analysis [6]. When the proteins in this fraction were incubated with insoluble cellulose, all the proteins were bound to the cellulose. N-Terminal amino acid sequence analysis of the cellulosomal proteins identified CipA and Cel48A as major components in the cellulosome.

Among the catalytic components, only Aga27A [7] and Cel8A [8] were biochemically characterized. Aga27A composed of a family 27 catalytic module of glycosyl hydrolases and a dockerin module involved in cellulosome assembly is the first α-galactosidase (EC 3.2.1.22) found in the cellulosomes from various origins. α-Galactosidases are substantially classified into three groups, families 4, 27 and 36 of glycosyl hydrolases on the basis of amino acid sequence homology [9]. Almost all of α-galactosidases from eukaryotic organisms belong to family 27 although an α-galactosidase of *Trichoderma reesei* is a member of family 36. On the other hand, most prokaryotic α-galactosidases have been classified into family 4 or 36. Therefore, *C. josui* Aga27A is the rare exception with *P. fluorescens* subsp. *cellulosa* Aga27A [10]. α-Galactosidase is an enzyme that hydrolyzes α-galactosidic linkages at non-reducing ends in galactose-containing oligosaccharides such as melibiose and raffinose, galactolipids, and/or galactomannan such as guar gum. On the basis of substrate specificity, there are two types of α-galactosidase: one is specific for small saccharides and another is extremely active on highly polymerized galactomannan. Some enzymes in family 27 are known to be more active on guar gum than on the small substrates [11]. *C. josui* Aga27A also prefers guar gum to the small substrates although bacterial α-galactosidases in family 4 or 36 show high activity toward the small substrates but negligible activity toward guar gum [12]. Western blot analysis using the anti-Aga27A antiserum detected Aga27A in the cellulosome fraction purified from the culture supernatant of *C. josui* grown on ball-milled cellulose. Since *C. josui* can not grow on galactomannan as the carbon source, it is likely that Aga27A contributes to degradation of galactomannan present in plant cell walls, allowing the cellulosome to access to cellulose chains that is buried in galactomannan and is not accessible unless galactomannan is hydrolyzed and removed. A similar phenomenon was observed in *C. thermocellum*, i.e., this bacterium is known to produce cellulosomal xylanases [13,14] and mannanases [15] although it uses neither xylan nor mannan as the carbon sources. The optimum temperature of this enzyme was determined to be about 58°C at pH 5.5.

Table 1. Molecular architecture of the cellulosomal enzymes of *C. josui*

Enzyme (Former name)	Module organization[1]
Aga27A	SP-CM27-Doc
Cel8A (CelB)	SP-CM8-Doc
Cel9A	SP-CM9-CBM3-Doc
Cel9B	SP-CBM4-Ig-CM9-Doc
Cel9C	SP-CM9-CBM3-Doc
Cel9D	SP-CM9-CBM3-Doc
Cel48A (CelD)	SP-CM48 -Doc
Man5A	SP-Doc-CM5

[1] SP, signal peptide; CM27, family-27 catalytic module; CBM3, family-3 CBM; Ig, Ig-like module; Doc, dockerin module.

Cel8A is an endoglucanase consisting of a family-8 catalytic module and a dockerin module [8]. The optima for the temperature and pH of the enzyme were 60°C and 5.5. The enzyme hydrolyzed cellotetraose, cellopentaose and cellohexaose but not cellobiose and cellotriose. The mode of degradation of cellooligomers (G4 → 2G2, G5 → G2 + G3, G6 → 2G3) of the enzyme suggested that it acts as a typical endoglucanase.

The temperature optima (about 60°C) of Aga27A and Cel8A appear to be higher than those of the enzymes from *C. cellulolyticum*, i.e., many enzymes from *C. cellulolyticum* have the temperature optima at about 50°C [16,17]. Difference in the temperature optima between

C. josui and *C. cellulolyticum* appears to reflect difference in optimal growth temperature between them.

1-4. Analysis of Cohesin/Dockerin Interactions within and Between *Clostridium Josui* and *Clostridium Thermocellum*

Cohesin and Dockerin Modules of C. Josui

The cellulosome components are assembled into the cellulosome complex by the interaction between one of the repeated cohesin modules of a scaffolding protein and the dockerin module of an enzyme component. Since *C. josui* CipA contains six cohesin modules, a molecule of CipA should anchor six molecules of enzyme components to form a cellulosome. Although amino acid sequences of the six cohesin modules are highly homologous with each other, the first and sixth cohesin modules seem to be less homologous to the others (Fig. 3B). The dockerin modules of *C. josui* are specified by the two pairs including an alanine in the first position and a hydrophobic residue (usually leucine or isoleucine) in the second (Fig. 4) like the dockerin modules of *C. cellulolyticum*.

```
Cj Cel8A   402 GDVNNDGAIDALDIAALKKAILTQSTSNINLTN
Cj Cel9A   674 GDVNSDKNVDALDLAALKKYLLGGTSSIDVKA
Cj Cel9B   818 GDVNGDGNKDAIDFALLKKALLSQDTSTINVAN
Cj Cel9C   683 GDINKDGNIDALDVALLKKYLLGNALEYDVSVADMN
Cj Cel9D   723 GDGSVDALDYALMKKYILYPTGDVDLNV
Cj Aga27A  405 GDVDGNGMIDALDYSLVKRYLLGQISDCPDSKGKLA
Cj Cel48A  663 GDVNGDETVDAIDLAMLKKYLLNSSTSIVAGN
Cc Cel5A   414 GDYNNDGNVDALDFAGLKKYIMAADHAYVKN
Ct Xyn10C  559 GDVNGDGRVNSSDLTLMKRYLLKSISDFPTPEGKIA
Ct Xyn11A  393 GDLNGDKQVNSTDYTALKRHLLNITRLSGTALAN
Ct Cel48A  675 GDVNDDGKVNSTDAVALKRYVLRSGISINTDN

Cj Cel8A   438 ADMNNDGNIDAIDFAQLKVKLLN*
Cj Cel9A   706 ADTYKDGNIDALDMATLKKYLLGTITQLPQG*
Cj Cel9B   851 SDLNQDGSIDAVDFALLKSFLLGKITL*
Cj Cel9C   715 ADMNSDKNIDALDFALLKKTLLSQ*
Cj Cel9D   751 WDMNKDGEINALDLALLKKTLLG*
Cj Aga27A  441 ADVDGDQQITALDFSLIKQYLLGTINKFPAQTASKIKP*
Cj Cel48A  695 ADMNGDGAIDAIDYALLKKALLANQ*
Cc Cel5A   445 LDVNLDNEVNAFDLAILKKYLLGMVSKLPSN*
Ct Xyn10C  595 ADLNEDGKVNSTDLLALKKLVLREL*
Ct Xyn11A  427 ADLNGDGKVDSTDLMILHRYLLGIISSFPRSNPQ
Ct Cel48A  712 ADLNEDGRVNSTDLGILKRYILKEIDTLPYKN*
```

Fig. 4. Alignment of amino acid sequences of *C. josui* cohesin modules. “AL” or “AI” motif in *C. josui* and *C. cellulolyticum* and “ST” or “SS” motif in *C. thermocellum* are boxed. Numbers refer to amino acid residues at the start of the respective lines; all sequences are numbered from Met-1 of the peptide.

Interactions Between Cognate Cohesin and Dockerin Polypeptides of C. Josui

In order to estimate the affinity of each cohesin module in *C. josui* CipA for the same dockerin module, four cohesin module polypeptides (the first, second, fifth, and sixth cohesin modules, abbreviated as rCoh1-Cj, rCoh2-Cj, rCoh5-Cj, and rCoh6-Cj), and two dockerin module polypeptides derived from Aga27A and Cel8A (rAga27ADoc and rCel8ADoc) were constructed and interactions between the cohesin and dockerin polypeptides were determined by surface plasmon resonance (SPR) analysis using a BIACore system [18]. All the cohesin polypeptides were covalently immobilized to carboxymethyl groups of the dextran on the surface of the CM5 sensor chip via the amine coupling method. The running buffer contained 10mM $CaCl_2$ since calcium ions are required for the binding of dockerin modules to cohesin modules [19]. Table 2 shows the affinity constants for the binding of cohesin and dockerin polypeptides from *C. josui*. In the case of the cohesin/dockerin interactions of *C. josui*, rAga27ADoc, which specifically recognized *C. josui* CipA but not *C. thermocellum* CipA in the affinity blotting analysis, associated with rCoh1-Cj, rCoh2-Cj, rCoh5-Cj and rCoh6-Cj with k_{on} values of 1.8 x 10^5 – 6.9 x 10^5 and k_{off} values of 1.3 x 10^{-4} – 3.4 x 10^{-4}, leading to K_D values of 1.9 x 10^{-9} – 2.4 x 10^{-10}, i.e., it showed the highest affinity for rCoh2-Cj (K_D, 2.4 x 10^{-10}) and the lowest affinity for rCoh1-Cj (1.9 x 10^{-9}), meaning that there is about 8-fold difference in the affinity between the cohesin modules in CipA and the dockerin module from Aga27A. rCel8ADoc also showed preference of rCoh2-Cj over rCoh1-Cj, i.e., rCel8ADoc associated with rCoh2-Cj with the K_D value of 1.3 x 10^{-10} and rCoh1-Cj with the K_D value of 4.4 x 10^{-9}, meaning 34-fold difference in the affinity between the cohesin modules in CipA and the dockerin module from Cel8A. Although the extreme difference in the k_{on} and k_{off} values was not observed between rAga27ADoc and rCel8ADoc of *C. josui* towards respective cohesin polypeptides, the product of relatively small differences in both of the k_{on} and k_{off} values resulted in the 34-fold difference. Both the two dockerin polypeptides, rCel8ADoc and rAga27ADoc, from *C. josui* showed the highest affinity for rCoh2-Cj and the lowest affinity for rCoh1-Cj, suggesting that a particular cohesin module has a higher affinity for the cognate dockerin modules than the other cohesin modules.

Table 2. Association and dissociation constants for the binding of dockerin polypeptides to immobilized cohesin polypeptides of *C. josui*. Numbers in parentheses indicate values relative to the smallest K_D value for each dockerin polypeptide. The smallest K_D value was defined as 1.0.

	rAga27ADoc			rCel8ADoc		
	k_{on} (s^{-1} M^{-1})	k_{off} (s^{-1})	K_D (M)	k_{on} (s^{-1} M^{-1})	k_{off} (s^{-1})	K_D (M)
rCoh1-Cj	1.8x10^5	3.4x10^{-4}	1.9x10^{-9} (7.9)	2.1x10^5	9.0x10^{-4}	4.4x10^{-9} (34.3)
rCoh2-Cj	6.9x10^5	1.7x10^{-4}	2.4x10^{-10} (1.0)	1.1x10^6	1.4x10^{-4}	1.3x10^{-10} (1.0)
rCoh5-Cj	3.9x10^5	2.5x10^{-4}	6.3x10^{-10} (2.6)	1.4x10^6	4.7x10^{-4}	3.5x10^{-10} (2.7)
rCoh6-Cj	2.1x10^5	1.3x10^{-4}	6.5x10^{-10} (2.7)	1.2x10^6	7.4x10^{-4}	6.3x10^{-10} (5.0)

Interactions Between Noncognate Combinations of Cohesin and Dockerin Polypeptides from C. Josui and C. Thermocellum

When interactions between the dockerin polypeptide of Xyn10C (rXyn10CDoc) and the cohesin polypeptides (rCoh1-Ct, rCoh2-Ct, rCoh3-Ct, rCoh4-Ct, and rCoh7-Ct) of CipA from C. thermocellum, k_{off} values exceeded the quantitatively measurable range (10^{-4}) of the BIACore system (Table 3). However, the affinities of the interactions between rXyn10CDoc and the cohesin polypeptides appear to be comparable to those obtained for the combinations of cohesin and dockerin polypeptides from *C. josui*. The k_{on} values of rXyn11ADoc toward the respective cohesin polypeptides were 5- to 14-fold larger than those of rXyn10CDoc toward the corresponding cohesin polypeptides, suggesting the possibility that there is the binding selectivity of cohesin and dockerin modules within *C. thermocellum*.

Table 3. Association and dissociation constants for the binding of dockerin polypeptides to immobilized cohesin polypeptides of *C. thermocellum*.

	rXyn11ADoc			rXyn10CDoc		
	k_{on} (s^{-1} M^{-1})	k_{off} (s^{-1})	KD (M)	k_{on} (s^{-1} M^{-1})	k_{off} (s^{-1})	KD (M)
rCoh1-Cj	5.4×10^{5}	$<10^{-4}$	$<10^{-9}$	5.2×10^{5}	$<10^{-4}$	$<10^{-9}$
rCoh2-Cj	6.9×10^{5}	$<10^{-4}$	$<10^{-9}$	5.1×10^{6}	$<10^{-4}$	$<10^{-9}$
rCoh3-Cj	3.2×10^{5}	$<10^{-4}$	$<10^{-9}$	5.3×10^{6}	$<10^{-4}$	$<10^{-9}$
rCoh4-Cj	4.8×10^{5}	$<10^{-4}$	$<10^{-9}$	8.0×10^{6}	$<10^{-4}$	$<10^{-9}$
rCoh7-Cj	1.8×10^{5}	$<10^{-4}$	$<10^{-9}$	3.8×10^{6}	$<10^{-4}$	$<10^{-9}$

In a previous paper of Pagès et al., they reported that *C. thermocellum* Cel48A did not interact with the first cohesin module of *C. cellulolyticum* CipC and *C. cellulolyticum* Cel5A did not recognize the second cohesin module of *C. thermocellum* CipA [20]. This phenomenon was recognized as "species specificity" of cohesin/dockerin interactions.

When we evaluated the species specificity of cohesin/dockerin interactions between *C. josui* and *C. thermocellum* by affinity blotting method, we found that no recombinant cohesin polypeptides derived from *C. thermocellum* CipA recognized cellulosomal proteins of *C. josui*, indicating the presence of the species specificity between *C. josui* and *C. thermocellum*. SPR analysis also showed that rAga27ADoc and rCel8ADoc of *C. josui* did not have any affinity for cohesin polypeptides from *C. thermocellum* CipA.

By contrast, recombinant cohesin polypeptides derived from *C. josui* CipA associated with two proteins in the *C. thermocellum* cellulosome. Immunological analysis of the *C. thermocellum* cellulosomal proteins using an anti-Xyn11A antiserum suggested that the two proteins reactive with the cohesin polypeptides were Xyn11A and Xyn11B. The affinity constants for the binding of rXyn11ADoc to the *C. josui* cohesin polypeptides determined by SPR are shown in Table 4. rXyn11ADoc interacted with all the cohesin polypeptides with affinities high enough to detect the interactions (K_D, 6.4 x 10^{-9} - 1.7 x10^{-8}) but with relatively low affinities compared with those observed in combinations of cognate cohesin and dockerin modules. Although the k_{on} values of rXyn11ADoc for the cognate and noncognate cohesin polypeptides were similar to each other, the k_{off} values of rXyn11ADoc for the *C. josui* cohesin polypeptides were larger than those for the *C. thermocellum* cohesin polypeptides, resulting in the increased K_D values. These results clearly indicated that there are exceptions in species specificity of cohesin/dockerin interactions.

Table 4. Association and dissociation constants for the binding of C. thermocellum Xyn11ADoc to immobilized cohesin polypeptides of *C. josui*. Numbers in parentheses indicate values relative to the smallest K_D value for each dockerin polypeptide. The smallest K_D value was defined as 1.0.

	rXyn11ADoc		
	k_{on} (s^{-1} M^{-1})	k_{off} (s^{-1})	K_D (M)
rCoh1-Cj	8.0x10^{5}	5.5x10^{-4}	6.9x10^{-9} (1.1)
rCoh2-Cj	2.5x10^{5}	2.3x10^{-4}	9.2x10^{-9} (1.4)
rCoh5-Cj	9.3x10^{5}	5.9x10^{-4}	6.4x10^{-9} (1.0)
rCoh6-Cj	3.6x10^{5}	4.4x10^{-4}	1.7x10^{-8} (2.7)

Mechaly et al. qualitatively and quantitatively studied the species specificity in cohesin/dockerin interactions using the dockerin modules of *C. thermocellum* Cel48A and *C. cellulolyticum* Cel5A [21]. Most dockerin modules, including Cel48A, from *C. thermocellum* contain an "ST" or "SS" motif in each of the duplicated segments, and most dockerin modules from *C. cellulolyticum* contain an "AL" or "AI" motif. The importance of these conserved amino acids in the species-specific recognition between the cohesin and dockerin modules was shown by site-directed mutagenesis, e.g., mutation of a single residue from threonine to leucine at the second segment of the Cel48A dockerin changed it from nonrecognition to high affinity recognition (K_D, 5 x 10^{-9}) by a cohesin from *C. cellulolyticum*, suggesting that the presence or absence of a single decisive hydroxyl group is critical to the observed biorecognition. However, complete displacement of these two amino acid residues in each segment, i.e., "ST" and "ST" to "AL" and "AL" in Cel48A and "AL" and "AF" to "ST" and "ST" in Cel5A, bestowed new affinities for noncognate cohesin modules but did not abolish their original preference for the cognate cohesins, suggesting that additional residues are also important as secondary determinants in the specificity of interaction. These observations are consistent with the phenomenon that rXyn11ADoc containing the "ST" motif can interact with the cohesin modules of *C. josui* CipA in addition to *C. thermocellum* CipA, although it is unclear why only the dockerin module of Xyn11A can recognize the noncognate cohesin modules as counterparts.

1-5. Noncellulosomal Catalytic Components as Cellulose-Binding Proteins

Cel5A (formerly CelA) [7] and Xyn10A [22] were identified as noncellulosomal components. These are typical modular enzymes: Cel5A consists of a family-5 catalytic module, a family-17 CBM, a family-28 CBM and three surface layer homologous (SLH) modules, and Xyn10A consists of two family-22 CBMs, a family-10 catalytic module, a family-9 CBM and two SLH modules from their N-termini. It is possible that extracellular enzymes such as Cel5A and Xyn10A having both CBM and SLH modules function as carbohydrate-binding proteins on the cell surface, i.e., the enzymes are anchored onto the cell surface by the interaction between SLH modules and secondary cell wall, and at the same time CBMs associate with insoluble polysaccharides, resulting in adsorption of the cells onto the polysaccharides. It was demonstrated that *Clostridium stercorarium* Xyn10B, which was highly homologous with *C. josui* Xyn10A, resided on the cell surface and mediated binding of *C. stercorarium* cells to insoluble cellulose [23]. Since *C. josui* Xyn10A showed affinities for both insoluble cellulose and cell wall fragments, it is likely that Xyn10A also play an important role for adsorption of *C. josui* cells to insoluble cellulose. In *C. thermocellum*, a type-2 dockerin module in CipA associates with a type-2 cohesin module in SdbA, a cell surface protein, and consequently it is anchored to the cell surface along with catalytic components as the cellulosome. Therefore, the cellulosome is thought to be responsible for binding of *C. thermocellum* cells to insoluble cellulose. On the other hand, since *C. josui* CipA includes a CBM and six cohesin modules but not a dockerin module as described above, the *C. josui* cellulosome must be fixed on the cell surface by a mechanism different from that in *C. thermocellum*.

1-6. Application of the Cohesin/Dockerin Interaction to Recombinant Protein Purification System in the Baculovirus Expression System

Need for Affinity Tags in the Baculovirus Expression System

High expression and purification of recombinant proteins facilitates production and detailed characterization of proteins of interest, and a wide variety of heterologous expression systems are currently being used for easy and rapid purification of recombinant proteins. The baculovirus expression system has been used to express various genes from bacteria, viruses, plants, and mammals. Post-translational modifications in insect cells closely resemble those in mammalian cells, making this system increasingly desirable for the expression of proteins destined to be post-translationally processed. In most cases, the procedure to infect insect cells with baculovirus is favorably used in the baculovirus expression system.

The use of affinity tags added to the proteins of interest can simplify the procedure of the purification of the recombinant fusion proteins by employing affinity chromatography. Although we first chose a combination of His-tag and nickel-nitrilotriacetic acid (Ni-NTA) metal-affinity chromatography for the silkworm-expression system, we found that this purification procedure was unsatisfactory for this expression system because of insufficient binding of the recombinant proteins and/or nonspecific binding of intrinsic proteins to Ni-NTA resin. Therefore, we attempted to use a combination of *C. josui* dockerin module as an affinity tag and the scaffolding protein immobilized to Sepharose as a resin for purification of the recombinant proteins. In this section, we describe the procedure for expression of recombinant proteins fused to a dockerin module in silkworm larvae or pupae and .for purification of the recombinant proteins by affinity chromatography on a CipA-immobilized-Sepharose column.

General Homologous Recombination and Expression of Recombinant Proteins in Insect Cell

When a recombinant transfer vector carrying a foreign gene and a linearized hybrid virus DNA are co-introduced into BmN cells with lipofection reagent, recombinant viruses are generated with homologous recombination in insect cells. Since the recombinant viruses are released into the culture medium, they can be harvested from the medium and used as virus suspension (Fig. 5). The virus suspension is injected into fifth instar larvae or pupae using a 1-cc syringe fitted with a 26-gauge needle. Since insects have an open circulatory system, hemolymph can be collected by piercing the proleg of larvae showing severe symptom of viral infection. Secreted proteins are collected from the hemolymph.

Expression of cell-associated proteins or trans-membrane proteins prefers pupae to larvae. The infected pupae were homogenized in PBS containing 10% glycerol, saturated phenyl-thiourea, 1 mM PMSF and 10 mM benzamidine, and further disrupted with an ultra-sonicator followed by centrifugation at 3,000 rpm for 15 min to remove the cell debris.

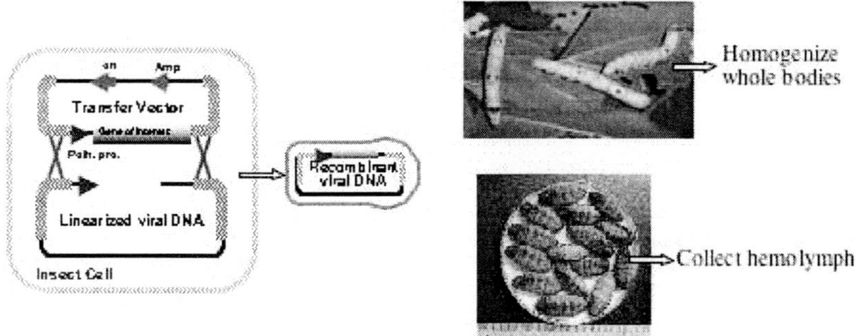

Fig. 5. Construction of a recombinant viral DNA by homologous recombination of a transfer vector and a linearized viral DNA for the expression of recombinant protein in pupae or fifth instar larvae.

Expression and Purification of Cohesin Proteins

We constructed recombinant baculoviruses for producing the full-length CipA of *C. josui* composed of a CBM and six cohesin modules (rCipA) and a polypeptide composed of the CBM and the first cohesin module (rCBM-Coh1) (Fig. 6A). Silkworm fifth instar larvae was infested with the virus and rCBM-Cohl and rCipA polypeptides were purified from hemolymph of the viral infected larvae by an anion-exchange column chromatography (Figs. 6B and 6C). rCipA was detected in the fraction eluted with 300 mM NaCl with the biotinylated dockerin polypeptide. The partially purified rCipA was coupled to CNBr-activated Sepharose and used for affinity chromatography. rCBM-Coh1 was biotinylated and used for the detection of the dockerin module in recombinant proteins.

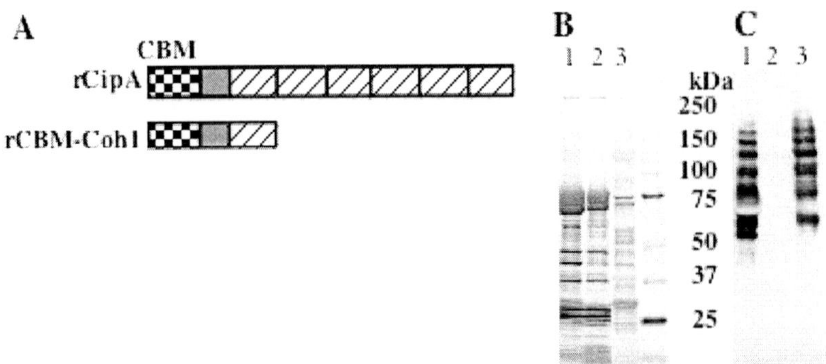

Fig. 6. Schematic model of rCipA and rCBM-Coh1 (A), the expression of rCipA detected by silver staining (B) and biotynilated dockerin polypeptide (C). Lane 1, hemolymph containing rCipA; lane 2, proteins not bound to an anion exchange column; lane 3, proteins eluted with 300 mM NaCl.

Dockerin-Cohesin Affinity Chromatography

Previously it was reported that the dockerin/cohesin interaction was calcium dependent [20], and therefore, that this interaction was inhibited by EDTA. In addition, amino acid sequence comparison between dockerin modules and EF-hand calcium-binding proteins strongly suggested that the dockerin module contained two calcium ion-binding sites, one per conserved segment. Tertiary structure analysis of dockerin modules in free form (*C. thermocellum* CelS) and in cohesin/dockerin complex form (*C. thermocellum* Xyn10B) confirmed that a dockerin module contains two calcium ion-binding sites. [24,25]. Therefore, we expected that dockerin-fused proteins would be bound to the rCipA-Sepharose column in a calcium-dependent manner and be eluted by removing calcium ion with ethylene glycol bis (β-aminoethylether)-N, N, N', N'-tetraacetic acid (EGTA).

Since the cohesin/dockerin interactions are highly specific (K_D of ca. 10^{-10}), the single step purification using the cohesion/dockerin interaction should be sufficient to recover homogeneous proteins. Purification of dockerin-tagged proteins can be done as follows: 1) diluted hemolymph or homogenate of pupae producing a foreign gene product is loaded onto an rCipA-Sepharose column to allow a dockerin-fused protein to bind specifically to rCipA-Sepharose; 2) the column is washed with suitable buffer containing calcium ions; 3) the column is filled with an elution buffer containing EGTA and incubated for several hours at 4°C; 4) the dockerin-fused protein is eluted with the elution buffer from the column.

Examples of Purification of Dockerin-Tagged Proteins

A new baculovirus transfer vector containing the dockerin module-encoding region of *C. josui cel8A* was constructed. The genes encoding GFP and mouse-interferon (mIFN) were inserted upstream and downstream of the dockerin-encoding region, respectively, and recombinant baculoviruses were obtained by general homologous recombination. The recombinant viruses carrying the genes encoding GFP-dockerin and dockerin-mIFN were allowed to infect pupae or fifth instar larvae, respectively. Dockerin-tagged GFP could be purified to almost homogeneous state on SDS-PAGE (Fig. 7) by one-step purification, i.e., rCipA-Sepharose affinity chromatography.

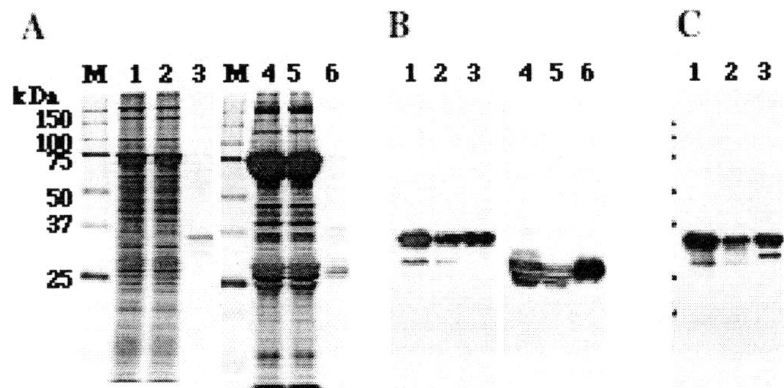

Fig. 7. Purification of dockerin-tagged proteins monitored by silver staining (A) and detected with biotinylated rCBM-Coh1 (B) and anti-GFP antibody. Lane M, molecular mass marker; lane 1, pupal homogenate containing dockerin-tagged GFP, proteins not bound to an rCipA-Sepharose column; lane 3, protein eluted with 5 mM EGTA; lane 4, pupal homogenate containing dockerin-tagged mIFN; lane 5, proteins not bound to an rCipA-Sepharose column; lane 6, protein eluted with 5 mM EGTA.

Aspargine-linked glycosylation (N-glycosylation) is always observed on silkworm-expression system. When recombinant proteins have two or more glycosylation sites, they are modified with heterogeneous oligosaccharide structures in larvae and pupae. Therefore, we often observe two or more molecular species with different molecular sizes even if they are encoded by a single gene. In fact, as shown in Fig.7, recombinant dockerin-tagged GFP was detected as two bands with biotinylated-rCBM-Coh1 and rabbit anti-GFP antiserum. This is plausible because the dockerin module of Cel8A has one possible glycosylation site, i.e., one molecular species with a carbohydrate chain and another without it. The fraction containing dockerin-tagged GFP showed green fluorescence, indicating that the purified GFP maintained its native structure and function. Dockerin-tagged mIFN was also almost completely separated from intrinsic proteins by rCipA-Sepharose affinity chromatography, suggesting that the position of the dockerin module in recombinant proteins does not affect purification efficiency, i.e., the dockerin module is active at either N- or C-terminal side of a recombinant protein.

We evaluated the antiviral activity of the recombinant dockerin-tagged mIFN purified from pupae according to the general method. Table 5 shows the antiviral activity of each fraction of the purification process. Hemolymph and the eluate containing dockerin-tagged mIFN from the rCipA-Sepharose column exhibited significant antiviral activity against Encephalomyocarditis (EMC) virus. This result showed that the dockerin-tagged mIFN exhibited strong antiviral activity against EMC virus comparable to the activity of a recombinant mIFN without dockerin-tag against the same virus.

Recombinant dockerin and cohesin polypeptides are N-glycosylated in both larvae and pupae. Since it is known that the scaffolding protein CipA and some catalytic components of *C. thermocellum* are naturally glycosylated in *C. thermocellum*, glycosylation of cohesin and dockerin polypeptides must not affect the interaction between them during the purification step.

Table 5. Antiviral activity of purified dockerin-tagged mouse Interferon.

Fraction	Dilution (fold)	IC50	IU
Hemolymph	100,000	$3.72x10^6$	$2.21x10^3$
Unbound to CipA column	1,000	$7.71x10^4$	$6.60x10^6$
Eluate from CipA column	1,000	$1.42x10^3$	$1.29x10^7$
	10,000	$7.84x10^3$	$4.83x10^7$
	100,000	$5.16x10^3$	$3.18x10^7$
MuIFN*	10,000	$9.80x10^3$	$6.62x10^7$
rDoc-GFP**	100	$<1.00x10^2$	$<7.52x10^3$

* Recombinant IFN without dockerin tag.
** Recombinant GFP protein fused to dockerin tag.

2. The Cellulolytic System of *Ruminococcus Albus*

Ruminococcus albus is a Gram-positive, anaerobic, cellulolytic microorganism belonging to cluster IV of the *Clostridium* subphylum, as shown by 16S rRNA sequence [26,27]. This bacterium commonly inhabits in rumen of herbivores such as cattle, sheep, and goats, and is one of major cellulolytic species, therefore has been studied largely for its ability to degrade plant cell wall. *R. albus* strain F-40 isolated from cow rumen grows on glucose, cellobiose, and cellulose to produce ethanol and short chain fatty acids, mainly acetate, which are essential nutrients for ruminants. It is well known that cellooligosaccharides produced from cellulosic materials are negligibly detected in the rumen contents, indicating that the newly produced oligosaccharides are consumed by rumen microorganisms immediately after their production. In other words, the degradation of cellulose is the rate-limiting step for the utilization of cellulosic materials by rumen microorganisms and consequently for the indirect utilization of them by ruminants. Hence fortification of the cellulose-degrading ability of rumen microorganisms might enhance the digestibility of cellulosic feedstuff. This motivated us to study cellulases and cellulase genes of rumen cellulolytic microorganisms with a view to enhancing forage digestibility and growing cattle or sheep at lower cost.

Some early studies suggested the occurrence of the cellulolytic complex, *e.g.*, Wood *et al.* [28] presumed that *R. albus* strain SY3 cellulases existed as an aggregate of low-molecular-weight cellulase components on the bacterial cell wall and the molecular mass of the aggregate depended on the culture conditions, although the concept of the cellulosome had not been established yet at the time; electron microscopic examinations revealed that the cells had specialized cell surface protuberant structures [28-30]. These observations were similar to those in the cellulosomes of *Clostridium* species. Nevertheless, it was not necessarily believed that *R. albus* produces cellulosomes as its cellulolytic system since early cloning experiments failed to show the presence of a dockerin module in enzymes. The *cel5A* (formerly *egI*) [31] and the *cel5B* (formerly *egIV*) [32] genes from *R. albus* F-40, *celA* and *celB* [33] from *R. albus* SY3, and *celA* [34] from *R. albus* 8, were cloned in *Escherichia coli* and their sequence analysis showed that these enzymes consisted of a catalytic module only, belonging to the family 5 of glycoside hydrolases.

With great effort we cloned some genes encoding modular endoglucanases from *R. albus* F-40 using a λZapII phage vector (Stratagene), including the *cel5C* (formerly *egV*), *cel5D* (*egVII*), and *cel9A* (*egVI*), because cloned genes are sometimes unstable and are deleted from recombinant plasmids in *E. coli* by unknown causes. These genes encode polypeptides with a dockerin-like module and a CBM or an unknown functional module [35,36], suggesting that *R. albus* F-40 has the cohesin/dockerin interaction to produces the cellulosome-like high molecular mass cellulolytic complex as observed by Wood *et al.* [28].

The occurrence of the cellulosome in *R. albus* was demonstrated as follows [37]. This bacterium was cultivated on ball-milled cellulose as the carbon source under anaerobic conditions whereby the cellulose was completely hydrolyzed. The cells were removed from the culture medium by centrifugation. Gel filtration chromatography on a Hiprep Sephacryl S-300HR column was used to isolate the cellulolytic complex preparation from the culture supernatant. As was expected, we obtained the high-molecular-weight cellulolytic complex (larger than 1.5×10^6 Da), containing at least 15 proteins with molecular weights in the range of 40 kDa to 250 kDa and these proteins were referred to as S1 to S15. Zymogram analysis using CMC indicated that eight of 15 proteins had CMC-degrading activity and four major activities corresponded to S2, S7, S11, and S14. On the other hand, nine proteins showed xylanolytic activity. These results were summarized in Table 6. Subunits S2, S8, S10, S11, S12, and S14 showed both cellulolytic and xylanolytic activities. By contrast, S1, S5, S13, or S15 did not show any detectable activity, indicating that there are at least 11 proteins with endoglucanase and/or xylanase activities on the zymograms [37].

Immunological analysis was done using the antiserum raised against Cel5D-dockerin module to detect the subunits having a dockerin module. At least seven of 15 subunits reacted with this antiserum, but some polypeptides such as S1, S6, S7, S11, S12, S13, S14, and S15 did not. This suggests that the complex from *R. albus* F-40 includes some components having a dockerin module. Since some subunits did not react with the antiserum, there might be different types of dockerins in other subunits or integrate into the complex with different mechanism. Rincón *et al.* showed phylogenic tree of dockerin modules in which dockerin modules of Cel5C (formerly EgV) and Cel5D (formerly EgVII) were located at different branches in their phylogenic trees [38,39]. When the Cel5D dockerin module was used as a probe to identify component(s) having an affinity for dockerin module, it reacted with only a 40-kDa protein (S15) of the complex in the presence of 15 mM $CaCl_2$. The molecular size of S15 was much smaller than those of usual scaffolding proteins such as CipA from *C. thermocellum* [40], CbpA from *Clostridium cellulovorans* [41], and CipC from *C. cellulolyticum* [18], which are more than 150 kDa. S15 might be one of the smaller adaptor scaffolding proteins like ScaC from *Ruminococcus flavefaciens* [38] and ScaB from *Acetivibrio cellulolyticus* [42], which bind to the primary scaffoldin, a large cellulosome-integrating scaffolding protein. The cellulosome-like cellulolytic complex of *R. albus* seems to be more complicate than the clostridial cellulosomes. It includes the 40-kDa dockerin-binding protein and catalytic components with a dockerin module having an affinity for the 40-kDa dockerin-binding protein, and in addition, it should include a different combination of a scaffolding protein and its counterparts of catalytic components.

Table 6. The enzymatic activities and immunoreactivity against the anti-Cel5D dockerin antiserum of the components of the cellulolytic complex from *Ruminococcus albus* F-40.

Subunit	MW(kDa)	Cellulase	Xylanase	Immunoreactivity*
S1	250	−	−	−
S2	110	+	+	+
S3	94	−	+	+
S4	88	−	+	+
S5	82	−	−	+
S6	76	−	+	−
S7	73	+	−	−
S8	67	+	+	+
S9	65	+	−	+
S10	62	+	+	+
S11	55	+	+	−
S12	51	+	+	−
S13	49	−	−	−
S14	44	+	+	−
S15	40	−	−	−

Cellulase and xylanase activities were detected by zymogram analysis using CMC and oat-spelt xylan, respectively.
* Immunoreactivity against the anti-Cel5D dockerin antiserum were measured by means of immunoblotting.

Morrison *et al.* [43] showed that *R. albus* cells adsorbed to cellulose not only with the cellulolytic complex but also with Pil-like proteins and several modular enzymes [44,45]. In this review, we discussed the possible function of *C. josui* modular enzymes containing CBMs and SLH modules as cellulose-binding proteins [22,23]. It is possible that modular enzymes play an important role to allow cells to bind to polysaccharides also in *R. albus*. Cellulolytic bacteria must have some different strategies to bind onto and degrade the targeted substrates. To understand the whole mechanism of degradation of plant cell wall materials, elucidation of the mechanism of adsorption of cells to them is as important as elucidation of the mechanism of adsorption of respective enzymes to their substrates.

Transmission electron microscope observations of thin sections of *R. albus* F-40 stained with cationized ferritin showed the presence of nodular structures on the cell surface, and scanning electron microscope observations identified these structures as surface protuberances in large numbers. The protuberant structures were present mainly on the bacterial surface and also bound to the cellulose substrate [46]. These observations are consistent with early observations that the cellulosome was released from cell surface to bind on cellulose substrate and suggest that the cellulosome-like complex exists not only at the cell surface of this bacterium but also on the substrate surface, detached from the cell surface.

Recently, genome sequence analysis of *R. albus* strain 8 has been completed and the sequence information is available on web site (TIGR Microbial Database: http://www.tigr.org/tdb/mdb/mdbinprogress.html). However, the nucleotide sequences of cellulase genes so far reported from *R. albus* F-40 do not show good match with *R. albus* 8 genomic sequences. The 16S rRNA gene sequence from *R. albus* F-40 as well as its physiological properties clearly showed that the strain F-40 belongs to *R. albus* species. This indicates that there is considerable variation in *R. albus* species. In case of *E. coli*, the enterohemorrhagic *E. coli* O157:H7 has 5.5 Mb in genome size while *E. coli* K-12 has only 4.6 Mb [47]. The genome projects of bacteria indicated that same species often had considerable diversity. Rumen microorganisms inhabit in the environment of higher population density. Such an environment in rumen might be one of reasons to horizontally transfer genetic materials among microorganisms, resulting in the formation of diversified microorganisms.

As described above, since some genes from *R. albus* F-40 are unfortunately unstable in routinely used *E. coli* strains, new *E. coli* strains or new host-vector systems in which *R. albus* genes are stably maintained are necessary for further genetic studies on the cellulolytic system of this bacterium.

ACKNOWLEDGEMENT

We thank Dr. Hidenori Hayashi of the Institute of RIKEN BioResource Center, Japan for his assistance in the construction of phylogenetic tree of cellulolytic clostridia and Ms. Makiko Sakka for helpful discussion.

REFERENCES

[1] Sukhumavasi, J., K. Ohmiya, S. Shimizu, and K. Ueno. 1988. *Clostridium josui* sp. nov., a cellulolytic, moderate thermophilic species from Thai compost. *Int. J. System. Bacteriol., 38*: 179-182.

[2] Cato, E. P., W. L. George, and M. Finegold. 1986. Genus Clostridium prazmowski 1880, 23 AL, p. 1141-1200. In P. H. A. Sneath, N. S. Mair, M. E. Sharpe, and J. G. Holt (ed.), *Bergey's manual of systematic bacteriology vol. 2.* Williams and Wilkins Co., Baltimore.

[3] Guglielmi G., P Béguin. 1998. Cellulase and hemicellulase genes of *Clostridium thermocellum* from five independent collections contain few overlaps and are widely scattered across the chromosome. *FEMS Microbiol. Lett. 161*: 209-215.

[4] Bélaich, J.-P., C. Tardif, A. Bélaich and C. Gaudin. 1997. The cellulolytic system of *Clostridium cellulolyticum. J. Biotechnol. 57*: 3-14.

[5] Fujino, T., S. Karita, and K. Ohmiya. 1993. Nucleotide sequences of the *celB* gene encoding endo-1,4-β-glucanase-2, ORF1 and ORF2 forming a putative cellulase gene cluster of *Clostridium josui*. *J. Ferment. Bioeng.* *76*: 243-250.

[6] Kakiuchi, M., A. Isui, K. Suzuki, T. Fujino, E. Fujino, T. Kimura, S. Karita, K. Sakka, and K. Ohmiya 1998. Cloning and DNA sequencing of the genes encoding *Clostridium josui* scaffolding protein CipA and cellulase CelD and identification of their gene products as major components of the cellulosome. *J. Bacteriol., 180*: 4303-4308, 1998.

[7] S. Jindou, S. Karita, E. Fujino. T. Fujino, H. Hayashi, T. Kimura, K. Sakka, and K. Ohmiya. 2002. α-Galactosidase Aga27A, an enzymatic component of the *Clostridium josui* cellulosome. *J. Bacteriol., 184*: 600-604.

[8] Fujino, T., and K. Ohmiya. 1991. Cloning of the *celB* gene encoding endo-1,4-β-glucanase-2 from *Clostridium josui* in *Escherichia coli* and the properties of the translated product. *J. Ferment. Bioeng. 72*: 422-425.

[9] Henrissat, B., and P. Coutinho. [Online] Glycoside hydrolase families. Architecture et Fonction des Macromolécules Biologaiques, CNRS, Marseille, France. http:// http://www.cazy.org/CAZY/

[10] Halstead, J. R., M. P. Fransen, R. Y. Eberhart, A. J. Park, H. J. Gilbert, and G. P. Hazlewood. 2000. α-Galactosidase A from *Pseudomonas fluorescens* subsp. *cellulosa*: cloning, high level expression and its role in galactomannan hydrolysis. *FEMS Microbiol. Lett. 192*: 197-203.

[11] Shibuya, H., H. Kobayashi, T. Sato, W. Kim, S. Yoshida, S. Kaneko, K. Kasamo, and I. Kusakabe. 1997. Purification, characterization, and cDNA cloning of a novel α-galactosidase from *Mortierella vinacea*. *Biosci. Biotechnol. Biochem. 61*:592-598.

[12] Suryani, T. Kimura, K. Sakka, and K. Ohmiya, Cloning. 2003. Sequencing, and Expression of the Gene Encoding the *Clostridium stercorarium* α-Galactosidase Aga36A in *Escherichia coli*. *Biosci. Biotechnol. Biochem., 67*: 2160-2166 (2003).

[13] Hayashi, H., K. Takagi, M. Fukumura, T. Kimura, S. Karita, K. Sakka, and K. Ohmiya. 1997. Sequence of *xynC* and properties of XynC, a major component of the *Clostridium thermocellum* cellulosome. *J. Bacteriol., 179*: 4246-4253.

[14] Hayashi, H., M. Takehara, T. Hattori, , T. Kimura, S. Karita, K. Sakka, and K. Ohmiya. 1999. Nucleotide sequences of two contiguous and highly homologous xylanase genes *xynA* and *xynB* and charactherization of XynA from *Clostridium thermocellum*, *Appl. Microbiol. Biotechnol., 51*: 348-357.

[15] J. Kurokawa, E. Hemjinda, T. Arai, S. Karita, T. Kimura, K. Sakka, and K. Ohmiya. 2001. Sequence of the *Clostridium thermocellum* mannanase gene *man26B* and characterization of the translated product. *Biosci. Biotechnol. Biochem., 65*: 548-554.

[16] Gaudin, C., A. Bélaich, S. Champ, and J.-P. Belaich. 2000. CelE, a multidomain cellulase from *Clostridium cellulolyticum*: a key enzyme in the cellulosome. *J. Bacteriol., 182*: 1910-1915.

[17] Gal, L., C. Gaudin, A. Bélaich, S. Pages, C. Tardif, and J.-P. Bélaich. 1997. CelG from *Clostridium cellulolyticum*: a multidomain endoglucanase acting efficiently on crystalline cellulose. *J. Bacteriol. 179*: 6595-6601.

[18] Jindou, S., A. Soda, S. Karita, T. Kajino, P. Béguin, J. H. D Wu, M. Inagaki, T. Kimura, K. Sakka, and K. Ohmiya. 2004. Cohesin-Dockerin Interactions within and

between *Clostridium josui* and *Clostridium thermocellum*. *J. Biol. Chem., 279*: 9867-9874.

[19] Yaron, S., E. Morag, E. A. Bayer, R. Lamed, and Y. Shoham. 1995. Expression, purification and subunit-binding properties of cohesins 2 and 3 of the *Clostridium thermocellum* cellulosome. *FEBS Lett., 360*: 121-124.

[20] Pagès, S., A. Belaich, J.-P. Belaich, E. Morag, R. Lamed, Y. Shoham, and E. A. Bayer. 1997. Species-specificity of the cohesin-dockerin interaction between *Clostridium thermocellum* and *Clostridium cellulolyticum*: prediction of specificity determinants of the dockerin domain. *Proteins 29*: 517-527.

[21] Mechaly, A., S. Yaron, R. Lamed, H. P. Fierobe, A. Bélaich, J. P. Bélaich, Y. Shoham, and E. A. Bayer. 2000. Cohesin-dockerin recognition in cellulosome assembly: experiment versus hypothesis. *Proteins 39*: 170-177.

[22] Feng, J.-X., S. Karita, E. Fujino, T. Fujino, T. Kimura, K. Sakka, and K. Ohmiya. 2000. Cloning, sequencing and expression of the gene encoding a cell-bound multi-domain xylanase from *Clostridium josui* and characterization of the translated product. *Biosci. Biotechnol. Biochem., 64*: 2614-2624.

[23] Ali, M. K., T. Kimura, K. Sakka, and K. Ohmiya. 2001. The multidomain xylanase Xyn10B as a cellulose-binding protein in *Clostridium stercorarium*. *FEMS Microbiol. Lett., 198*: 79-83.

[24] Lytle, B. L., B. F. Volkman, W. M. Westler, M. P. Heckman, and J. H. D. Wu. 2001. Solution structure of a type I dockerin domain, a novel prokaryotic, extracellular calcium-binding domain. *J. Mol. Biol., 307*: 745-753.

[25] Carvalho, A. L., F. M. V. Dias, J. A. M. Prates, T. Nagy, H. J. Gilbert, G. J. Davies, L. M. A. Ferreira, M. J. Romão, and C. M. G. A. Fontes. 2003. Cellulosome assembly revealed by the crystal structure of the cohesin–dockerin complex. *Proc. Natl. Acad. Sci., 100*: 13809-13814.

[26] Krause, D. O., B. P. Dalrymple, W. J. Smith, R. I. Mackie, and C. S. McSweeney. 1999. 16S rDNA sequencing of *Ruminococcus albus* and *Ruminococcus flavefaciens*: design of a signature probe and its application in adult sheep. *Microbiology, 145*: 1797-1807.

[27] Tajima, K., R. I. Aminov, T. Nagamine, K. Ogata, M. Nakamura, H. Matsui, and Y. Benno. 1999. Rumen bacterial diversity as determined by sequence analysis of 16S rDNA libraries. *FEMS Microbiol. Eco., 29*: 159-169.

[28] Wood, T. M., C. A. Wilson, and C. S. Stewart. 1982. Separation of the cellulase from the cellulolytic anaerobic rumen bacterium *Ruminococcus albus* and its release from the bacterial cell wall. *Biochem. J., 105*: 129-137.

[29] Lamed, R., J. Naimark, E. Morgenstern, and E. A. Bayer. 1987. Specialized cell surface structure in cellulolytic bacteria. *J. Bacteriol., 169*: 3792-3800.

[30] Miron, J., M. T. Yokoyama, and R. Lamed. 1989. Bacterial cell surface structures involved in lucerne cell wall degradation by pure cultures of cellulolytic rumen bacteria. *Appl. Microbiol. Biotechnol., 32*: 218-222.

[31] Ohmiya, K., T. Kajino, A. Kato, and S. Shimizu. 1989. Structure of *Ruminococcus albus* endo-1,4-β-glucanase gene. *J. Bacteriol., 171*: 6771-6775.

[32] Karita, S., K. Morioka, T. Kajino, K. Sakka, K. Shimada, and K. Ohmiya. 1993. Cloning and sequencing of a novel endo-1,4-β-glucanase gene from *Ruminococcus albus*. *J. Ferment. Bioeng., 76*: 439-444.

[33] Poole, D. M., G. P. Hazlewood, J. I. Laurie, P. J. Barker, and H. J. Gilbert. 1990. Nucleotide sequence of the *Ruminococcus albus* SY3 endoglucanase genes *celA* and *celB*. *Mol. Gen. Genet., 223*: 217-223.

[34] Attwood, G. T., F. Herrera, L. A. Weissenstein, and B. A. White. 1996. An endo-β-1,4-glucanase gene (*celA*) from the rumen anaerobe *Ruminococcus albus* 8: cloning, sequencing, and transcriptional analysis. *Can. J. Microbiol., 42*: 267-278.

[35] Ohara, H., J. Noguchi, S. Karita, T. Kimura, K. Sakka, and K. Ohmiya. 2000. Sequence of *egV* and properties of EgV, a *Ruminococcus albus* endoglucanase containing a dockerin domain. *Biosci. Biotechnol. Biochem., 64*: 80-88.

[36] Taguchi, H., D. Hagiwara, T. Genma, S. Karita, T. Kimura, K. Sakka, and K. Ohmiya. 2004. Cloning of the *Ruminococcus albus cel5D* and *cel9A* genes encoding dockerin module-containing endoglucanases and expression of *cel5D* in *Escherichia coli*. *Biosci. Biotechnol. Biochem., 68*: 1557-1564.

[37] Ohara, H., S. Karita, T. Kimura, K. Sakka, and K. Ohmiya. 2000. Characterization of the cellulolytic complex (cellulosome) from *Ruminococcus albus*. *Biosci. Biotechnol. Biochem., 64*: 254-260.

[38] Rincón, M. T., S.-Y. Ding, S. I. McCrae, J. C. Martin, V. Aurilia, R. Lamed, Y. Shoham, E. A. Bayer, and H. J. Flint. 2003, Novel organization and divergent dockerin specificities in the cellulosome system of *Ruminococcus flavefaciens*. *J. Bacteriol., 185*: 703-713.

[39] Rincón, M. T., J. C. Martin, V. Aurilia, S I. McCrae, G. J. Rucklidge, M. D. Reid, E. A. Bayer, R. Lamed, and H. J. Flint. 2004. ScaC, an adaptor protein carrying a novel cohesin that expands the dockerin-binding repertoire of the *Ruminococcus flavefaciens* 17 cellulosome. *J. Bacteriol., 186*: 2576-2585.

[40] Gerngross, U., M. P. M. Romaniec, T. Kobayashi, N. S. Huskisson, and A. L. Demain. 1993. Sequencing of a *Clostridium thermocellum* gene (*cipA*) encoding the cellulosomal SL-protein reveals an unusual degree of internal homology. *Mol. Microbiol., 8*: 325-334.

[41] Shoseyov, O., M. Takagi, M. A. Goldstein, and R. H. Doi. 1992. Primary sequence analysis of *Clostridium cellulovorans* cellulose binding protein A. *Proc. Natl. Acad. Sci. USA. 89*: 3483-3487.

[42] Xu, Q., W. Gao, S.-Y. Ding, R. Kenig, Y. Shoham, E. A. Bayer, and R. Lamed. 2003. The cellulosome system of *Acetivibrio cellulolyticus* includes a novel type of adaptor protein and a cell surface anchoring protein. *J. Bacteriol., 185*: 4548-4557.

[43] Morrison, M. and J. Miron. 2000. Adhesion to cellulose by *Ruminococcus albus*: a combination of cellulosomes and Pil-proteins. *FEMS Microbiol. Lett., 185*: 109-115.

[44] Devillard, E., D. B. Goodheart, S. K. R. Karnati, E. A. Bayer, R. Lamed, J. Miron, K. E. Nelson, and M. Morrison. 2004. *Ruminococcus albus* 8 mutants defective in cellulose degradation are deficient in two processive endocellulases, Cel48A and Cel9B, both of which possess a novel modular architecture. *J. Bacteriol., 186*: 136-145.

[45] Xu, Q., M. Morrison, K. E. Nelson, E. A. Bayer, N. Atamna, and R. Lamed. 2004. A novel family of carbohydrate-binding modules identified with *Ruminococcus albus* proteins. *FEBS Lett., 566*: 11-16.

[46] Kim, Y.-S., A. P. Singh, S. G. Wi, K. H. Myung, S. Karita, and K. Ohmiya. 2001. Cellulosome-like structures in ruminal cellulolytic bacterium *Ruminococcus albus* F-40 as revealed by electron microscopy. *Asian-Aust. J. Anim. Sci., 14*: 1429-1433.

[47] Ohnishi, M., J. Terajima, K. Kurokawa, K. Nakayama, Y. Ogura, H. Tamura, H. Watanabe, and T. Hayashi. 2002. Genomic diversity of enterohemorrhagic *Escherichia coli* O157 revealed by whole genome PCR scanning. *Proc. Natl. Acad. Sci. USA, 99*: 17043-17048.

In: Cellulosome
Eds: V. Uversky and I. A. Kataeva, pp. 191-209
ISBN 1-59454-950-8

Chapter IX

CELLULOSOME CHIMERAS: A POWERFUL TOOL TO INVESTIGATE THE FUNCTIONING OF CELLULOSOMES WITH POTENTIAL BIOTECHNOLOGICAL APPLICATIONS

Henri-Pierre Fierobe[1,*], Sandrine Pagès[1,2], Chantal Tardif[1,2], Anne Bélaïch[1] and Jean-Pierre Bélaïch[1,2]
[1]BIP-CNRS 31, Chemin Joseph-Aiguier, Marseille, France,
[2]Université de Provence, Marseille, France.

ABSTRACT

Cellulosome chimeras designate hybrid minicellulosomes composed of a miniscaffoldin containing two (or more) cohesins of different specificities that anchors two (or more) engineered cellulolytic enzymes appended with the corresponding dockerin. Two important observations made it possible for the construction of such minicomplexes: the bacterial cohesins and dockerins are genuine modules keeping their biological properties when they are produced alone or grafted to a heterologous protein, and the high affinity cohesin / dockerin interaction is species-specific between some cellulosome-producing bacterial species. Compared to natural cellulosomes, cellulosome chimeras exhibit three major differences: they contain a limited number of enzymes, they are homogeneous, and the position of each enzyme along the miniscaffoldin is controlled. These characteristics make the cellulosome chimeras a convenient tool to dissect the functioning of natural cellulosomes, and this strategy was applied to investigate the cellulosomes produced by *Clostridium cellulolyticum*. It was observed that organization of most tested *C. cellulolyticum* enzyme pairs into cellulosome chimeras induced enhanced activities on

* Correspondence concerning this article should be addressed to Henri-Pierre Fierobe, 1BIP-CNRS 31, Chemin Joseph-Aiguier, F-13402 Marseille cedex 20, France. E-mail: hpfierob@ibsm.cnrs-mrs.fr.

recalcitrant cellulose, assigned to both enzyme proximity on the resulting complex that triggers additional synergies, and to the cellulose-binding module of the scaffoldin which anchors the whole complex at the surface of the substrate. The most efficient hybrid minicellulosomes remain, however, much less active than natural cellulosomes on crystalline cellulose. For this reason, future prospects involve the construction of new hybrid scaffoldins which are able to bind onto specific locations three or more different enzymes. Besides these fundamental studies, the cellulosome chimeras may also be exploited for biotechnological applications in various fields.

Keywords: Cellulosome chimeras, controlled incorporation, synergy, proximity effect, CBM effect

Introduction

In the early 90s the molecular basis of the cellulosome assembly was discovered. Tokatlidis and co-workers [39] unambiguously showed that the cellulosomal enzymes from *Clostridium thermocellum* bind to the scaffoldin (called at that time SL or S1) of the complex *via* two conserved stretches of 22 residues later called dockerins, whereas Fujino and co-workers identified in 1992 [10] the receptor domains in the scaffoldin, subsequently named "cohesins". The same year, Shoseyov and co-workers [38] published the first entire primary sequence of a scaffoldin (CbpA from *C. cellulovorans*) which contains nine cohesins, in addition to a Cellulose Binding Module (CBM) and four domains of unknown function. More than a decade after this breakthrough, the complete primary structures of scaffoldins issued from seven bacteria have become available (*C. thermocellum*, *C. cellulolyticum*, *C. Josui*, *C. acetobutylicum*, *Ruminococcus flavefaciens*, *Acetivibrio cellulolyticus*, *Bacteroides cellulosolvens*). Meanwhile the number of dockerin-containing coding sequences from various organisms keeps growing. For instance, the draft genome of *C. thermocellum* provided by the Joint Genome Institute at the US Department of Energy encompasses over 60 dockerin-containing coding sequences. Sequence comparisons lead to the classification of both cohesins and dockerins into several distinct groups [44], whereas cohesins (and dockerins) belonging to different groups may coexist within the same species [35, 44, 45].

Concomitantly, a picture of the bacterial cellulosomes raised, from molecular biology, genetic, biochemical and structural studies. The bacterial cellulosomes can be divided into two groups. The cellulosomes produced by the mesophilic *Clostridia* (*C. cellulolyticum*, *C. cellulovorans*, *C. josui*, *C. acetobutylicum*) are the simplest since they only contain one scaffoldin bearing 6 to 9 type I cohesins. Their molecular masses range from 0.5 to 1 MDa. It has not been clearly established whether the cellulosomes are bound to the cell surface at an early stage of the growth, but they are found in the supernatant at the late exponential or stationary phases. In *C. cellulovorans*, however, one cellulosomal cellulase was found to contain three repeated domains with partial homology to those of bacterial S-layer proteins [18]. This enzyme may connect the cellulosomes to the cell surface to some extent. To date, only type I cohesins and dockerins have been discovered in these bacteria.

In the second group are found the cellulosomes produced by *C. thermocellum*, *A. cellulolyticus*, *R. flavefaciens* and *B. cellulosolvens* which display a higher degree of complexity with several interacting scaffoldins. They are transitorily or permanently attached to the cell surface. In the case of *C. thermocellum*, the main scaffoldin, CipA, contains 9 type I cohesins that bind to the dockerins of the catalytic sub-units. CipA also bears a type II dockerin. It has been showed that this dockerin binds to type II cohesins borne by three outer layer proteins [11,17] thus anchoring the cellulosomes at the cell surface. *A. cellulolytic* produces at least three different scaffoldins, and three types of specific cohesin/dockerin devices have been found, suggesting a complex network of cohesin/dockerin mediated interactions among cellulolytic enzymes, scaffoldins and anchoring proteins [44].

Despite this diversity and complexity, bacterial cellulosomes share many common features in addition to their high specific activity on cellulose and plant cell wall polysaccharides. The same recurrent building blocks are found: cohesin, dockerin, CBM, and catalytic domains. The cohesin/dockerin interaction is calcium dependent and appears to be generally strong since the dissociation of the cellulosomes requires harsh treatments (SDS, EDTA, elevated temperatures etc.) Quantification of the type I interaction has been performed in the case of *C. cellulolyticum*, *C. thermocellum* and *C. Josui*, and $K_A \geq 10^9\ M^{-1}$ was reported [7,16,36]. Recently, the crystal structure of a cohesin/dockerin complex from *C. thermocellum* has been determined [3], leading to the identification of the interacting residues in both modules. It has also been shown for *C. thermocellum* and *C. cellulolyticum* that the intra-species cohesin/dockerin interaction is relatively non specific; a given cohesin appears to recognize all of the dockerin-containing enzymes within the same species [27,46]. Similar observations were reported concerning the cellulosomes produced by *C. josui*, though Jindou and co-workers [16] showed that the first cohesin domain has a slightly lower affinity (in average 10 fold) for the dockerins of two cellulosomal enzymes (Aga27A and Cel8A), compared to cohesins 2, 5 and 6. The calculated K_A, however, remains in the 2-5 $10^8\ M^{-1}$ range, and is very similar for both dockerins, thus suggesting a lack of selectivity for a given dockerin even for this particular cohesin. Therefore, it is very likely that dockerin-borne enzymes are principally incorporated at random onto the scaffoldin in these species. In the case of *C. cellulolyticum*, this hypothesis has recently been confirmed using the genetic approach [30]. Random incorporation may also partly explain the heterogeneity of the cellulosomes: SDS-Polyacrylamide gel electrophoresis performed on purified cellulosomes and followed, by far, western blot using cohesins as probes have shown that the cellulosomes from either *C. thermocellum* [46], *C. cellulolyticum* [22,27] or *C. josui* [16] contain many more dockerin-borne proteins than expected from the number of cohesins displayed by their scaffoldin. Furthermore, the number of different polypeptides appended with a dockerin is probably underestimated with this technique, since several cellulosomal enzymes may have very similar molecular masses. The heterogeneity of the cellulosomes may therefore be more considerable, and in this context the discovery of approximately 60 dockerin containing coding sequences in the draft genome of *C. thermocellum* is not surprising.

A number of cellulosomal cellulases, hemicellulases, and pectinases from various bacteria have been overproduced in *Escherichia coli*, and characterized. Among the cellulases participating to the cellulosomes are found exo- and endo-processives enzymes (Glycoside Hydrolase families 9 and 48), as well as endoglucanases (GH families 5, 8 and 9). These enzymes efficiently degrade substituted soluble cellulose and/or amorphous cellulose but display much lower activities towards crystalline celluloses. In the free state, some of these enzymes were found to act synergistically on crystalline cellulose, especially in the case of family 48/family 9 enzyme pairs, but the specific activity of the resulting enzyme mixture remains very low compared to that of natural cellulosomes. For instance, in the case of *C. cellulolyticum*, the most efficient free enzyme pair (Cel48F + Cel9G) on Avicel is 40-fold less active than purified cellulosomes at the same concentration [9].

Most probably, the synergies between the catalytic sub-units are maximized in the cellulosomes, but the cooperation between enzymes within the complexes was not studied for obvious reasons: Contrary to the known multi-proteic complexes of similar size synthesized by living cells such as the proteasome, or the ribosome, the cellulosomal proteins do not occupy a unique specific position in the complex. The bacteria thus concurrently secrete a whole population of cellulosomes with various enzyme composition, stoichiometry and arrangement. The heterogeneity, size and complexity of natural cellulosomes naturally hinder the exploration of their catalytic efficiency by conventional biochemical approaches. Furthermore, genetic tools were not available for any cellulosome-producing bacterium until very recently [28,30,41].

In this context, the possibility to build simplified, homogeneous and defined cellulosomes by means of cellulosome chimeras was explored to gain new insight on the functioning of natural cellulosomes.

The Concept of the Cellulosome Chimeras

The lack of specificity of the cohesin/dockerin interaction within a species is an obstacle to the construction of homogeneous minicellulosomes. As shown in Figure 1, if a miniscaffoldin containing two cohesins of the same species and type is mixed with stoichiometric amounts of two enzymes appended with the cognate dockerins, four different complexes are likely to be obtained. On the other hand, many reports have shown that cohesins and dockerins from various species and types, are genuine modules keeping their binding properties when they are produced alone or grafted to a carrier protein [7,16,19,20]. It was thus assumed that defined and homogeneous minicellulosomes could be obtained by combining several cohesins of different specificities within the same polypeptide chain and by grafting the corresponding dockerins to the target enzymes. Theoretically, a mixture of the various components in stoichiometric proportions should yield to 100% of homogeneous complex.

The first application of this concept was reported by Leibovitz and Beguin [20], who constructed a miniscaffoldin containing the seventh type I cohesin from CipA (*C. thermocellum*) and the type II cohesin from SdbA (*C. thermocellum*). The miniscaffoldin was mixed with wild type Cel9D (containing its native type I dockerin domain), and an engineered Cel5C bearing the type II dockerin domain of CipA. Nevertheless, analysis by non denaturing gel electrophoresis showed that complexation was not total and only 56% of the resulting complexes contain both enzymes. Since the type I cohesin/dockerin interaction in *C. thermocellum* is of high affinity, the authors concluded that this reduced amount of expected complex is due to the low affinity of the type II cohesin/dockerin interaction [19,20].

In 1997, Pagès and co-workers [26] showed that despite rather high sequence homologies, particularly among the dockerins, the (type I) cohesin/dockerin interaction is strictly species-specific between *C. thermocellum* and *C. cellulolyticum*. The residues of the dockerin involved in the species-specificity were subsequently identified by site directed mutagenesis [23]. Interestingly, the species-specificity between *C. thermocellum* and *C. josui* which is very close to *C. cellulolyticum* appeared to be less strict, since at least the dockerin modules of two *C. thermocellum* enzymes were found to bind strongly to *C. josui* cohesins [16]. Such "cross reactivity" was never observed between *C. thermocellum* and *C. cellulolyticum* type I cohesin/dockerin domains, which therefore constitute promising building blocks for the construction of cellulosome chimeras (Figure 1), both interactions being highly specific and very strong.

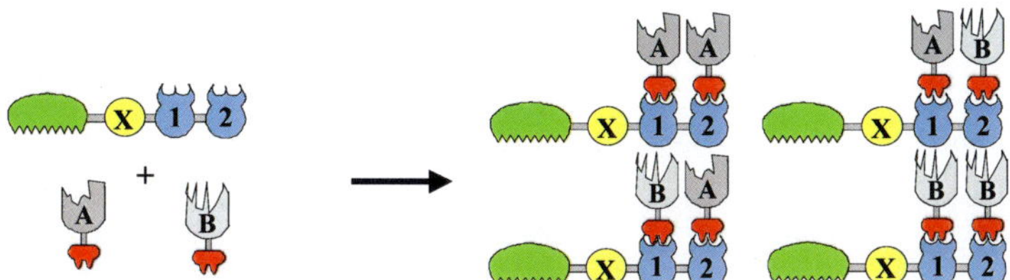

Same species and type of cohesin/dockerin devices

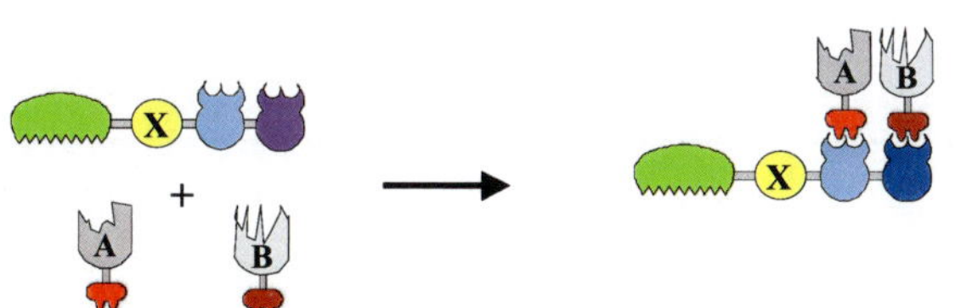

Different species or types of cohesin/dockerin devices

Key to symbols

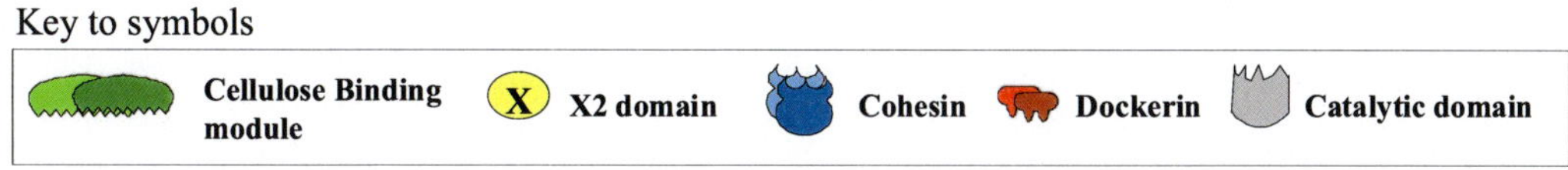

Figure 1: The cellulosome chimera concept.

In a first study [8], four different hybrid miniscaffoldins with different topologies but containing one cohesin from each species were engineered, overproduced in *E. coli*, and purified. As shown in figure 2, the cohesins are either separated by an extra domain (Scaf1 and 2) or adjacent (Scaf3 and 4). In parallel, the native dockerins of Cel48F and Cel5A from *C. cellulolyticum* were replaced by the dockerin of Cel48S from *C. thermocellum*. The engineered cellulases were named 48Ft and 5At and overproduced in *E. coli* as well as wild-type Cel48F (48Fc) and Cel5A (5Ac). The binding properties of the various partners and the ability to achieve nearly 100% complexation were investigated using three different but complementary techniques. Surface Plasmon Resonance (SPR) experiments (Figure 3) with immobilized scaffoldins, and various order of cellulase injections demonstrated that each type of cohesin binds in an independent manner to the appropriate dockerin-containing enzyme, even when the cohesins are adjacent in the hybrid scaffoldin, irrespective of the order of incorporation. The affinity constants of 5Ac or 48Fc (2-4 10^9 M^{-1}) for the *C. cellulolyticum* cohesin contained in the hybrid miniscaffoldins was in accordance with previously determined values for the cohesin-dockerin interaction in *C. cellulolyticum* [7]. The affinity of 5At or 48Ft, however, was too high (>10^{11} M^{-1}) to be determined using the BIAcore system, suggesting a stronger cohesin-dockerin interaction in *C. thermocellum* as also observed by others [16]. Furthermore, the amount of enzyme bound at saturation corroborated a stoichiometry of 1:1:1 for the ternary complexes. Total or nearly total complexation and 1:1:1 stoichiometry was subsequently confirmed by nondenaturing polyacrylamide gel electrophoresis and gel filtration HPLC. Therefore, this article showed that the cohesin-dockerin modules from *C. thermocellum* and *C. cellulolyticum* are suitable devices for the construction of cellulosome chimeras: both interactions are of high affinity, and full complexation is simply achieved by mixing stoichiometric amounts of the three partners in presence of calcium.

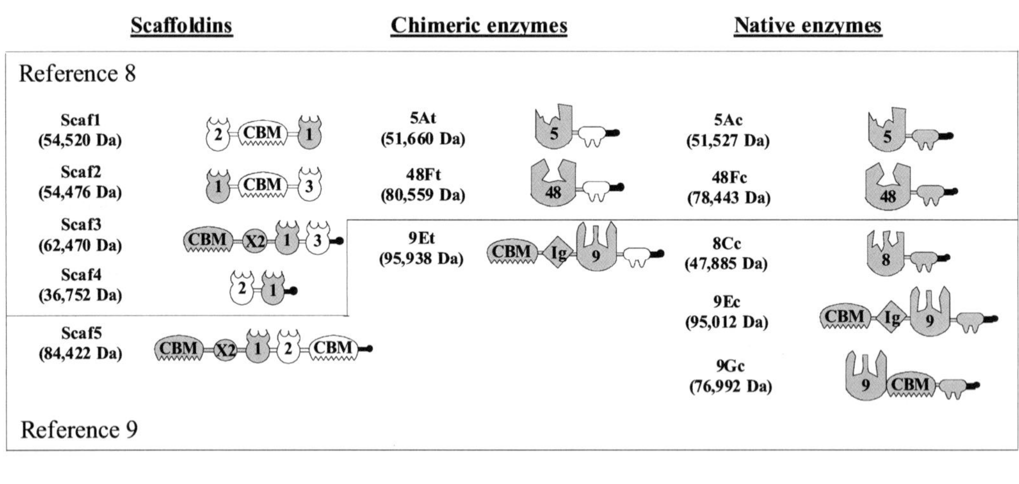

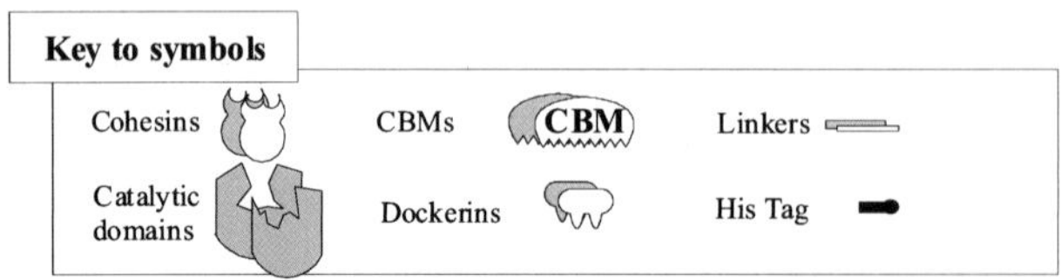

Figure 2: Schematic representation of the recombinant proteins used in this study. *White* (*C. thermocellum*) and *grey* (*C. cellulolyticum*) symbols denote the source of the respective domains (see « *Key to symbols* »). The cohesin domains are numbered according to their original position in the respective native cellulosomal scaffoldin. A hydrophylic domain (*X*) of unknown function is part of the *C. cellulolyticum* scaffoldin. An immunoglobulin-like domain (*Ig*) is part of *9Ec* (and *9Et*). In the shorthand notation for the enzymes, numbers indicate the corresponding family of the catalytic domain ; *A*, *C*, *E*, *F*, and *G* refer to the original name of the enzymes (CelA, CelC, CelE, CelF, and CelG, repectively) ; *c* and *t* indicate the source of the dockerin domain, *C. cellulolyticum* or *C. thermocellum*, respectively.

CELLULOSOME CHIMERAS USED TO INVESTIGATE THE CELLULOSOMES PRODUCED BY *C. CELLULOLYTICUM*

When this study commenced [9], the genes encoding five different cellulases from *C. cellulolyticum* had already been overexpressed in *E. coli*, and the corresponding enzymes had been purified and characterized. The available recombinant cellulases include two typical endoglucanases, 5Ac and 8Cc [5,6], and two endo-processive cellulases, 48Fc [32] and 9Ec [14], which were found to be the two major cellulosomal enzymes [12]. Finally, 9Gc [13] was described as an unusual endo-cellulase, the activity pattern of which resembles that of the homologous enzyme Cel9A from *Thermobifida fusca* [15]. Among these cellulases, The most active enzyme on crystalline cellulose was found to be 9Ec [9,14], but it is approx. 100 fold less active than purified cellulosomes at the same concentration. Several studies have shown that synergy occurs between these enzymes in the free state on Avicel, especially between 48Fc and 9Gc, and between 9Ec and 9Gc [9,14], but the reported synergy factors are ≤ 2. Therefore the specific activity of free *C. cellulolyticum* enzyme mixtures also remains much lower compared to that of purified natural cellulosomes.

To get some clues on the elevated activity of the cellulosomes towards crystalline cellulose, the cellulosome chimeras were used to dissect their functioning. Our goal was to study both the impact of the family 3A CBM of the scaffoldin, and the enzyme proximity within the complex on the overall activity. Compared to the original article [8], the library of native and engineered cellulases, as well as hybrid miniscaffoldins to incorporate in the cellulosome chimeras was enlarged (Figure 2). A new miniscaffoldin was designed and produced in *E. coli*. This protein named Scaf5 derives from Scaf3 but contains a second family 3A CBM stemming from CipA of *C. thermocellum* at the C-terminus. Thus five hybrid scaffoldins were available with 0 (Scaf4), 1 (Scaf1-3), or 2 (Scaf5) family 3A CBMs. In addition to 5At and 48Ft, a third *C. cellulolyticum* cellulase, 9Ec, was engineered to bear a *C. thermocellum* dockerin, and named 9Et. 5At, 9Et and 48Ft were combined with one of the five available native cellulases (5Ac, 8Cc, 9Ec, 48Fc and 9Gc) and incorporated onto each of the five hybrid scaffoldins leading to 75 different cellulosome chimeras. Total or nearly total complexation was checked by non-denaturing polyacrylamide gel electrophoresis (PAGE). The stability of some cellulosome chimeras at 37°C was investigated: the half life of the complexes varied from approx. 24h (Scaf4-based complexes) to several days (Scaf2-based complexes). In the case of the Scaf4-based complexes, the relatively low half life was found to be due to spontaneous proteolysis of the scaffoldin.

The activity of the 75 cellulosome chimeras (0.1 μM) was determined and compared to that of corresponding free enzyme pairs on Avicel and ribbons of bacterial cellulose (BC) at a substrate concentration of 3.5 g/L. Kinetics were performed for 24h at 37°C, and the activity was monitored by measuring the released soluble sugars.

Avicel is a microcrystalline cellulose extracted from wood pulp, with a degree of polymerization (DP) of 100-250 glucopyranose units, and is 50-60% crystalline. The data collected on Avicel (summarized in Table I), showed that most ternary complexes were found to be more active (up to 7.3-fold) than the corresponding free enzyme systems. The lowest activity enhancements were, as expected, observed for homogeneous enzyme pairs such as

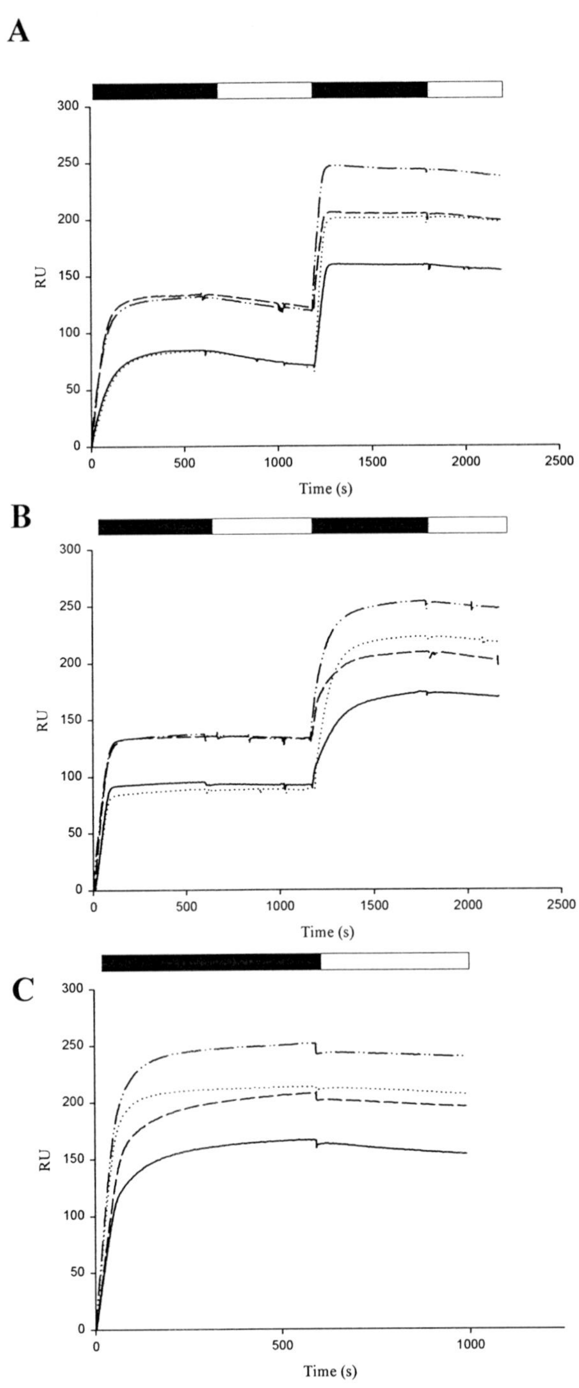

Figure 3: Sensograms for the binding of various combinations of cellulases (5 nM) to immobilised Scaf4. (A) Initial injection of the cellulases bearing a *C. cellulolyticum* dockerin, followed by the binding of the cellulases harboring a *C. thermocellum* dockerin. (B) Initial injection of the cellulases bearing a *C. thermocellum* dockerin, followed by the binding of the cellulases harboring a *C. cellulolyticum* dockerin. (C) co-injection of cellulases containing both types of dockerin. Solid lines, Ac + At ; dotted lines Ac + Ft ; dashed lines, Fc + At, dotted and dashed lines, Fc + Ft.

48Fc + 48Ft, but for a given enzyme pair, the same "hierarchy" between the scaffoldins in terms of activity enhancement was found: Scaf1 ≈ Scaf2 > Scaf3 > Scaf5 ≥ Scaf4. These experiments revealed that incorporation of cellulase pairs into a cellulosome-like particle can induce two major effects: i) increased activity due to the presence of the family 3A CBM of the scaffoldin and ii) enhanced synergy due to the proximity of the enzymes in the complex.

Table I

Released soluble sugars from Avicel (3.5 g/L) after 24 h at 37°C by cellulosome chimeras, free enzyme pairs and free enzymes alone[a]

Pairs	Scaf1		Scaf2		Scaf3		Scaf4		Scaf5		Free enzyme pairs	
	µM	SF[b]	µM	S F	µM	SF	µM	SF	µM	SF	µM	Synergy[c]
5At/5Ac	70[d]	2.5	84	3	65	2.3	37	1.3	50	1.8	28	0.9
5At/8Cc	93	3.3	100	3.6	79	2.8	41	1.5	45	1.6	28	1
5At/9Ec	220	2.9	238	3.1	147	1.9	113	1.5	123	1.6	76	1.2
5At/48Fc	175	3.6	193	4	152	3.1	92	1.9	116	2.4	49	1.2
5At/9Gc	277	7.2	281	7.3	225	5.8	134	3.5	181	4.7	39	0.9
9Et/5Ac	201	2.4	225	2.7	155	1.9	116	1.4	133	1.6	83	1.4
9Et/8Cc	202	2.9	229	3.3	152	2.2	92	1.3	150	2.2	69	1.3
9Et/9Ec	156	1.8	161	1.9	127	1.5	98	1.1	123	1.4	86	1
9Et/48Fc	171	2.2	169	2.2	153	2	140	1.8	122	1.5	77	1.2
9Et/9Gc	419	4.2	448	4.5	336	3.4	220	2.2	249	2.5	99	1.5
48Ft/5Ac	158	3	160	3	126	2.4	85	1.6	100	1.9	53	1.2
48Ft/8Cc	197	4.1	201	4.1	140	2.9	58	1.2	137	2.8	48	1.2
48Ft/9Ec	203	2.2	199	2.2	152	1.7	146	1.6	147	1.6	91	1.3
48Ft/48Fc	85	1.4	82	1.4	74	1.2	66	1.1	79	1.3	60	1.1
48Ft/9Gc	413	3.8	441	4	376	3.4	230	2.1	271	2.5	109	2

[a]

Enzyme	5Ac	5At	8Cc	9Ec	9Et	48Fc	48Ft	9Gc
µM	17	15.5	13	45.5	40.2	25.6	26	27.8

[b] SF (Stimulation Factor) is the impact of the complexation of an enzyme pair on a given scaffoldin : SF = (released soluble sugars by ternary complex)/(released soluble sugars by the corresponding free enzyme pairs). [c] Synergy between enzymes in the free state : Synergy factor = (released soluble sugars by a free enzyme pair)/(sum of released soluble sugars by corresponding enzymes alone). [d] Averages of 2 to 6 experiments varying within ± 10%.

The "CBM effect" was highlighted by the fact that single-CBM scaffoldins (Scaf1-3) gave the best enhancement of activities, compared with CBM-less scaffoldin (Scaf4). The Family 3A CBMs are known to be powerful crystalline cellulose binding modules [2], but the lower activities observed with Scaf5, compared to Scaf1-3, probably reflect superfluous levels of interaction with the substrate preventing optimal activity of the associated enzymes. Alternatively, it may also be hypothesized that upon binding to the substrate, Scaf5 adopts a more rigid conformation than single-CBM scaffoldins thus limiting the mobility of the associated catalytic domains within the complex and reducing their activity toward adjacent cellulose chains. This hypothesis is consistent with the fact that the known bacterial scaffoldins [1] never contain two or more family 3A CBMs. It may also be correlated with the fact that *C. cellulolyticum* cellulosomal cellulases, which have been discovered to date, lack an efficient cellulose binding module. Among single-CBM scaffoldins, Scaf1 and 2 systematically displayed higher activity enhancement compared to Scaf3. This difference may reflect the fact that the CBM of Scaf3 stems from *C. cellulolyticum*, whereas Scaf1 and 2 contains the CBM of CipA from *C. thermocellum*, although both domains are highly homologous, share almost identical three dimensional structures and were both found to bind strongly to crystalline cellulose [37,40]. Alternatively, the lower activity enhancement which was observed with Scaf3 may be due to its different domain organization: an X2 domain lays between the CBM and the cohesins, whereas in Scafl and 2, both cohesins are adjacent to the CBM.

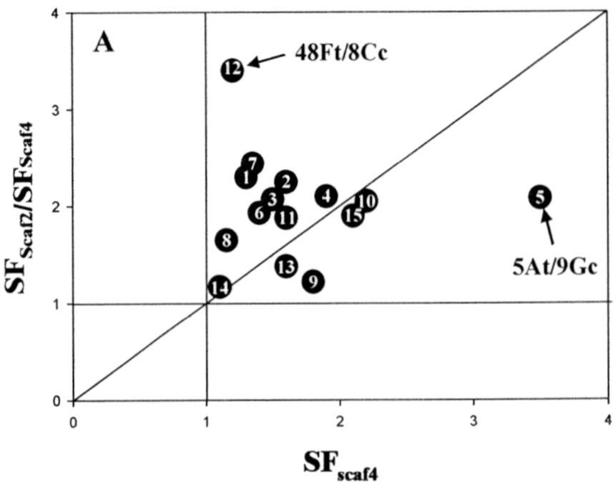

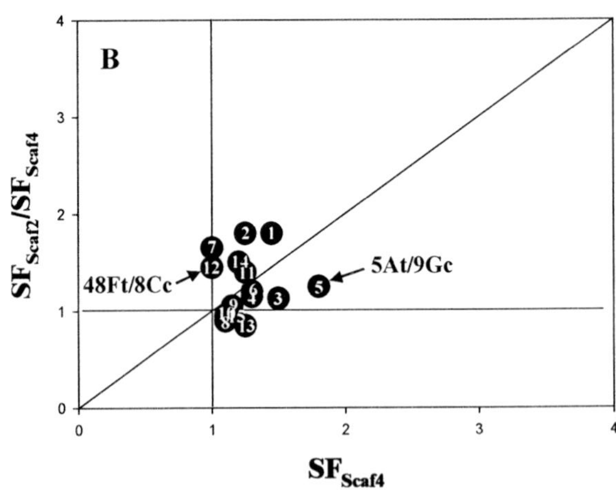

Figure 4: Discrimination between CBM and proximity effects on the hydrolysis of Avicel (*A*) and Bacterial cellulose (*B*). Data are from Table I (*A*) and Table II (*B*). On the X-axis are reported S/F ratios for Scaf4 (S/F_{Scaf4}), and on the Y-axis are reported $(S/F_{Scaf2})/(S/F_{Scaf4})$. *Vertical line* (X = 1) discriminates enzyme pairs displaying enhanced synergy in Scaf4-based complexes (X > 1) and enzyme pairs which are not stimulated (or are inhibited) by the binding to this scaffoldin (X ≤ 1). *Horizontal line* (Y = 1) distinguishes enzyme pairs in the case of which the binding to Scaf2 induces additional activity compared to Scaf4 (Y > 1). The *diagonal line* (Y = X) symbolises an equal contribution between proximity (S/F_{Scaf4}), and CBM effect ($(S/F_{Scaf2})/(S/F_{Scaf4})$). *White* numbers refer to the enzyme pairs : *1*, 5At + 5Ac ; *2*, 5At + 8Cc ; *3*, 5At + 9Ec ; *4*, 5At + 48Fc ; *5*, 5At + 9Gc ; *6*, 9Et + 5Ac ; *7*, 9Et + 8Cc ; *8*, 9Et + 9Ec ; *9*, 9Et + 48Fc ; *10*, 9Et + 9Gc ; *11*, 48Ft + 5Ac ; *12*, 48Ft + 8Cc ; *13*, 48Ft + 9Ec ; *14*, 48Ft + 48Fc ; *15*, 48Ft + 9Gc.

The effect of enzyme proximity was reflected by the Scaf4-based complexes. The activity of most heterogeneous enzyme pairs was significantly increased by complexation onto the CBM-less Scaf4. The highest level of enhancement by Scaf4-based complexes (3.5-fold) was observed for the enzyme pair 5At + 9Gc, which displays no synergy in the free state. Additional investigations showed that this enhancement was not due to a stabilization of the enzymes by complexation, or to a modification of their activity pattern upon binding to the cognate cohesin [9]. The physical proximity of these two enzymes probably enables them to hydrolyze additional sites of the substrate. On the other hand, 48Ft + 9Gc, the enzyme pair that displays the highest synergy in the free state was also strongly stimulated when complexed onto Scaf4. Furthermore, the enhancement of activity due to enzyme proximity is likely to be underestimated, because after 24h of incubation at 37°C, about 50% of Scaf4 undergoes spontaneous proteolysis, which was not observed with Scaf2 for instance.

CBM and proximity effects are indeed not exclusive and occurred simultaneously for most enzyme pairs upon complexation onto CBM-scaffoldins. To estimate the relative contribution of the CBM and the proximity in this type of complexes, the data obtained with Scaf4 and 2 were plotted in Figure 4A: on the abscissa are reported the enhanced synergy exclusively due to enzyme proximity (SF_{Scaf4}), and the ordinates represent the calculated contribution of the CBM in Scaf2-based complexes (SF_{Scaf2}/SF_{Scaf4}). As show in Figure 4A, the majority of the spots are located near the diagonal line, indicating that the contribution of the CBM and proximity effects to the enhanced synergy observed for the Scaf2-based complexes are in the same range. For two enzyme pairs, however, one effect seemed to predominate over the other. Incorporation of 48Ft and 8Cc onto Scaf4 has no impact on the overall activity, and the enhancement observed when these cellulases are bound to Scaf2 seems exclusively due to the CBM of the scaffoldin. On the other hand, the proximity effect is predominant when the cellulase pair 5At + 9Gc is attached to Scaf2, although the CBM induces an additional 2-fold increase. As expected, the CBM effect was found inversely proportional to the substrate concentration, whereas the activity enhancement due to proximity (Scaf4-based complexes), was constant at all Avicel concentrations [9].

On Avicel, the enzyme pair 5At+9Gc displayed the highest activity enhancements on all tested scaffoldins, but for a given scaffoldin, the most active ternary complexes always contain 9Gc associated with 48Ft or 9Et, the two major cellulases of the cellulosomes produced by *C. cellulolyticum* [12]. Therefore these experiments also highlighted the role of these two endo-processive cellulases which are probably key enzymes of the cellulosomes. The fact that the family-48 enzyme is a critical cellulosomal enzyme was recently confirmed by *in vivo* studies [28].

The kinetics performed on ribbons of BC at 3.5 g/L, revealed another interesting information about the cellulosomes. Before this study [9], the cellulases from *C. cellulolyticum* had never been tested on BC. The ribbons have a degree of polymerization of approx. 2000 glucopyranose units, along which highly crystalline regions (~60%) are interrupted by amorphous or less ordered regions. This crystalline cellulose of bacterial origin was chosen, since several reports showed that the synergy between endo and exo (or processive) cellulases was optimal on this model substrate [42]. Many enzyme pairs displayed unexpectedly high levels of synergy in the free state (Figure 5). Indeed, the highest synergies were observed for pairs containing one endoglucanase (Cel5A, Cel8C, or Cel9G) and one endoprocessive cellulase (Cel48F or Cel9E). For instance, the synergy reached a value of approx. 10 when Cel48F is mixed with Cel5A or Cel8C, and these free enzyme pairs were found 8-12 times more active on BC than on Avicel at the same concentration. In general, the complexation onto hybrid scaffoldins induced much lower activity enhancement than that found with Avicel. The compilation of the data in Figure 4B (SF_{Scaf2}/SF_{Scaf4} graphed as a function of SF_{Scaf4}) shows that in the case of BC the spots are less dispersed and closer to coordinates 1,1, indicating that the CBM and proximity effects are much lower than on Avicel. In fact most significant enhancements (above 2) were only observed for endoglucanase pairs that showed low activity and no synergy in the free state such as 5At and 9Gc. Furthermore, complexation on the double-CBMs scaffoldin Scaf5 induced a modest but significant inhibition of half of the tested enzyme pairs.

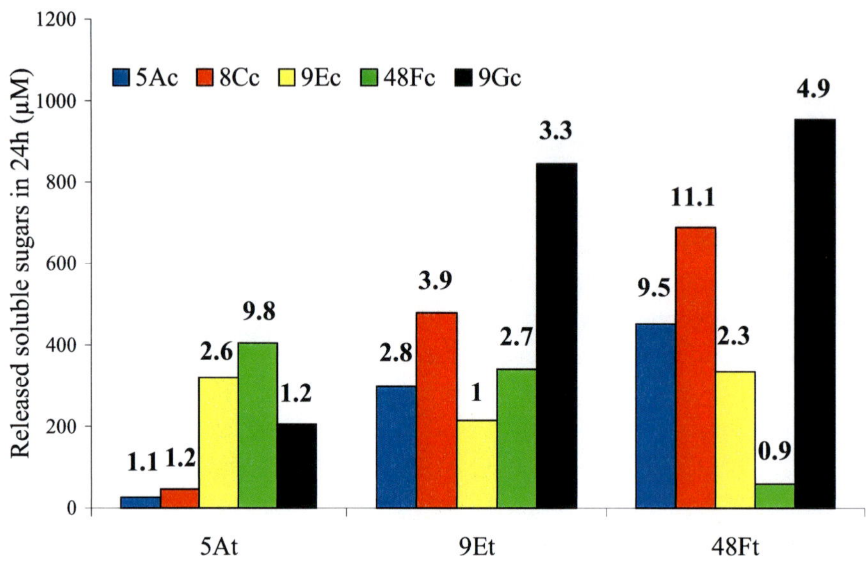

Figure 5: Released soluble sugars from Bacterial Cellulose (3.5 g/L) after 24 h at 37°C by free enzyme pairs. Numbers on top of the bars indicate the synergy between enzymes in the free state: Synergy = (released soluble sugars by a free enzyme pair)/(sum of released soluble sugars by corresponding enzymes alone).

These large variations between the two types of cellulose cannot be explained by the crystallinity, because the crystallinity index of BC is even slightly higher than that of Avicel. Nevertheless, the two substrates display very different ultrastructures: BC is described as long, thin and well dispersed ribbons along which crystalline regions are interrupted by amorphous regions. In contrast, Avicel is composed of large aggregates of microcrystals that have a low surface area. Thus, in BC, the crystalline and less ordered regions of the ribbons are much more accessible to an enzyme or an enzyme complex, and reactive and accessible sites are likely to be much more abundant in BC than in Avicel. This hypothesis was confirmed by two observations: at lower BC concentration (0.35 g/L), complexation of 48Ft and 9Gc onto Scaf2 and Scaf4 induced a 13- and a 2.7-fold activity enhancement, respectively, whereas at 3.5 g/L the gain in activity due to the binding onto either scaffoldin was in the 10% range. The same cellulosome chimeras were also assayed on Bacterial MicroCrystalline Cellulose (BMCC) which is obtained from BC by moderate hydrolysis with HCl. Such treatment acts selectively on the disordered regions, leading to a reduction in DP to 100-150 residues and to an aggregation of the crystalline domains. BMCC is slightly more crystalline than BC, and the reactive sites are less accessible. The ultrastructure of BMCC thus resembles that of Avicel more than BC, and indeed complexation of 48Ft + 9Gc on either Scaf2 or Scaf4 resulted in significant increase in activity towards BMCC, similar to that observed in the case of Avicel.

The use of cellulosome chimeras showed that incorporation of the enzymes into cellulosomes induce two major effects: an enhancement of the activity due to the strong family 3A CBM of the scaffoldin which anchors the complex at the surface of the substrate, and enhanced synergy due to the proximity of the catalytic domains within the complex. This study also confirmed the importance of the two major cellulosomal enzymes, Cel48F and Cel9E, since the most active cellulosome chimeras on Avicel or BC always contain one of these two endoprocessive cellulases. It was also demonstrated that cellulosomes are optimized for degradation of recalcitrant cellulose substrates, whereby recalcitrance reflects both the intrinsic crystallinity and the low accessibility of the reactive sites, such as that found in native plant cell walls.

Nevertheless, on Avicel, the most efficient cellulosome chimera (Scaf2:48Ft/9Gc), remained approx. 10–fold less active compared to purified cellulosomes from *C. cellulolyticum* at the same concentration of 0.1 μM. Thus, even on pure crystalline cellulose substrates, two-enzymes minicellulosomes cannot compete with natural cellulosomes, and a larger number of complementary cellulases is required for optimal degradation of this substrate.

FUTURE PROSPECTS

Extension of the Cellulosome Chimeras Concept

To increase the capacity of the cellulosome chimeras new specific cohesin/dockerin devices that do not cross react with type I cohesins and dockerins from *C. thermocellum* and *C. cellulolyticum* have to be used. Type II cohesin/dockerin from *C. thermocellum* were shown not to be suitable for this purpose because the interaction is rather weak [20] whereas Jindou and co-workers [16] have demonstrated that the species-specificity of the (type I) cohesin/dockerin interaction between *C. thermocellum* and *C. josui* is not as strict as observed between *C. thermocellum* and *C. cellulolyticum*. Furthermore, *C. josui* cohesins and dockerins are highly homologous to that from *C. cellulolyticum*, and inter-species recognition is likely to occur. On the other hand, the new scaffoldins from other cellulolytic bacteria which gene have recently been sequenced, may provide a number of suitable cohesins and dockerins to be exploited in the construction of cellulosome chimeras. Sequence analyses have shown that their cohesin and dockerin domains are quite divergent from those found in *C. thermocellum* and *C. cellulolyticum*. Furthermore, several specific cohesin/dockerin types of high affinity coexist within the same bacterium. For instance, in *A. cellulolyticus* [44], at least three different specific cohesin/dockerin devices were found, thus providing new promising interacting building blocks. Alternatively, new artificial specificities may be generated by random or site directed mutagenesis based on the available crystal structure of the type I cohesin/dockerin complex from *C. thermocellum* [3].

A second generation of cellulosome chimeras able to contain three different enzymes, as shown in Figure 6, would constitute a valuable tool to further explore the functioning of natural cellulosomes. With this second generation of complex, it would be possible to gather the three types of cellulases discovered to date in *C. cellulolyticum*: a genuine endoglucanase (such as Cel5A or Cel8C), an atypical endocellulase Cel9G, and a processive cellulase (Cel48F or Cel9E). It would also be of interest to create a chimera that contains both major cellulosomal processive enzymes (Cel48F and Cel9E) and an endoglucanase. Furthermore, other parameters could be explored: in the former study with two-enzymes complexes [9], only cellulosome chimeras with a 1:1 enzyme stoichiometry could be built. With a hybrid scaffoldin containing three specific cohesins, it should be possible to build cellulosome chimeras containing only two different cellulases with a 2:1 stoichiometry. The influence of the location of the enzymes onto the scaffoldin may also be explored, using only two enzymes appended with the dockerins corresponding to the most distant or adjacent cohesins. Finally, incorporation of one hemicellulase with the two most active cellulases (Cel48F and Cel9G) may be planned, and the activity of the resulting cellulosome chimera could be directly tested on plant cell walls. Alternatively, a complementary cellulase from an other microorganism belonging to a GH-family that has not been found to date in *C. cellulolyticum* (such as GH family 6), may also be incorporated with Cel48F and Cel9G, to obtain more efficient cellulosome chimeras on crystalline cellulose.

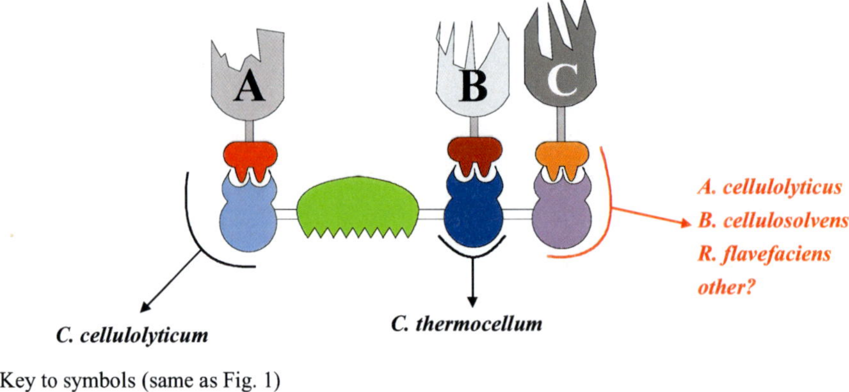

Figure 6: Schematic representation of the second generation of cellulosome chimeras.

In vivo Studies

If the studies performed *in vitro* with cellulosome chimeras [8,9], provided new insights on the functioning of natural cellulosomes, this type of approach, however, has two types of limitations: i) substantial saccharification of crystalline cellulose requires prolonged incubations at 37°C of more than one day, whereas some cellulosome chimeras were found to have a half life of 24 h. ii) the main product of cellulose degradation by cellulase mixture is cellobiose which may cause feed back inhibition of the enzymes. For these reasons, the expression of the genes encoding cellulosome chimeras in a heterologous host for *in vivo* studies may give new insights on the functioning of the cellulosomes in this context.

Many genes encoding cellulosomal components from various species have been successfully cloned and expressed in *E. coli*. The corresponding proteins were usually easily overproduced in the cytoplasmic compartment of the bacterium. Nevertheless, Only very few examples of successful heterologous production and secretion in the medium of intact cellulosomal components have been reported [17,24].

To perform *in vivo* studies of the cellulosome chimeras, the suitable host should be non cellulolytic, able to grow on and consume rapidly cellobiose, and related to cellulolytic *Clostridia*. That is why, our group has selected *C. acetobutylicum* for these experiments. This bacterium is not cellulolytic though its chromosome contains a whole set of cellulolytic genes including the gene encoding a scaffoldin, but interestingly, it has been shown that *C. acetobutylicum* produces very low amounts of cellulosome which are inactive towards crystalline cellulose [25,34]. *C. acetobutylicum* grows on cellobiose, and has a codon usage similar to that of *C. cellulolyticum*. Furthermore, this bacterium produces large amounts of solvents of industrial interest such as butanol, and the construction of an engineered strain capable to convert cellulose into organic solvents would have biotechnological applications. The first step towards the production and secretion of cellulosome chimeras was recently achieved in our laboratory, with the cloning of the gene encoding the hybrid miniscaffoldin Scaf3 (Figure 2) downstream of the strong constitutive *thl* promoter [29]. The heterologous protein was secreted in the medium at approx. 10 mg/L, and the family 3A CBM as well as the cohesins from *C. cellulolyticum* and *C. thermocellum* were found to be fully operational. Attempts to co-produce in *C. acetobutylicum*, cellulosomal enzymes appended with a suitable dockerin are currently underway.

Alternatively, *in vivo* studies of cellulosome chimeras may also be directly performed in a mutant strain of *C. cellulolyticum* which has lost the cellulolytic phenotype. In *C. cellulolyticum*, as well as in other mesophilic cellulolytic *Clostridia*, most of the known genes encoding cellulosomal components are gathered in a 26-kb cluster beginning with the gene encoding the scaffoldin CipC. We have isolated a mutant strain in which the *cipC* gene is disrupted with an insertion sequence, and the 10 downstream genes encoding cellulosomal enzymes are not expressed [21,22]. This mutant strain was shown to produce a truncated CipC and to secrete complexes that contain about 12 dockerin-borne enzymes. Nevertheless, the mutant strain is no longer able to degrade and grow on crystalline cellulose [22]. This strain is therefore suitable to perform *in clostridio* studies of the artificial minicellulosomes and determine whether the secretion of cellulosome chimeras containing selected cellulases can restore the cellulolytic phenotype.

POTENTIAL BIOTECHNOLOGICAL APPLICATIONS

To date, only cellulosomal cellulases have been incorporated into cellulosome chimeras, and their complexation resulted in enhanced activity towards recalcitrant cellulose. Is this technology inspired by bacterial cellulosomes applicable to other categories of enzymes from prokaryotic or eukaryotic origins that do not participate to complexes *in vivo*? Such demonstration is definitely required. On the other hand, it is worth noticing that the various types of cellulosomal cellulases, hemicellulases and pectinases discovered to date are not restricted to cellulosomes, since similar enzymes lacking the dockerin but belonging to the same GH families are also produced and secreted, in the free state, by other micro-organisms. It thus seems that the only difference between cellulosomal enzymes and their counterparts in free enzyme systems resides in the presence of a dockerin module. Their sequence and specific activity in the free state resembles that of the homologous enzymes that do not form complexes. The application of this technology by grafting appropriate dockerin modules to selected complementary enzymes that do not participate to cellulosomes *in vivo* is therefore likely to be successful. Furthermore, Riedel and Bronnenmeier [33] have shown that the fusion of two synergistic free cellulases, Cel48Y and Cel9Z from *C. stercorarium*, in a single peptide chain resulted in a hybrid enzyme with enhanced activity towards cellulose, compared to the activity of the corresponding free enzyme pair.

Once the demonstration that non-cellulosomal enzymes can be incorporated into cellulosome chimeras to boost their activity is established, these artificial complexes should have considerable biotechnological value for a broad variety of applications.

In the field of biomass utilization, one can design "cellulosome" chimeras containing the synergistic fungal esterases, xylanases and/or laccases [4,31] bound onto a hybrid scaffoldin to remove hemicellulose and lignin from paper pulp and improve the bleaching process. In a similar field of application, the powerful cellulolytic system of the fungus *T. reesei* has been selected in European and American "Bio-ethanol" projects for the saccharification of biomass prior to ethanolic fermentation by yeast(s). Nevertheless, it has been shown that at high substrate concentration, the synergy between the exo- and endo-glucanases secreted by *T. reesei* and their overall activity is reduced due to a dispersion of the complementary cellulases at the surface of the substrate [43]. It may therefore be possible to prevent the phenomenon of "substrate inhibition" by combining some of the *T. reesei* complementary cellulases into cellulosome chimeras.

The potential applications described above concern the hydrolysis of large insoluble polysaccharides or polymers, similar to those degraded by natural cellulosomes. But the possibility to incorporate selected enzymes at specific locations into defined complexes may be exploited in more distant biotechnological applications. Tandem acting enzymes which are used in many processes from the food industry to drug synthesis would constitute suitable candidates to be incorporated into cellulosome chimeras. The proximity of the enzymes within the complex may prevent the diffusion of the intermediate substrate(s)/product(s) thus increasing the velocity of the overall enzymatic reaction. Furthermore, such complexes could be easily immobilized onto the inexpensive cellulose thanks to the family 3A CBM of the scaffoldin.

CONCLUSION

Cellulosome chimeras have been shown to constitute valuable tools for the exploration of natural cellulosomes. This technology, which has only recently been set up, can be extended and the next generation of chimeras should help to understand further the functioning of these sophisticated enzymatic machineries. Besides, simplified and homogeneous artificial cellulosomes constitute suitable objects for structural studies involving, for instance, small angle X-ray scattering. Furthermore, the capacity to combine several enzymes or proteins at specific locations onto a miniscaffoldin is expected to have biotechnological applications in various fields.

REFERENCES

Bayer E. A., Belaich J. P., Shoham Y. and Lamed R. (2004) *Annu. Rev. Microbiol. 58*, 521-54.

Carrard G., Koivula A., Soderlund H. and Beguin P. (2000) *Proc. Natl. Acad. Sci. U S A. 97*, 10342-7.

Carvalho A. L., Dias F. M., Prates J. A., Nagy T., Gilbert H. J., Davies G. J., Ferreira L. M., Romao M. J. and Fontes C. M. (2003) *Proc. Natl. Acad. Sci. U S A. 100*, 13809-14.

de Vries R. P., Kester H. C., Poulsen C. H., Benen J. A. and Visser J. (2000) *Carbohydr. Res. 327*, 401-10.

Fierobe H. P., Gaudin C., Belaich A., Loutfi M., Faure E., Bagnara C., Baty D. and Belaich J. P. (1991) *J. Bacteriol. 173,* 7956-62.

Fierobe H. P., Bagnara-Tardif C., Gaudin C., Guerlesquin F., Sauve P., Belaich A. and Belaich J. P. (1993) *Eur. J. Biochem. 217*, 557-65.

Fierobe H. P., Pages S., Belaich A., Champ S., Lexa D. and Belaich J. P. (1999) *Biochemistry 38*, 12822-32.

Fierobe H. P., Mechaly A., Tardif C., Belaich A., Lamed R., Shoham Y., Belaich J. P. and Bayer E. A. (2001) *J. Biol. Chem. 276,* 21257-61.

Fierobe H. P., Bayer E. A., Tardif C., Czjzek M., Mechaly A., Belaich A., Lamed R., Shoham Y. and Belaich J. P. (2002) *J. Biol. Chem. 277*, 49621-30.

Fujino T., Beguin P. and Aubert J. P. (1992) *FEMS Microbiol. Lett. 73*, 165-70.

Fujino T., Beguin P. and Aubert J. P. (1993) *J. Bacteriol. 175*, 891-9.

Gal L., Pages S., Gaudin C., Belaich A., Reverbel-Leroy C., Tardif C. and Belaich J. P. (1997) *Appl. Environ. Microbiol. 63*, 903-9.

Gal L., Gaudin C., Belaich A., Pages S., Tardif C., Belaich J. P. (1997) *J. Bacteriol. 179*, 6595-601.

Gaudin C., Belaich A., Champ S., Belaich J. P. (2000) *J. Bacteriol. 182*, 1910-5.

Irwin D., Shin D. H., Zhang S., Barr B. K., Sakon J., Karplus P. A. and Wilson D. B. (1998) *J. Bacteriol. 180*, 1709-14.

Jindou S., Soda A., Karita S., Kajino T., Beguin P., Wu J. H., Inagaki M., Kimura T., Sakka K. and Ohmiya K. (2004) *J. Biol. Chem. 279*, 9867-74.

Kim A. Y., Attwood G. T., Holt S. M., White B. A. and Blaschek H. P. (1994) *Appl. Environ. Microbiol. 60*, 337-40.

Kosugi A., Murashima K., Tamaru Y. and Doi R. H. (2002*) J. Bacteriol. 184*, 884-8.

Leibovitz E, Ohayon H, Gounon P, Beguin P. (1997) *J. Bacteriol. 179*, 2519-23.

Leibovitz E. and Beguin P. (1998) *Enz. Microbial Technol. 22*, 588-593.

Maamar H., de Philip P., Belaich J. P. and Tardif C. (2003) *J. Bacteriol. 185*, 714-25.

Maamar H., Valette O., Fierobe H. P., Belaich A., Belaich J. P. and Tardif C. (2004) *Mol. Microbiol. 51*, 589-98.

Mechaly A., Fierobe H. P., Belaich A., Belaich J. P., Lamed R., Shoham Y. and Bayer E. A. (2001) *J. Biol. Chem. 276*, 9883-8.

Murashima K., Chen C. L., Kosugi A., Tamaru Y., Doi R. H. and Wong S. L. (2002) *J. Bacteriol. 184*, 76-81.

Nolling J., Breton G., Omelchenko M. V., Makarova K. S., Zeng Q., Gibson R., Lee H. M., Dubois J., Qiu D., Hitti J., Wolf Y. I., Tatusov R. L., Sabathe F., Doucette-Stamm L., Soucaille P., Daly M. J., Bennett G. N., Koonin E. V. and Smith D. R. (2001) *J. Bacteriol. 183*, 4823-38.

Pagès S., Belaich A., Belaich J. P., Morag E., Lamed R., Shoham Y. and Bayer E. A. (1997) *Proteins 29*, 517-27.

Pages S., Belaich A., Fierobe H. P., Tardif C., Gaudin C. and Belaich JP. (1999) *J. Bacteriol. 181*, 1801-10.

Perret S., Maamar H., Belaich J. P., Tardif C. (2004) *Mol. Microbiol. 51,* 599-607.

Perret S., Casalot L., Fierobe H. P., Tardif C., Sabathe F., Belaich J. P. and Belaich A. (2004) *J. Bacteriol. 186,* 253-7.

Perret S., Belaich A., Fierobe H. P., Belaich J. P. and Tardif C. (2004) *J. Bacteriol.* in press.

Record E., Asther M., Sigoillot C., Pages S., Punt P.J., Delattre M., Haon M., van den Hondel C. A., Sigoillot J. C., Lesage-Meessen L. and Asther M. (2003) *Appl. Microbiol. Biotechnol. 62*, 349-55.

Reverbel-Leroy C., Pages S., Belaich A., Belaich J. P. and Tardif C. (1997) *J. Bacteriol. 179*, 46-52.

Riedel K. and Bronnenmeier K. (1998) *Mol. Microbiol. 28*, 767-75.

Sabathe F., Belaich A. and Soucaille P. (2003) *FEMS Microbiol. Lett. 217*, 15-22.

Salamitou S., Raynaud O., Lemaire M., Coughlan M., Beguin P. and Aubert J. P. (1994) *J. Bacteriol. 176*, 2822-7.

Schaeffer F., Matuschek M., Guglielmi G., Miras I., Alzari P. M. and Beguin P. (2002) *Biochemistry 41*, 2106-14.

Shimon L. J., Pages S., Belaich A., Belaich J. P., Bayer E. A., Lamed R., Shoham Y. and Frolow F. (2000) *Acta Crystallogr. D Biol. Crystallogr. 56*, 1560-8.

Shoseyov O., Takagi M., Goldstein M. A. and Doi R. H. (1992) *Proc. Natl. Acad. Sci. USA. 89*, 3483-7.

Tokatlidis K., Salamitou S., Beguin P., Dhurjati P. and Aubert J. P. (1991) *FEBS Lett. 291*, 185-8.

Tormo J., Lamed R., Chirino A. J., Morag E., Bayer E. A., Shoham Y. and Steitz T. A. (1996) *EMBO J. 15,* 5739-51.

Tyurin M. V., Desai S. G. and Lynd L. R. (2004) *Appl. Environ. Microbiol.* 70, 883-90.

Väljamäe P., Sild V., Nutt A., Pettersson G. and Johansson G. (1999) *Eur. J. Biochem. 266*, 327-34.

Väljamäe P., Pettersson G. and Johansson G. (2001) *Eur. J. Biochem. 268*, 4520-6

Xu Q., Gao W., Ding S. Y., Kenig R., Shoham Y., Bayer E. A. and Lamed R. (2003) *J. Bacteriol. 185*, 4548-57.

Xu Q., Bayer E. A., Goldman M., Kenig R., Shoham Y. and Lamed R. (2004) *J. Bacteriol.* 186, 968-77.

Yaron S., Morag E., Bayer E. A., Lamed R. and Shoham Y. (1995) *FEBS Lett.* 360, 121-4.

In: Cellulosome
Eds: V. Uversky and I. A. Kataeva, pp. 211-229
ISBN 1-59454-950-8

Chapter X

CELLULOSOME ORGANISATION IN *RUMINOCOCCUS FLAVEFACIENS*

Harry J. Flint** and Marco T. Rincon
Rowett Research Institute, Bucksburn, Aberdeen, UK

ABSTRACT

Ruminococcus flavefaciens is a cellulolytic anaerobic bacterium that is one of the main agents of plant cell wall breakdown in the gut of herbivorous mammals, notably in the rumen. Plant cell wall degrading enzymes in this species display an elaborate form of cellulosome organisation involving at least four distinct dockerin:cohesin specificities. Three cohesin-containing structural proteins (ScaA, B, C) are encoded by linked genes within the *sca* cluster. The primary scaffoldin ScaA binds to one type of dockerin that is found in many enzymes as well as in the structural protein ScaC. ScaC apparently functions as an adaptor, since its unusual cohesin binds an entirely different group of polypeptides. Connection of the ScaA complex to the cell surface occurs via binding of ScaA to the larger ScaB, which in turn appears to bind to a component of the bacterial cell wall. Not all enzyme dockerins bind either ScaA or ScaC, however, suggesting the existence of further interactions and scaffolding proteins. ScaA lacks an identified cellulose binding module. Cellulose-binding modules are however present in a number of cellulosomal enzymes and in at least one specialised binding protein; cellulose binding may also involve the yellow affinity substance after which *R. flavefaciens* is named. ScaA and ScaB homologues have recently been detected in the related strain *R. flavefaciens* FD1, and information from the ongoing genome sequencing project on FD1 promises to expand our picture of cellulosomal complexes in this species in the near future.

Keywords: Rumen; *Ruminococcus*; dockerin specificity; cohesin phylogeny; adaptor

INTRODUCTION

Most detailed knowledge of microbial cellulolytic enzyme systems has come from the study of aerobic fungi and thermophilic and mesophilic anaerobic bacteria that are found in soils, compost or similar environments. The digestive tract of herbivores is however one of the most active sites for the degradation of lignocellulosic plant material in nature. The rumen microbial community, and in turn the ruminant animal, derives most of its energy from the activities of a few primary degraders which include strictly anaerobic cellulolytic fungi, bacteria and protozoa. Only two groups of ruminal bacteria have been commonly reported as cellulolytic, the gram-negative *Fibrobacter succinogenes* and the gram-positive Ruminococci. Recent evidence indicates that at least in *Ruminococcus flavefaciens* an elaborate and unusual form of cellulosome organisation plays a key role in the ability to degrade plant cell wall material. *Ruminococcus* spp. can be found in two major lineages among the low G+C content gram-positive bacteria based on 16S rRNA sequencing – Clostridial clusters XIVa and IV (Collins et al 1994, Rainey and Janssen 1995). Cellulolytic *Ruminococcus* species (*R. flavefaciens* and *R. albus*) have however been reported only from cluster IV (also known as the *C. leptum* group) (Fig.1). It is worth noting that cellulolytic Clostridia studied from non-gut habitats belong to other Clostridial clusters (eg. *C. thermocellum* belongs to cluster III) and thus share limited relatedness with the ruminococci. Bacteria related to *R. flavefaciens* are readily detected in many gut habitats (Suau et al 1999, Krause et al 1999,Tajima et al 1999, Leser et al, 2002, Daly et al 2001) and may play an important role in fibre breakdown in the large intestine in man (Robert & Bernalier 2003), pigs (Varel et al 1984) and horses (Julliand et al 1999).

CELLULOSOME ORGANISATION IN *R. FLAVEFACIENS*

In their early work, Ed Bayer and Raphael Lamed noted protuberant structures on the surface of a number of cellulolytic bacteria, including *Clostridium thermocellum* and *Ruminococcus albus* (Lamed et al 1987). Early work on cellulolytic enzymes from Ruminococci suggested the presence of high molecular weight enzyme complexes (Pettipher and Latham 1979, Wood et al 1982) and individual enzymes including an exoglucanase were successfully purified (Gardner et al 1987, Doerner and White 1990). The instability of some cloned inserts however complicated the molecular analysis of cellulolytic enzymes from ruminococci. For example, the cellulase EndA from *R. flavefaciens* 17 was originally reported to be a single domain enzyme (Cunningham et al 1990) but subsequent genomic PCR extension demonstrated multiple domains including a C terminal dockerin (Kirby et al 1997). Catalytic domains implicated in plant cell wall breakdown have nevertheless been described that belong to glycoside hydrolase families 3, 5, 9, 10, 11, 16, 44 and 48 (reviewed Selinger et al, 1996, Chesson and Forsberg 1997) and to carbohydrate esterase families 1 and

* Correspondence concerning this article should be addressed to; Harry Flint (h.flint@rowett.ac.uk) Microbial Ecology Group, Rowett Research Institute, Greenburn Road, Bucksburn, Aberdeen AB21 9SB, UK

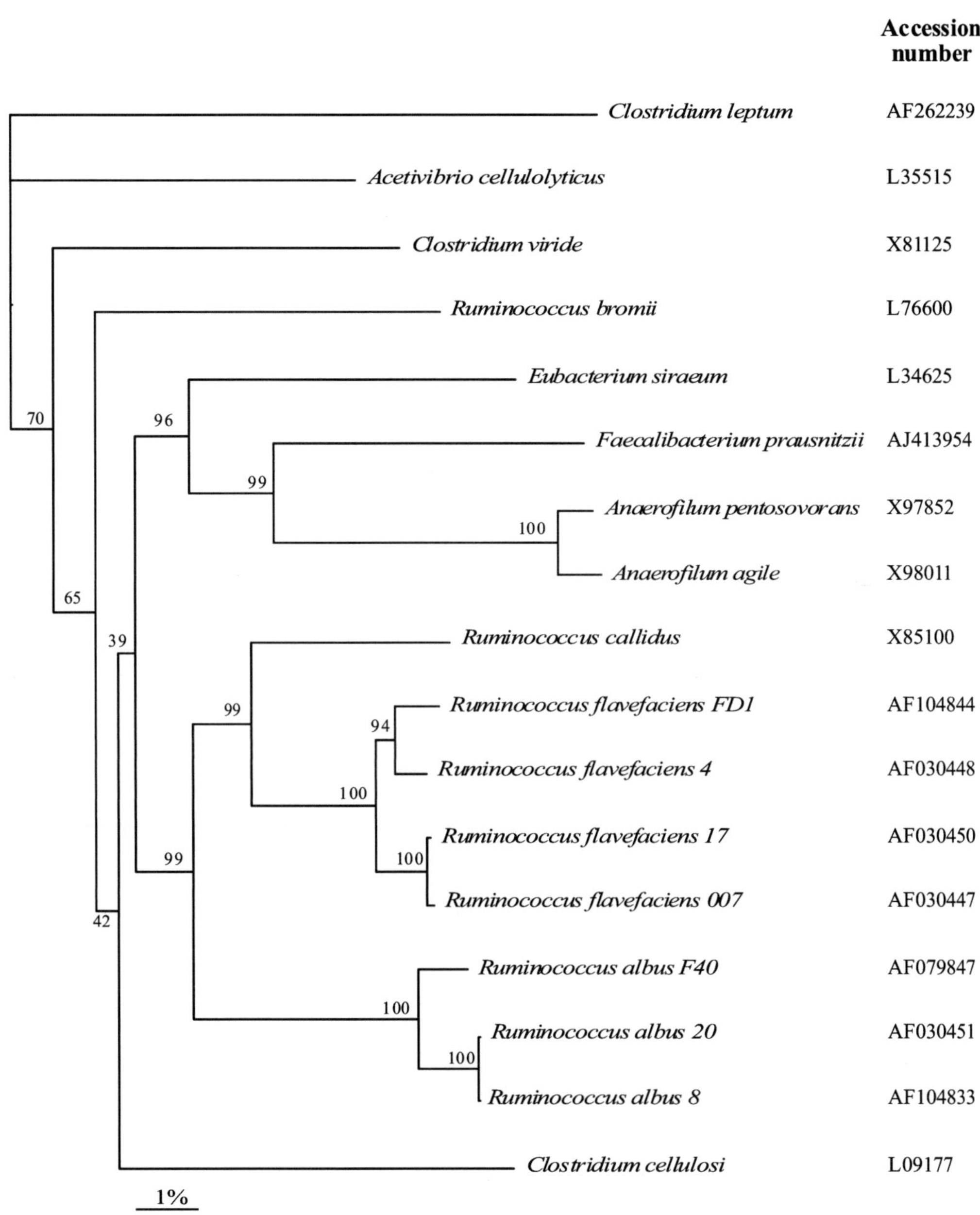

Figure 1. Unrooted phylogenetic tree based on the 16S rRNA gene sequences of *Ruminococcus flavefaciens* strains and other representatives of Clostridial cluster IV (*Clostridium leptum* group). The tree was constructed using the program ClustalX (ftp://ftp-igbmc.u-strasbg.fr/pub/ClustalX/) using the Neighbour-Joining method based on the alignment of approximately 1300 nt and edited using the program Treeview (http://taxonomy.zoology.gla.ac.uk/rod/treeview.-html). Bootstrap values, expressed as percentage of 1000 replications, are given at node points.

3 (Aurilia et al 2000a); genome sequencing (Antonopoulos et al 2004ab) can be expected to greatly expand this list. The clearest evidence for cellulosome organisation has so far emerged from work on *R. flavefaciens* 17, as discussed below. Comparative work suggests

that the main features will be found to apply in other related *R. flavefaciens* strains including the type strain FD1.

Domain Organisation of Plant cell wall Polysaccharidases in R. Flavefaciens 17

Complete sequences are available for seven extra-cellular plant cell wall degrading enzymes from *R. flavefaciens* strain 17. All of these enzymes are multi-modular; four carry catalytic domains that belong to different enzyme families within the same polypeptide (eg. Flint et al 1993) while a fifth (CesA) includes an unidentified C terminal domain (Fig. 2). There is some indication of synergistic action, for example between the acetyl xylan esterase and xylanase domains of XynB in the breakdown of acetylated xylans, although it is not known whether this effect is enhanced when the two domains reside in the same polypeptide (Aurilia et al 2000a).

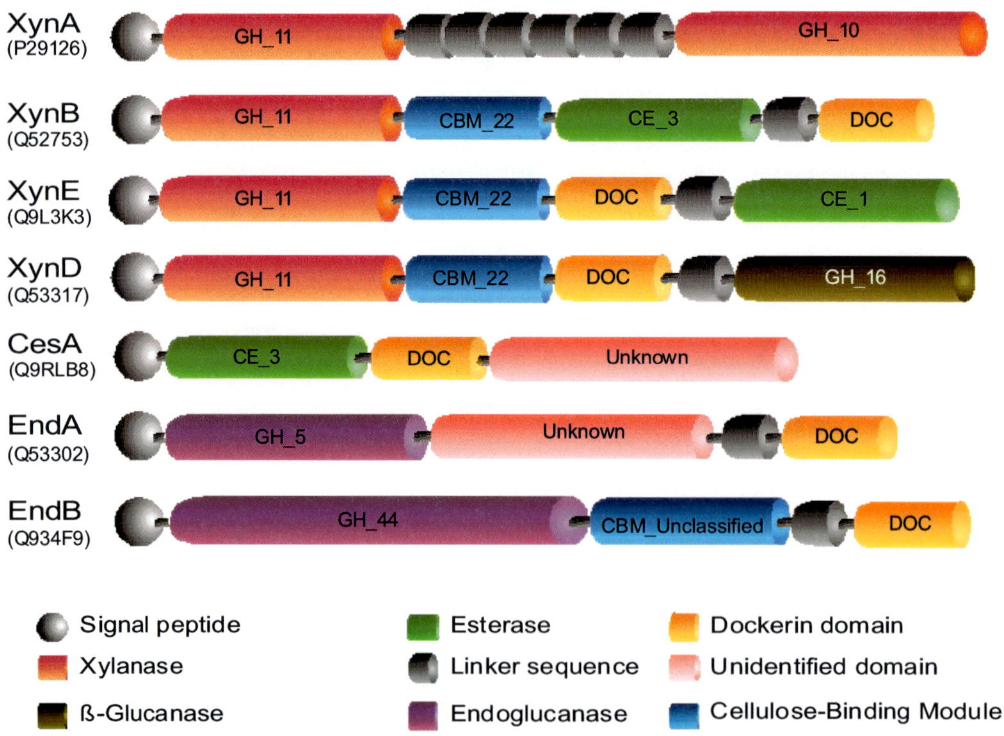

Figure 2. Diagram showing the domain structure organization of some glycosyl hydrolases isolated and characterized from *R. flavefaciens* 17. Accession numbers are given below the name of each enzyme. Families of glycoside hydrolase (GH), carbohydrate-binding module (CBM) and carbohydrate esterase (CE) domains are indicated.

Six of the seven polypeptides carry dockerin-like sequences, either at the C terminus or located internally (Kirby et al 1997, Aurilia et al 2000a), typically adjacent to threonine-rich linker sequences. Four of these dockerins are closely related, comprising what will be referred to as the EndB group. The remaining two (XynE and CesA) are only distantly related in sequence to each other and to the EndB group (Aurilia et al 2000a). The phylogenetic analysis shown in Fig. 3 reveals that all of these *R. flavefaciens* dockerins diverge significantly in sequence from dockerins previously identified in cellulolytic Clostridia, although they retain the essential feature of at least one copy of a calcium-binding motif (Fig. 4).

One extra-cellular polysaccharidase, XynA, from *R. flavefaciens* 17 (Zhang & Flint 1992) is currently known that lacks a dockerin sequence (Fig. 2). This unusual polypeptide carries family 11 and 10 xylanase domains connected by a lengthy (374 residue) asparagine/glutamine rich linker. Interestingly, a *xynA* homologue was recently reported from *R. flavefaciens* FD1 in which the family 10 and 11 domains are inverted (Antonopoulos et al 2004a).

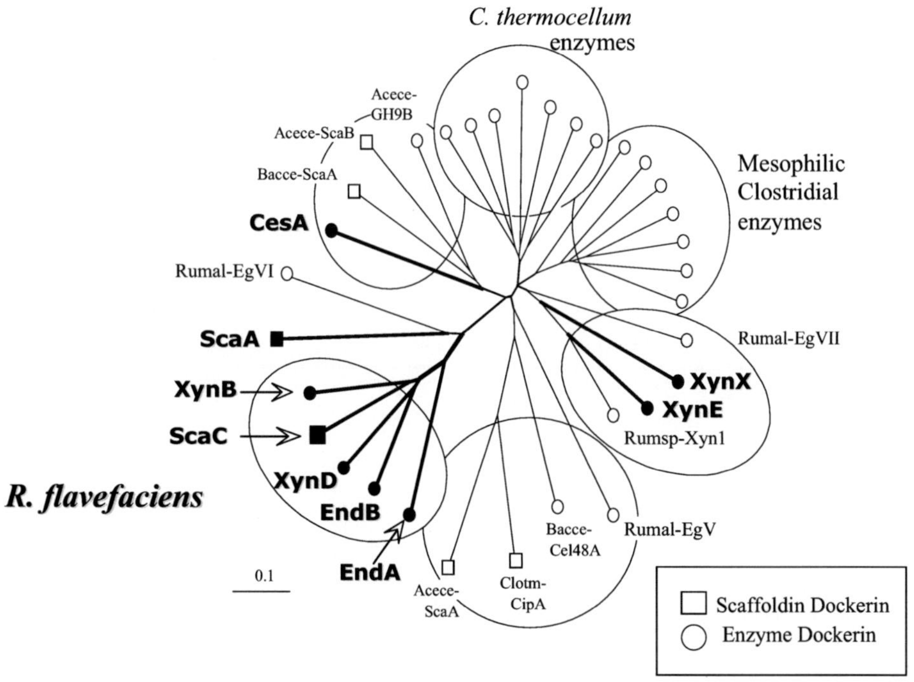

Figure 3. Unrooted phylogenetic tree of dockerin sequences from *R. flavefaciens* 17 (in bold) by comparison with dockerins from other bacteria. The known *R. flavefaciens* 17 dockerins fall into four distinct phylogroups. Enzyme dockerins are represented by a circle whereas dockerins present in scaffoldin proteins are represented by a square. The XynX dockerin is from a fifth, partially sequenced, xylanase gene (AJ586790) (Rincon et al 2004). The figure is from Rincon et al (2004) by kind permission.

The *sca* Gene Cluster

Firm evidence for cellulosome organisation in *R. flavefaciens* 17 has come from the identification of proteins carrying cohesin sequences that are able to interact with enzyme dockerins (Ding et al 2001) (Fig. 5). These are encoded by the *sca* gene cluster, which is now known to encode at least three cellulosome-related proteins. The roles of the ORFs that lie immediately downstream of *scaC*, *A* and *B* are still under investigation.

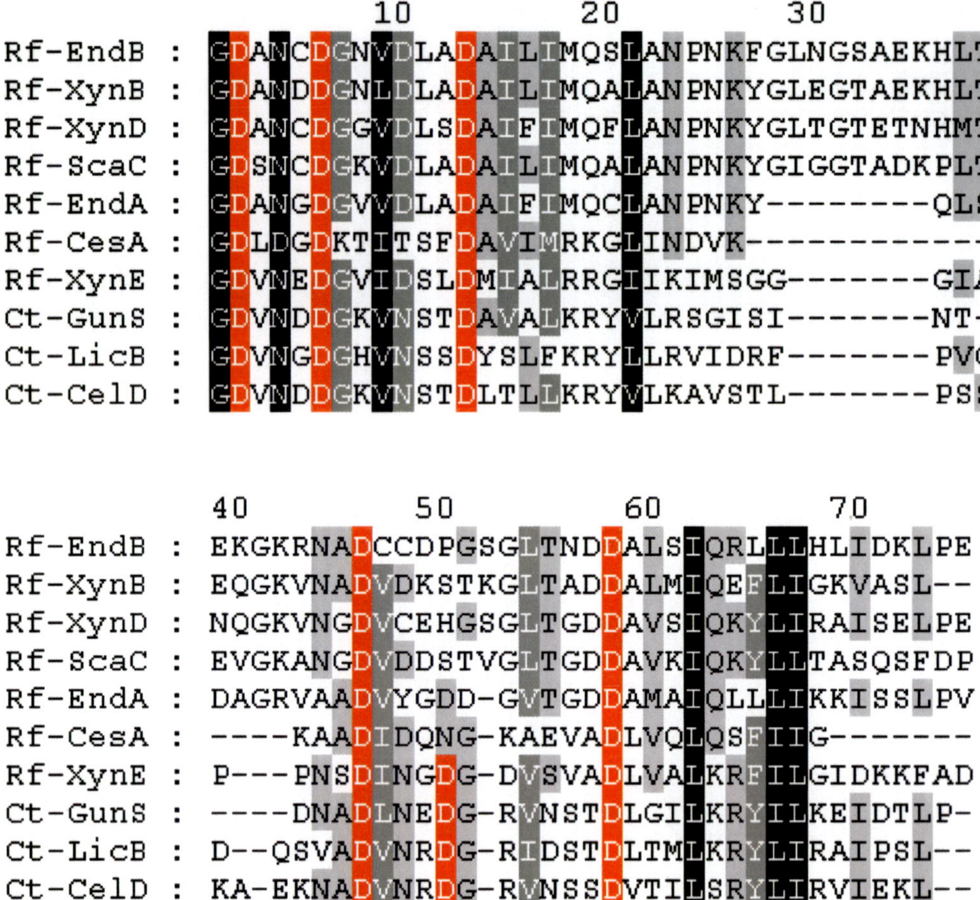

Figure 4. Multiple sequence alignment of dockerins. *R. flavefaciens* dockerins (Rf) are aligned with representative dockerins from the *Clostridium thermocellum* (Ct) GunS, LicB and CelD proteins (accessions P38686, P29716 and P04954 respectively. Aspartic acid residues that form the calcium-binding loop (positions 2, 6, 13 and 46, 50, 58 for the first and second loop respectively) are highlighted in red.

Roles of ScaA and ScaB in Cellulosome Assembly

ScaA is a 130kDa glycosylated protein in its native form. It carries three very similar cohesin domains that recognise dockerins present in the EndB group of enzymes. This dockerin:cohesin interaction has been demonstrated through the specific binding of recombinant enzyme fragments carrying dockerins to the native ScaA protein, and the binding of recombinant ScaA-cohesin2 to native proteins and to recombinant dockerin containing fragments (Ding et al 2001, Rincon et al 2001, 2003). ScaA can therefore be regarded as a scaffolding protein, by analogy with those found in cellulolytic Clostridia. As with the larger *C. thermocellum* CipA scaffoldin, ScaA carries a C terminal dockerin sequence that is distinct from dockerins found in cellulosomal enzymes. Just as the dockerin in *C. thermocellum* CipA mediates binding to cohesins within a cell wall anchoring protein, OlpA (Fujino et al 1993), the C terminal dockerin of ScaA binds to a second cohesin-containing protein, ScaB in *R. flavefaciens* (Ding et al 2001).

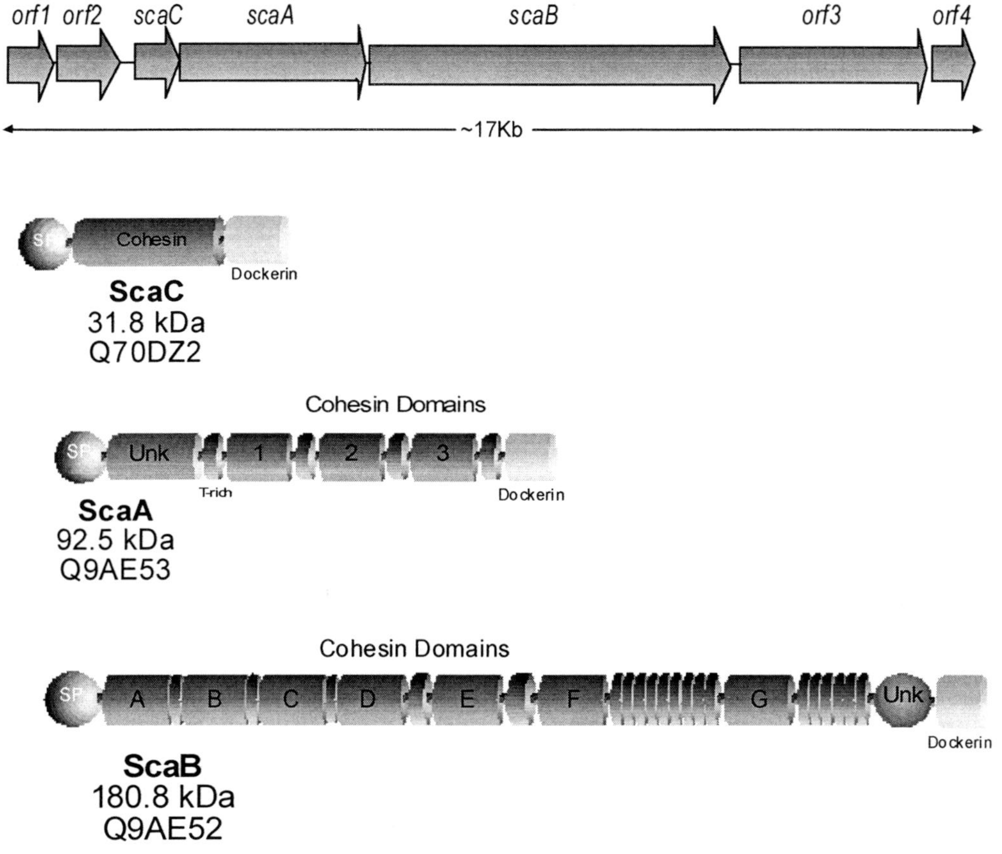

Figure 5. Diagram showing the gene organization of the scaffoldin cluster in *R. flavefaciens* 17. The domain organization of the gene products is indicated, along with their estimated molecular weights and the sequence accession numbers.

There is no evidence that ScaA possesses a cellulose-binding module, although this is a feature of all other scaffolding proteins reported to date. The role of the unique N terminal domain of ScaA has yet to be established, but an over-expressed recombinant protein fragment carrying this domain did not bind detectably to cellulose or xylan. Substrate binding in *R. flavefaciens* must therefore be a function of other proteins, or groupings, that are part of the cellulosome complex, or that are attached to the cell surface.

ScaB is a heavily glycosylated protein that migrates at around 300kDa in its native form. Two very long threonine-rich linkers in ScaB, as well as one similar sequence in ScaA, contain numerous potential glycosylation sites, but the nature of the carbohydrate decorations has not been established. ScaB carries seven cohesin domains that exhibit considerable sequence divergence from ScaA cohesins. With the exception of two similar cohesins, 5 and 6, the ScaB cohesins also differ from each other in sequence. Recombinant forms of six divergent ScaB cohesins all gave specific recognition of the same 130kDa native protein, which was identified by N terminal amino acid sequencing as ScaA, in *R. flavefaciens* 17 (Rincon et al 2003). Furthermore, the recombinant C terminal dockerin of ScaA was shown to recognise a protein co-migrating with the native ScaB (Ding et al 2001). There was no evidence to suggest avid binding of enzyme dockerins to ScaB cohesins. Thus ScaB can provide the platform for anchoring up to seven ScaA subunits, and up to 21 enzyme subunits, within a single large enzyme complex. ScaB can therefore be regarded as a putative cell wall anchoring protein, and shows a closer association to the cell surface than ScaA, which can be detected by EM attached to substrate complexes as well as on the cell surface in cellulose-grown *R. flavefaciens* cultures (Fig. 6).

The C terminal 241 residues of ScaB show little close similarity to other sequences found in the database, and may represent more than one functional domain. They include a sequence that shows some similarity to a dockerin module, including one likely calcium-binding motif.

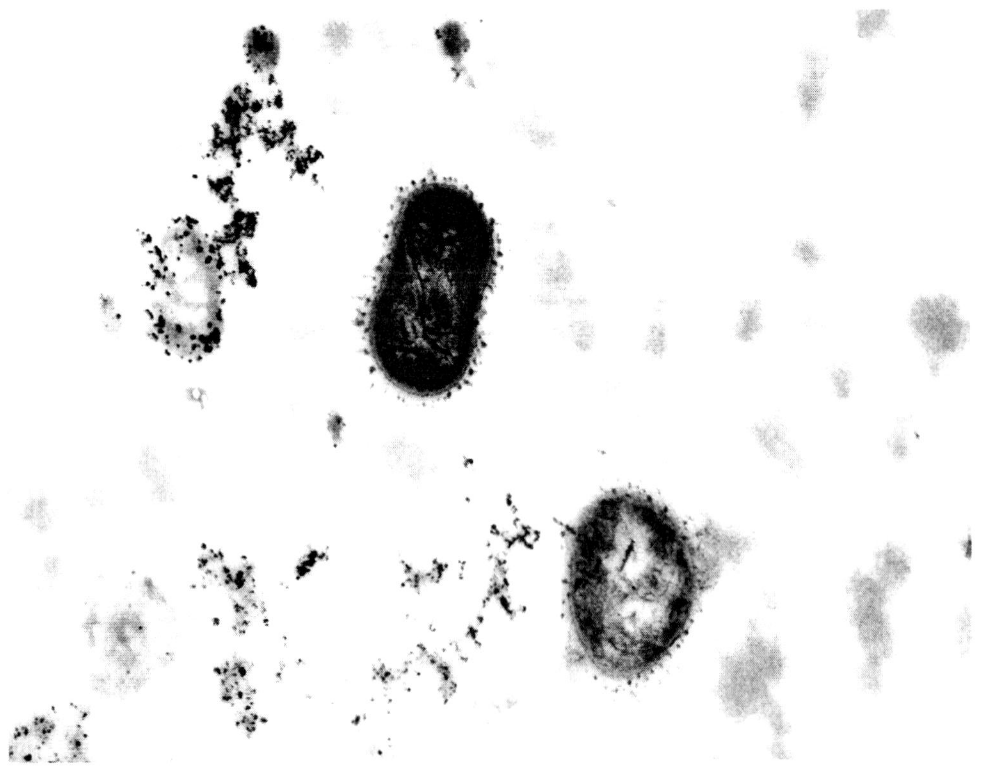

Figure 6. Transmission electron microscopy of *R. flavefaciens* 17 cells. The specimen was incubated with rabbit polyclonal antibody against ScaB-cohesin and then with gold-labelled anti-rabbit antibody. Gold particles are seen surrounding the cell surface as well as sloughed off material. Reproduced from Rincon et al (2003) by kind permission.

The ScaC Adaptor

The cohesins of ScaA and ScaB clearly diverge in sequence from those found in other cellulolytic bacteria (Fig. 7). The third cohesin-containing protein encoded by the *sca* cluster,

ScaC, carries a single cohesin that is again quite distinct in its sequence from any other known cohesins, including those from ScaA and ScaB. ScaC also appears unusual for a structural protein in carrying at its C terminal a dockerin that is closely related to the EndB group of enzyme dockerins. It was predicted that ScaC would bind ScaA via this dockerin sequence, and this was confirmed by showing that recombinant ScaA-coh2 binds to a 29kDa native *R. flavefaciens* protein whose MALDI profile matches that of ScaC (Rincon et al 2004).

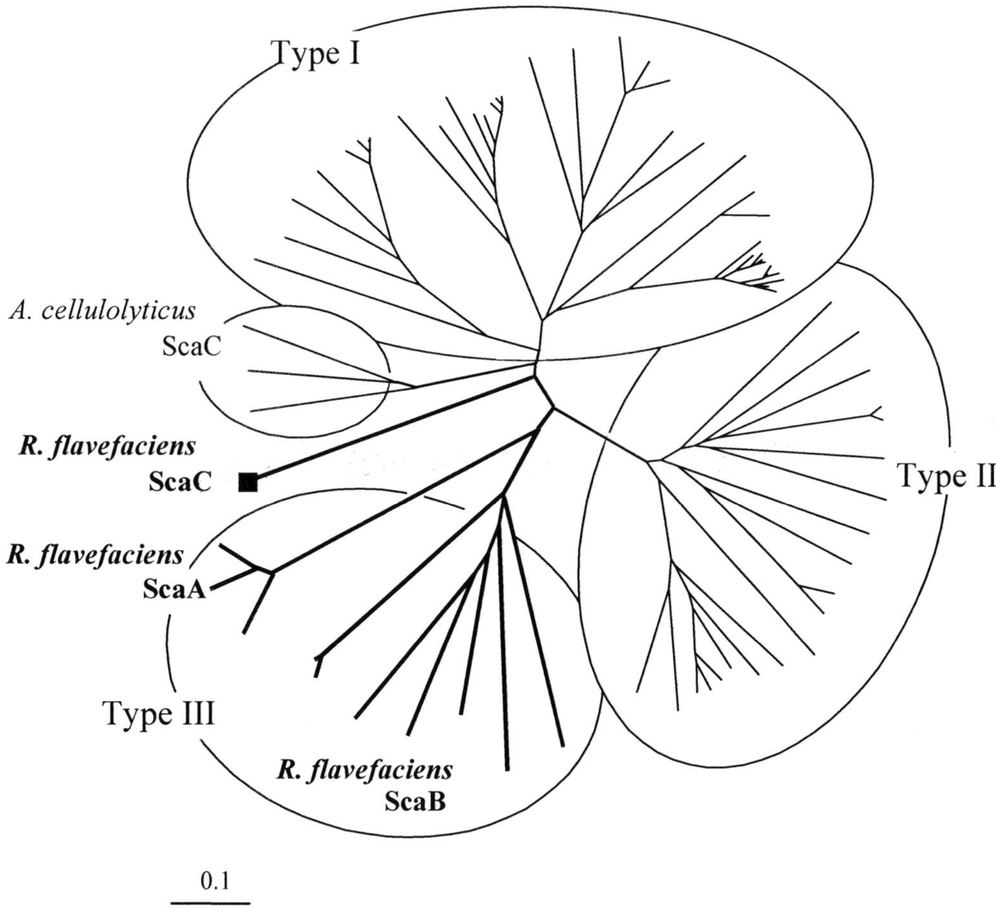

Figure 7. Phylogenetic analysis of cohesin domains from *R. flavefaciens* 17 scaffoldin proteins in relation to the known type I and II cohesins. *R. flavefaciens* ScaA and ScaB cohesins comprise a distinct phylogenetic group (type III). The ScaC cohesin maps outside the group in a separate branch distinct from all other cohesins. Reproduced from Rincon et al (2004) by kind permission.

The binding capability of the novel ScaC cohesin was investigated by probing two-dimensional separations of *R. flavefaciens* 17 proteins with recombinant ScaC-coh (Fig. 8). This revealed binding of ScaC-coh to multiple cell proteins ranging from 30 to 100 kDa in size. Notably, there was no overlap apparent with the set of proteins recognised by ScaA-coh2 (Rincon et al 2004). Therefore ScaC appears to recognise a new set of proteins that are hypothesised to carry a distinct, as yet undescribed, type of dockerin. ScaC apparently acts as a type of adaptor that greatly expands the range of proteins that can be incorporated into the ScaA-based complex (Fig. 9).

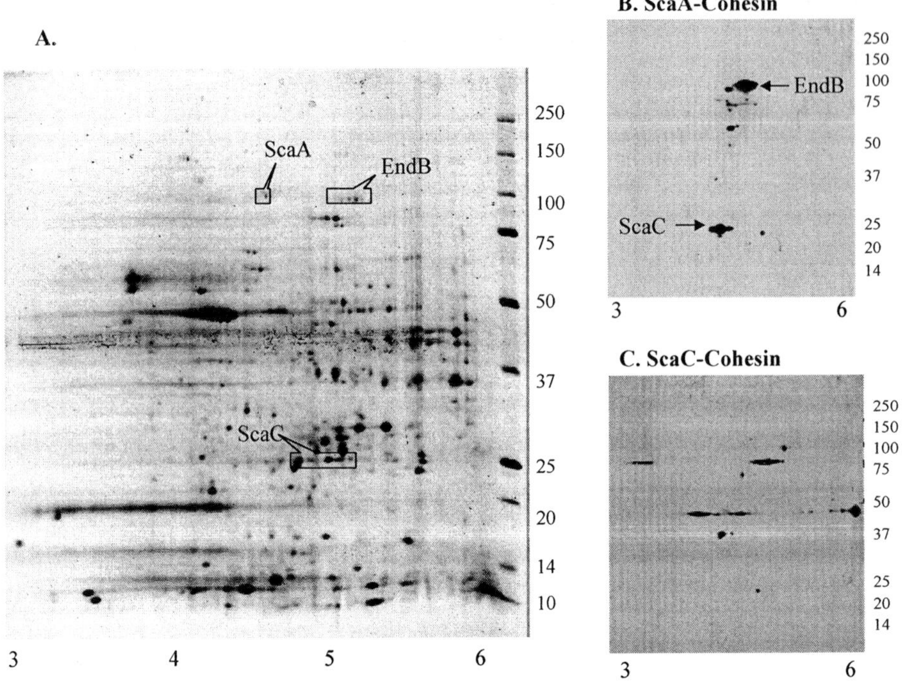

Figure 8. 2D gel electrophoresis of cell surface-associated protein fraction of *R. flavefaciens* 17 grown on birchwood xylan. (A) Colloidal Coomassie blue-stained gel. The positions of the enzyme EndB and the structural cellulosomal proteins ScaA and ScaC are indicated. The numbers at bottom indicate the pH gradient of the first separation. The numbers on the right indicate the positions of the protein markers for the second separation. (B) 2D Western blot analysis performed with a similar gel after blotting onto a PVDF membrane and probing with His6-ScaA-Coh. The positions of EndB and ScaC are indicated. (C) 2D Western blot analysis performed with a similar gel after the protein spots were blotted onto a PVDF membrane and probed with His6-ScaC-Coh. Cohesin-dockerin interactions were exposed using a Ni^{2+}-conjugated peroxidase that interacts specifically with the polyhistidine tag and enhanced chemiluminescence. Reproduced from Rincon et al (2004) by kind permission.

Multiplicity of Dockerin:Cohesin Binding Specificities

As with CipA from *C. thermocellum*, the interactions of the *R. flavefaciens* ScaA scaffolding protein involve two distinct dockerin:cohesin specificities, one connecting EndB-type enzyme subunits to ScaA cohesins, and a second connecting the distinct ScaA dockerin to cohesins in the ScaB putative anchoring protein. The situation in *R. flavefaciens* is however more complex (Table 1). As discussed above, the existence of ScaC as a cohesin-containing adaptor subunit attaching to ScaA creates a third specificity between an unknown dockerin type and the divergent ScaC cohesin.

In addition the enzymes CesA and XynE carry dockerins that are not closely related in sequence to each other or to the EndB group (Fig. 3). No binding was detected between these dockerins and ScaA-coh2, although CesA and XynE fragments do appear to bind native *R. flavefaciens* proteins other than ScaA (Rincon et al 2003). It is therefore assumed that at least one other type of cohesin, perhaps present in an alternative scaffolding protein, must be responsible for binding these divergent enzyme dockerins. This cohesin is apparently not the one present in ScaC, which gave no binding to recombinant XynE or CesA fragments (Rincon et al 2004).

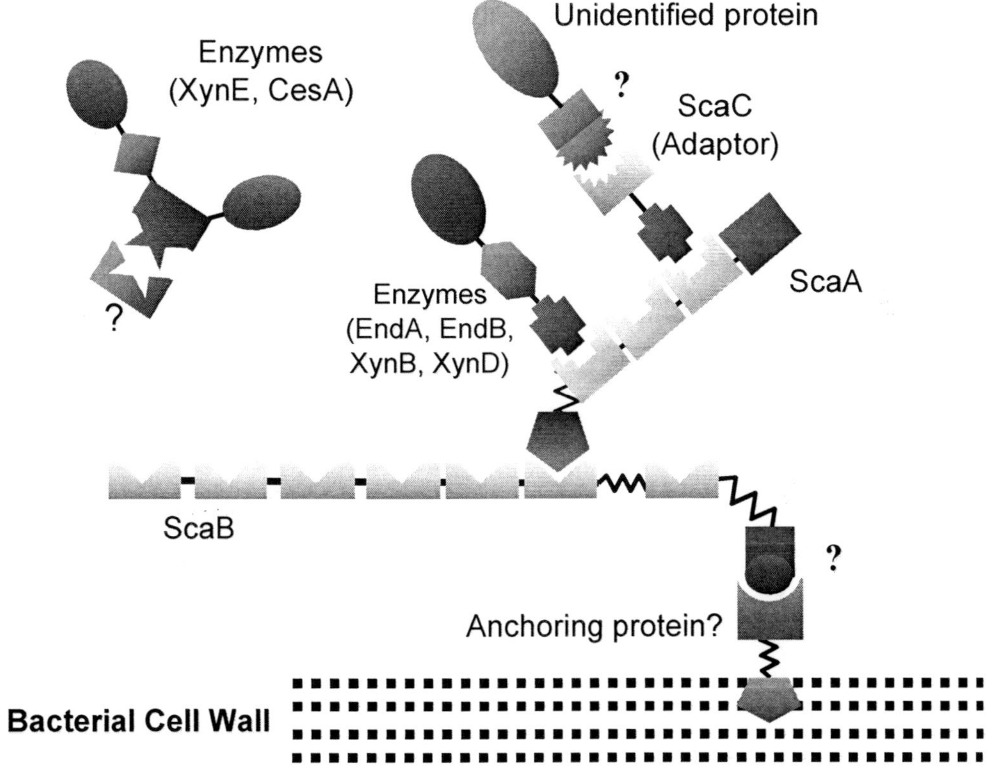

Figure 9. Postulated cellulosome organization in *R. flavefaciens* 17. Proteins coloured in grey are unidentified or inferred from protein-protein binding experiments.

In summary (Table 1) two dockerin:cohesin specificities have been defined in *R. flavefaciens* 17 for which both partners are known, but at least two others are inferred where only one partner has been identified (the cohesin in the case of ScaC, the dockerin in the case of XynE and CesA).

Cellulose Binding

As in other cellulolytic species, many enzymatic polypeptides from *R. flavefaciens* carry putative or proven carbohydrate binding modules. These include family 22 CBMs that are adjacent to family 11 xylanase domains in several gene products (XynD, XynB, XynE), and a new CBM sequence next to the family 44 cellulase domain in EndB (Fig. 2, Rincon et al

2001). Since most of these proteins can bind ScaA via dockerin:cohesin interactions, they may contribute to the attachment of the enzyme complex to its substrate. The newly identified EndB CBM is able to bind insoluble cellulose, and has a binding constant for Avicel of approximately 6μM (McCrae, Rincon and Flint, unpublished data).

Three large, abundant and tightly adherent proteins can be recovered from cellulose following growth of R. flavefaciens 17 on Avicel and extensive washing of the residual substrate (Kirby 1996, Rincon et al unpublished). These include an unidentified protein of around 130kDa and a second polypeptide of around 110kDa that is thought to carry a family 48 catalytic domain and an EndB-type dockerin. The third large, unidentified protein is a possible candidate for an additional scaffolding type protein that might for example bind XynE type and/or CesA-type dockerins.

Table 1. Summary of dockerin:cohesin pairings in *R. flavefaciens* 17.

Dockerin:	*Polypeptide*	*Cohesin:*			
		ScaA	***ScaB***	***ScaC***	***unknown***
EndB-type	(EndA,B; XynB,D; ScaC)	+	-	-	(-)
ScaA	(ScaA)	-	+	-	(-)
XynE-type	(XynE,X)	-	-	-	? (+)
CesA	(CesA)	-	-	-	? (+)
Unknown	Unknown	-	-	+	?

Regulation of Cellulosome Components

Very different profiles of dockerin-containing proteins are detected when the ScaA-cohesin is used to probe total cell proteins from cultures grown on xylan compared with cellulose (Rincon et al 2004). Expression of *xynD* and *xynB* was previously shown to be up-regulated at the mRNA level in cultures grown with xylan (Flint et al 1991). In addition two cellulases in *R. flavefaciens* FD1 showed increased expression in cellulose- compared to cellobiose-grown cultures (Doerner et al 1992).

These observations point to the likelihood of substantial changes in the expression of cellulosomal subunits, and in cellulosome composition, with different substrates and growth conditions. Whether the diversity of dockerin-cohesin specificities seen in *R. flavefaciens* plays any role in coordinating the expression of different complexes under different growth conditions is currently unknown.

OTHER ASPECTS OF PLANT CELL WALL BREAKDOWN IN *R. FLAVEFACIENS*

Xylan Utilization Genes

Cellulolytic *R. flavefaciens* strains produce a range of xylanases and other hemicellulases that are required to degrade the matrix material in the plant cell wall. Some strains of *R. flavefaciens*, including *R. flavefaciens* 17, can utilize the products of xylan degradation for growth whereas others cannot (Dehority & Scott 1967, Aurilia et al 2000b). *R. flavefaciens* 17 possesses a xylose isomerase gene (*xsi*) as part of a cluster that lies downstream from the gene for the cellulosome-associated enzyme XynD (Aurilia et al 2000b, Fig. 10). Upstream of *xsi* lies a family 3 glycoside hydrolase that functions as a β-xylosidase and is assumed to be involved in the intracellular processing of transported xylo-oligosaccharides. Downstream of *xsi* are elements of an ABC transporter system that is likely to be responsible for xylo-oligosaccharide transport. Two AraC-type regulatory domains flank *xynD*, and these may be involved in the observed up-regulation of *xynD* mRNA and of β-xylosidase activity (due to XylA) (Aurilia et al 2000b) in response to xylan.

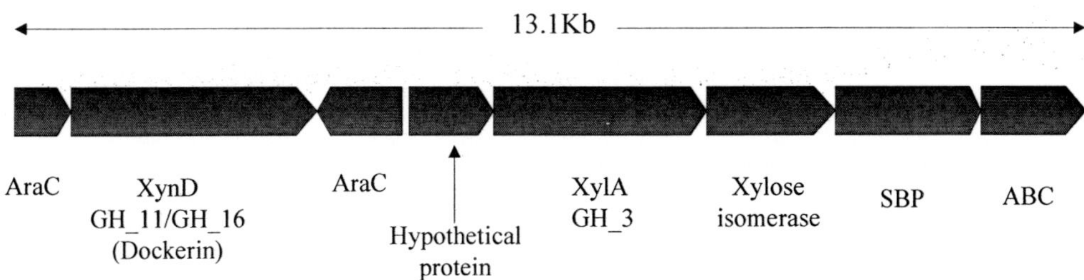

Figure 10 . Diagram showing the gene organization of the *R. flavefaciens* 17 xylan utilization operon. AraC: regulatory protein; XynD: multidomain GH_11/GH_16 cellulosomal enzyme; XylA GH_3: family 3 glycosyl hydrolase; SBP: substrate-binding protein; ABC: ABC transport protein. (Aurilia et al, 2000b). The open reading frames are drawn to scale using the program Artemis (http://www.sanger.ac.uk/Software/Artemis/)

DNA from two *R. flavefaciens* strains (C1a and B34b) that are unable to grow on xylan failed to hybridise with *xsi* or *xylA* probes from *R. flavefaciens* 17, whereas two xylan-utilising strains (C94 and B1a) did show cross-hybridization (Aurilia et al 2000b). This suggests the existence at least two functionally distinct groups of *R. flavefaciens* strains that presumably occupy somewhat different niches in the gut ecosystem.

Adaptation

Some laboratory cultures of *R. flavefaciens* 007 (a strain that is very closely related to *R. flavefaciens* 17 based on its proteome and 16S rRNA sequence) have been found to lose the ability to degrade dewaxed cotton cellulose following prolonged cultivation on cellobiose, although retaining the ability to breakdown Avicel (Stewart et al 1989, 1990). Cotton degrading ability could, however, be recovered upon repeated sub-cultivation on cotton. These events were ascribed to mutational changes. Recent proteomic analyses have revealed key differences between the non-cotton-degrading strain 007S and the strains 007C and 17, both of which can degrade cotton cellulose (Rincon et al unpublished). This should help to define the molecular requirements for attachment and enzymatic attack on this substrate. Morphological differences between 007C and 007S include chain length and cell surface architecture (Stewart et al 1991, Fig. 11) and are suggestive of significant changes in the cell surface.

Adaptation was also observed in *R. flavefaciens* 17 when this strain was repeatedly cultured with ryegrass as the sole energy source (Saluzzi et al 2001). Increased ryegrass degradation was accompanied by a 60% increase in total culture xylanase activity per unit cell protein. Again, an undefined regulatory mutation was suspected to be involved.

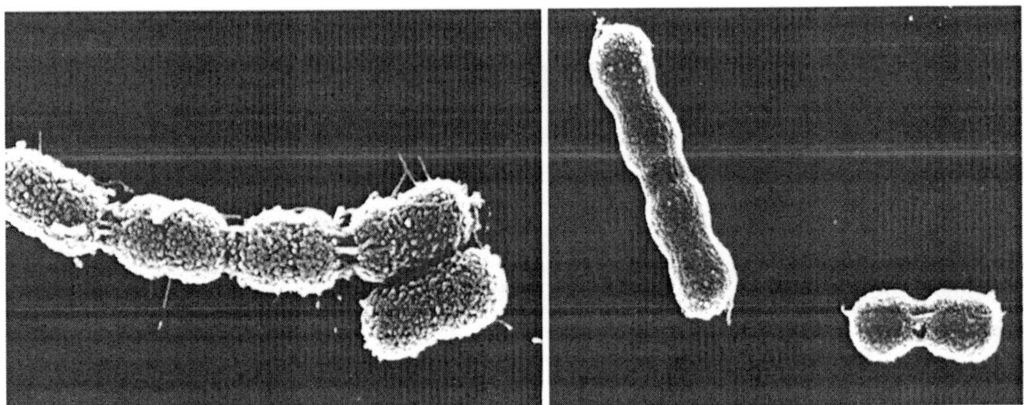

Figure 11. Scanning electron micrograph of cells of *R. flavefaciens* strains 007C (A) and 007S (B). Note that the surface of strain 007S appears smoother compared with strain 007C.

Yellow Affinity Substance

R. flavefaciens derives its name from the production of a distinctive yellow pigment (YAS) during growth on cellulose (Sijpestein 1948, 1951) Two YAS peaks have been recovered from *R. flavefaciens* 17; partially purified YAS showed affinity for crystalline cellulose, and appeared to promote the binding of *R. flavefaciens* cellulases to cellulose (Kopecny and Hodrova 1997). YAS was sensitive to oxidation by air and H_2O_2. Yellow pigments with similar properties are also produced by other cellulose degrading bacteria, including *C. thermocellum* (Hon Nami et al 1985).

The presence of a functional cysteine desulphurase gene downstream of the *endA* cellulase gene is also intriguing, leading to the speculation that this activity might act on a cellulosome component (Kirby et al 1998). There is however no direct evidence that this juxta-positioning has any significance for cellulose breakdown.

The extensive cell surface glycocalyx is also proposed to play a role in adhesion of *R. flavefaciens* to plant material (Latham et al, 1978).

Conclusion

It should be apparent that the number of different dockerin:cohesin specificities involved in *R. flavefaciens* cellulosome organisation is likely to be larger than in previously studied cellulolytic Clostridia. This might in part reflect the occurrence of extensive horizontal gene transfer within the dense communities of plant cell wall degrading micro-organisms that colonise gut habitats. Thus new sets of degradative enzymes might have been acquired from other species that bring with them divergent dockerins requiring to be attached via alternative cohesins to the cell surface. It seems likely that possession of a particularly wide range of degradative enzymes, and the ability to vary the composition of degradative enzyme complexes, would be important in responding to changes in the type of plant material that is ingested by the host animal.

A partial genome sequence has been determined for *R. flavefaciens* strain FD1 (Antonopoulos et al 2004ab). This strain is quite closely related to *R. flavefaciens* 17, showing 97% identity in its 16S rRNA sequence and giving detectable cross-hybridization between homologous genes (Flint et al 1992). Close homologues of ScaA and ScaB have now been detected in FD1, and exhibit the same essential features as reported for strain 17 (Antonopoulos et al 2004b). An analysis of the dockerin sequences present in this strain, together with their flanking regions, is in progress. This can be expected ultimately to reveal the full range of dockerin:cohesin interactions as well as the range of protein domains involved in cellulosomal complexes in this species. It will be of particular interest to define the range of other protein domains and modules that are associated with each type of dockerin, and thus to establish the likely functional significance of the different dockerin:cohesin specificities. Potentially this diversity of specificities might help in regulating the expression of different enzymatic activities as part of the cellulosome complex, and in the formation of alternative types of enzyme complex, perhaps in response to plant cell wall substrates of differing macromolecular composition.

While the evidence for cellulosome organisation in *R. flavefaciens* is clear, not all enzymes involved in plant cell wall breakdown belong to such complexes. XynA provides an example of a non-cellulosomal extracellular hydrolase from *R. flavefaciens* 17 (Zhang & Flint 1992). Meanwhile in R. albus 8 two major processive endocellulases do not possess dockerin motifs (Devillard et al 2004) and evidence for cellulosomal organisation is currently lacking in this strain, although proposed for R. albus F-40 (Ohara et al, 2001, Morrison and Miron 2000). It appears likely that there are some genuine inter-strain and inter-species differences in the organisation of plant cell wall degrading enzymes between *R. albus* and *R. flavefaciens*.

Acknowledgements

The Rowett Research Institute is largely funded by the Scottish Executive Environment and Rural Affairs department. MTR is supported by the EU framework V project GEMINI. We would particularly like to acknowledge the contribution of Ed Bayer and Raphael Lamed to the progress of work on the *R. flavefaciens* cellulosome; other contributions are reflected in the authorship of the original papers.

Note added in Proof: Attachment of the ScaB-based cellulosome complex to the bacterial cell surface in *R. flavefaciens* 17 was shown recently to be mediated by another product of the *sca* gene cluster, ScaE (Rincon MT, et al (2005) J. Bacteriol. 187, 7569-7578).

REFERENCES

Antonopoulos, D.A., Nelson, K.E., Morrison, M. and White, B.A. 2004 (a). Strain-specific genomic regions of *Ruminococcus flavefaciens* FD-1 as revealed by combinatorial random-phase genome sequencing and suppressive subtractive hybridization. *Environ. Microbiol. 6*, 335-346.

Antonopoulos, D.A., Bayer, E.A, and White, B.A. 2004 (b). Identification of cellulase components from a random phase genome sequence of *Ruminococcus flavefaciens* FD1. P.309-317 In: *Biotechnology of Lignocellulose Degradation and Biomass Utilization*. Ed. Ohmiya, K., Sakka, K., Karita, S., Kimura, T., Sakka, M., Onishi, Y. Uni Publishers Co Ltd, Tokyo.

Aurilia, V., Martin, J.C., McCrae, S.I., Scott, K.P., Rincon, M.T. and Flint, H.J. 2000 (a). Three multidomain esterases belonging to the plant cell wall degrading enzyme system of the rumen cellulolytic bacterium *Ruminococcus flavefaciens* 17 carry divergent dockerin sequences. *Microbiology 146*, 1391-1397.

Aurilia, V., Martin, J.C., Munro, C.A., Mercer, D.K. and Flint, H.J. 2000 (b). Organisation and strain distribution of genes responsible for the utilization of xylans by the rumen cellulolytic bacterium *Ruminococcus flavefaciens*. *Anaerobe 6*, 333-340.

Chesson, A., and Forsberg, C.W. 1997. Polysaccharide degradation by rumen micro-organisms. Pp. 329-381. In: *"The Rumen Microbial Ecosystem" 2nd edition* Ed. Hobson, P.N. and Stewart, C.S. Blackie, London.

Collins, M. D., Lawson, P. A., Willems, A., Cordoba, J. J., Fernandez-Garayzabal, J., Garcia, P., Cai, J., Hippe, H. and Farrow, J. A. E. 1994. The phylogeny of the genus *Clostridium*: proposal for five new genera and eleven new species combinations. *Int J Syst Bacteriol 44*, 812-826.

Cunningham, C., McPherson, C.A., Martin, J., Harris, W.D. and Flint H.J. 1991. Sequence of a cellulase gene from the rumen anaerobe *Ruminococcus flavefaciens* 17. *Mol. Gen. Genet. 228*, 320-323.

Daly, K., Stewart, C.S., Flint, H.J. and Shirazi-Beechey, S.P. 2001. Bacterial diversity within the equine large intestine as revealed by molecular analysis of cloned 16S rRNA genes. *FEMS Microbiol. Ecol. 38*, 141-151.

Dehority, B.A. and Scott H.W. 1967. Extent of cellulose and hemicellulose degradation in various forages by pure cultures of rumen bacteria. *J. Dairy Sci. 50*, 1136-1141.

Devillard, E., Goodheart, D.B., Karnati, S.K.R., Bayer, E.A., Lamed, R., Miron, J., Nelson, K.E., Morrison, M. 2004. *Ruminococcus albus* 8 mutants defective in cellulose degradation are deficient in two processive endocellulases Cel48A and Cel9B, both of which possess a novel modular architecture. *J. Bacteriol. 186*, 136-145.

Ding, S.Y., Rincon, M.T., Lamed, R., Martin, J.C., McCrae, S.I., Aurilia, V., Shoham, Y., Bayer, E.A. and Flint, H.J. 2001. Cellulosomal scaffoldin-like proteins from *Ruminococcus flavefaciens*. *J. Bacteriol. 183,* 1945-1953.

Doerner, K.C. and White, B.A. 1990. Assessment of endo-1,4-β-glucanase components of *Ruminococcus flavefaciens* FD1. *Appl. Environ. Microbiol. 56*, 1844-1850.

Doerner, K.C., Howard, G.T., Mackie, R.I. and White, B.A. 1992. β-glucanase expression by *Ruminococcus flavefaciens* FD-1. *FEMS Microbiol. Letts. 93*, 147-154.

Flint, H.J., McPherson, C.A. & Martin, J. 1991. Expression of two xylanase genes from the rumen cellulolytic bacterium *Ruminococcus flavefaciens* 17 cloned in pUC13. *J. Gen. Microbiol. 137*, 123-129

Flint, H.J., Zhang, J-X., Martin, J.C., Garcia-Campayo, V., Daniel, A., and Wood, T.M. 1992. Relationships of xylanase genes from rumen bacteria. Pp 188-196. In *"Lignocellulose degradation"* ed. K.Shimada, K. Ohmiya, Y. Kobayashi, S. Hoshino, K. Sakka and S. Karita. Uni Publishers Ltd.

Flint, H.J., Martin, J., McPherson, C. A., Daniel, A. and Zhang, J-X. 1993. A bifunctional enzyme having separate xylanase and ß (1,3-1,4) glucanase domains, encoded by the *xynD* gene of *Ruminococcus flavefaciens*. *J. Bacteriol. 175*, 2943-2951.

Fujino, T., P. Béguin, P., and J.P. Aubert 1993. Organisation of a *Clostridium thermocellum* gene cluster encoding the cellulosomal scaffolding protein CipA and a protein possibly involved in attachment of the cellulosome to the cell surface. *J. Bacteriol. 175*, 1891-1899.

Gardner, R.M., Doerner, K.C. and White, B.A. 1987. Purification and characterization of an exo-β-1,4-glucanase from *Ruminococcus flavefaciens* FD-1. *J. Bacteriol. 169*, 4581-4588.

Hon-Nami, H., Coughlan, M.P., Hon-Nami, K., and Lungdahl, L. 1985. The time course production of the yellow affinity substance and of the bound and free cellulase complexes by *Clostridium thermocellum* JW20. *Proc. Irish Academy 87B*, 83-92.

Julliand V., de Vaux A., Millet L and Fonty, G. 1999. Identification of *Ruminococcus flavefaciens* as the predominant cellulolytic bacterial species of the equine cecum. *Appl. Environ. Microbiol. 65*, 3738-3741.

Kirby J. 1996. Multiplicity and organisation of plant cell wall degrading enzymes in *Ruminococcus flavefaciens*. PhD thesis, U. Aberdeen.

Kirby, J. and Flint, H.J. 1998. A cysteine desulphurase gene from the cellulolytic rumen anaerobe *Ruminococcus flavefaciens. Biochim. Biophy. Acta 1386,* 233-237.

Kirby, J., Martin, J.C., Daniel, A.S. and Flint, H.J. 1997. Dockerin-like sequences in cellulases and xylanases from the rumen cellulolytic bacterium *Ruminococcus flavefaciens*. *FEMS Microbiol. Letts. 149*, 213-219.

Kopecny, J. and Hodrova, B. 1997. The effect of yellow affinity substance on cellulases of *Ruminococcus flavefaciens*. *Lett. Appl. Microbiol. 25*, 191-196.

Krause, D.O., Dalrymple, B.P., Smith, W.J., Mackie, R.I. and McSweeney, C.S. 1999. 16S rDNA sequencing of *Ruminococcus albus* and *Ruminococcus flavefaciens*: design of a signature probe and its application in adult sheep. *Microbiology 145*, 1797-1807.

Lamed, R., Naimark, R.J., Morgenstern, E. and Bayer, E.A. 1987. Specialised cell surface structures in cellulolytic bacteria. *J. Bacteriol. 169*, 3792-3800.

Latham, M.J., Brooker, B.E., Pettipher, G.L. and Harris, P.J. 1978. *Ruminococcus flavefaciens* cell coat and adhesion to cotton cellulose and to cell walls in leaves of perennial rye grass (*Lolium perenne*). *Appl. Environ Microbiol. 35*, 156-165.

Leser, T.D., Amenuvor, J.Z., Jensen, T.K., Lindecrona, R.H., Boye, M., and Moller, K. 2002. Culture-independent analysis of gut bacteria: the pig gastrointestinal tract revisited. *Appl. Environ. Microbiol. 68*, 673-690.

Morrison, M. and Miron J. 2000. Adhesion to cellulose by *Ruminococcus albus*: a combination of cellulosomes and Pil proteins? *FEMS Microbiol. Lett. 185*, 109-115.

Ohara, H., S. Karita, T. Kimura, K. Sakka and K. Ohmiya 2000. Characterisation of the cellulolytic complex (cellulosome) from *Ruminococcus albus*. *Biosci. Biotech Biochem. 64*, 254-260.

Pettipher, G.L. and Latham, M.J. 1979. Characteristics of enzymes produced by *Ruminococcus flavefaciens* which degrade plant cell walls. *J. Gen Microbiol. 110*, 21-27.

Rincon, M.T., McCrae, S.I., Kirby, J., Scott, K.P. and Flint, H.J. 2001. EndB, a multidomain cellulase from the rumen cellulolytic bacterium *Ruminococcus flavefaciens* 17, binds cellulose via a novel cellulose binding domain and to a 130 Kda *R. flavefaciens* protein via a dockerin domain. *Appl. Environ. Microbiol. 67,* 4426-4431.

Rincon, M.T., Ding, S-Y., McCrae, S.I., Martin, J.C., Aurilia, V., Lamed, R., Shoham, Y., Bayer, E.A. and Flint, H.J. 2003. Novel organisation and divergent dockerin specificities in the cellulosome system of *Ruminococcus flavefaciens*. *J. Bacteriol. 185*, 713-713.

Rincón M.T, Martin J.C., Aurilia V., McCrae S.I., Rucklidge G., Reid M., Bayer E.A., Lamed R., and Flint H.J. 2004. ScaC, an adaptor protein carrying a novel cohesin that expands the dockerin-binding repertoire of the *Ruminococcus flavefaciens* 17 cellulosome. *J. Bacteriol. 186*, 2576-2585.

Rainey, F.A. and Janssen, P.H. 1995 Phylogenetic analysis by 16S ribosomal DNA sequence comparison reaveals two unrelated groups of species within the genus *Ruminococcus*. *FEMS Microbiol. Letts. 129*, 69-74.

Robert C. and Bernalier-Donadille, A. 2003. The cellulolytic microflora of the human colon: evidence of microcrystalline cellulose-degrading bacteria in methane excreting subjects. *FEMS Microbiol. Ecol. 46*, 81-89.

Saluzzi, L., Flint, H.J. and Stewart, C.S. 2001. Adaptation of *Ruminococcus flavefaciens* to degradation of ryegrass cell walls. *FEMS Microbiol. Ecol. 36*, 131-137.

Selinger, L.B., Forsberg, C.W. and Cheng, K.L. 1996. The rumen: a unique source of enzymes for enhancing livestock production. *Anaerobe 2*, 263-284.

Sijpestein, A.K. 1948. Cellulose-decomposing bacteria from the rumen of cattle. PhD thesis, U. Leiden.

Sijpestein, A.K. 1951. On *Ruminococcus flavefaciens*, a cellulose decomposing bacterium from the rumen of sheep. *J. Gen Microbiol. 5*, 317-325.

Stewart, C.S., Duncan, S.H., McPherson, C.A., Richardson, A.J. & Flint, H.J. 1989. The implications of the loss & regain of cotton degrading activity for the degradation of straw by *Ruminococcus flavefaciens* 007. *J. Appl. Bacteriol. 68*, 349-356.

Stewart, C.S., Duncan, S.H. and Flint, H.J. 1990. The properties of forms of *Ruminococcus flavefaciens* which differ in their ability to degrade cotton cellulose. *FEMS Microbiol. Letts 72*, 47-50.

Suau, A., R. Bonnet, M. Stutren, J.-J. Godon, G.R. Gibson, M.D. Collins and J. Dore. 1999. Direct analysis of genes encoding 16S rRNA from complex communities reveals many novel molecular species within the human gut. *Appl. Environ. Microbiol. 65*, 4799-4807.

Tajima, K., R. I. Aminov, T. Nagamine, K. Ogata, M. Nakamura, H. Matsui and Y. Benno. 1999. Rumen bacterial diversity as determined by sequence analysis of 16S rDNA libraries. *FEMS Microbiol. Ecol. 29*, 159-169.

Varel, V.H., Fryda, S.F. and Robinson, I.M. 1984. Cellulolytic bacteria from the pig large intestine. *Appl. Environ. Microbiol. 47*, 219-221.

Wood, T.M., Wilson, C.A. and Stewart, C.S. 1982. Preparation of the cellulase from *Ruminococcus albus* and its release from the bacterial cell wall. *Biochem. J. 205*, 129-137.

Zhang, J-X. and Flint, H.J. 1992. A bifunctional xylanase encoded by the *xynA* gene of the rumen cellulolytic bacterium *Ruminococcus flavefaciens* 17 comprises two dissimilar domains linked by an asparagine/glutamine rich sequence. *Mol. Microbiol. 6*, 1013-1023.

In: Cellulosome
Eds: V. Uversky and I. A. Kataeva, pp. 231-259
ISBN 1-59454-950-8

Chapter XI

CLOSTRIDIUM CELLULOLYTICUM: CELLULOSOMES AND CELLULOLYSIS

Chantal Tardif[1,2,*], *Anne Bélaïch*[1], *Henri-Pierre Fierobe*[1], *Sandrine Pagès*[1,2], *Pascale de Philip*[1,2] *and Jean-Pierre Bélaïch*[1,2]
[1]BIP-CNRS, 31 chemin Joseph Aiguier, Marseille, France
[2]Université de Provence, Marseille, France

ABSTRACT

The enzymatic system of *Clostridium cellulolyticum* that hydrolyzes plant cell wall polysaccharides comprises multi-enzyme complexes, named cellulosomes, and non-cellulosomal components. Each cellulosome consists of a scaffolding protein CipC complexed with a maximum of eight enzymes. CipC is composed of a cellulose binding module at the N terminus followed by one hydrophilic X2 module, seven highly homologous cohesin modules, a second X2 module, and a final cohesin module. Individual cellulosomes range in mass from about 600,000 to 700,000 Da. Cellulosomes prepared from cellulose culture revealed three major components, which were identified to be CipC, the processive endocellulase Cel48F, and the cellobiohydrolase Cel9E, in addition to numerous other components with molecular masses ranging from 100 to 30 kDa. Enzymatic components of the complexes were all found to contain one specialized docking module (called dockerin) able to specifically interact with the CipC cohesins. *Clostridium cellulolyticum* synthesized at least 22 different dockerin-bearing proteins. Thus the complexes may vary widely in their composition. This composition is regulated by the relative amounts of the enzymes available to be incorporated into the complexes. The composition of the cellulosomes was efficiently engineered *in vivo* by structural gene or regulatory gene expression. The gene encoding CipC is the first gene of a large cluster spanning 26 kb. It is immediately followed by *cel48F*, and then by nine other

* Correspondence concerning this article should be addressed to Chantal Tardif, BIP-CNRS, 31 chemin Joseph Aiguier, F-13402 Marseille cedex 20, France. E-mail :tardif@ibsm.cnrs-mrs.fr.

genes (*cel8C-cel9G-cel9E-orfX-cel9H-cel9J-man5K-cel9M-rgl11Y-cel5N*) coding for seven cellulases, one mannanase and one pectinase, also found to be components of the cellulosomes. The cellulases encoded by the genes located in this cluster were found to be essential for the building of cellulosomes efficient in crystalline cellulose degradation. Studies focused on the metabolism of *C. cellulolyticum* in culture with cellulose or cellobiose as substrate indicated that the strain is well adapted, and even restricted to a low carbon flow, which is characteristic of growth on its natural substrate, cellulose. The strain is not able to regulate soluble sugar entry and thus the carbon flow through glycolysis. Under these conditions, the nutrients or products of its metabolism accumulate intracellularly to toxic levels. Improvement of the wild-type strain was obtained by heterologous expression of an artificial operon to eliminate excess intracellular pyruvate.

Keywords: cellulose degradation, *Clostridium cellulolyticum*, cellulases, cellulosome, enzymatic complexes, growth inhibition, oligotrophy, metabolic engineering.

1. INTRODUCTION

The conversion of cellulose to fuels and chemical feedstocks by microbial fermentation might represent a partial answer to waste accumulation problems and the depletion of fossil fuel reserves. Indeed, cellulose is a major biological compound on earth. Around 40 gigatonnes are synthesized per year, essentially by plants that use it as the major component in their cell wall, embedded within other polymers. Cellulose molecules, composed of long unbranched polymer of glucose which are linked by β-1,4 bonds, are associated together to form microfibrils, and this insoluble substrate is so compact and highly organized that it is resistant to degradation, the majority of β-1,4 bonds being not accessible to enzymes in cellulose fibers. During the degradation of very complex organic matter such as the organic fraction of municipal solid waste, cellulose hydrolysis is the rate-limiting step of the whole process. Due to its recalcitrant, durable nature, cellulose accumulates in terrestrial environments. In aerobic niches, biodegradation of cellulose is carried out primarily by fungi, whereas, in anaerobic environments, bacteria are the major cellulose hydrolysers. Around 5 to 10% of cellulosic materials were estimated to be degraded anaerobically. In these biotopes, cellulolytic clostridia play a prominent part in such processes [11]; numerous *Clostridium* strains were thus isolated from many hot and ambient temperature biotopes [1,3,40,41,50,55,72,73,85,86,89,93-96,98,99]. Among them, only a few were extensively studied.

Clostridium cellulolyticum, a nonruminal mesophilic cellulose-degrading bacterium was isolated from decayed grass twenty years ago [12,72]. As found with other cellulolytic clostridia, due to early inhibition of metabolism and growth, this strain classically uses low quantities of sugar [33]. Taking into account that many natural ecosystems rarely contain all nutrients in high quantities, this is especially the case in cellulose-rich biotopes in which cellulose hydrolysis is a slow phenomenon, the cellulolytic bacteria are adapted to grow under conditions of low levels of soluble sugars. *C. cellulolyticum* cultivated in complex medium with a high concentration of soluble substrate was shown to encounter great difficulties for regulating its metabolism; under these conditions, the nutrients and products of its metabolism accumulate into the cells to toxic levels. It can be argued that during the course of evolution, *C. cellulolyticum* has evolved to optimize cellobiose (the major product of cellulolysis) catabolism and nitrogen anabolism under nutrient-poor conditions. It can thus be considered as an oligotrophic micro-organism [44,45]. Besides cellulose and the soluble products of cellulolysis, cellobiose and glucose, *C. cellulolyticum* is able to fermentate some other sugar polymers and their hydrolysis products, such as xylan and xylose, mannane and mannose. Several other monosaccharides as arabinose (a component of arabinoxylan) and with a lesser efficiency, fructose, galactose and ribose [72] can also be used by this strain. This indicates that in addition to cellulose, *C. cellulolyticum* is able to degrade and use as carbon source some of the cellulose-associated polymers of the plant cell wall.

Cellulolytic micro-organisms produce a battery of cellulases, with different specificities, which are supposed to cooperate in the cellulose hydrolysis. They often also produce other enzymes, such as xylanases, mannanases, pectinases, etc..., that degrade the cellulose associated polymers. These enzymes, as cellulases, are secreted by the micro-organisms. Anaerobic strains assemble most of the enzymes onto scaffolding proteins (scaffoldins) leading to the formation of cellulolytic/hemicellulolytic/pectinolytic complexes, named "cellulosomes". In these complexes, the proteins are tightly bound; the dissociation of the complexes requires harsh conditions that lead to component denaturation. Preliminary experiments carried out to purify the various cellulolytic enzymes from *C. cellulolyticum* culture supernatants showed the presence of many endocellulases and at least one processive cellulase [4,49]. Nevertheless, extensive purification of these components was not complete and seemed to be difficult to achieve with biochemical methods. Studies of *C. cellulolyticum* cellulosome components have been facilitated by the use of gene cloning technology [6,10,21,22,68,80,82]. T7 promotor-based expression vector were essentially used to overproduce several of the representative enzymes and cellulosomal scaffolding proteins in *Escherichia coli* as histidine-tagged proteins. The purified recombinant proteins were subjected, in most cases, to extensive studies to determine their individual characteristics and to reconstitute mini-cellulosomes. *In vivo* modifications of the *Clostridium* strain was also chosen as a mean to obtain information on the role of targeted components in the entire cellulolytic system context.

Besides the characterization of the enzymes and the study of the cellulosomes, the studies focused on carbon metabolic pathways gave very interesting information about biomass formation and metabolic features which control rates, yields, and types of fermentation end-products. From a basic point of view and to achieve biotechnological applications, it is indeed essential to understand the conditions that direct the bacterial growth.

2. Cellulosomal Enzymes

A large set of genes and proteins related to the cellulolytic system of *C. cellulolyticum* has been studied. These genes are in majority located into a large cluster spanning 26 kb ("*cel*" cluster), which encodes many hydrolases belonging to various structural families of glycosyl hydrolases (GH) (Fig. 1). Eight genes coding for cellulases, including Cel48F and

Cel9E, the two major cellulases of the complexes, one gene coding for a mannanase (Man5K) and one gene coding for a rhamnogalacturonan lyase (Rgl11Y) are located into the "*cel*" cluster. In addition, three genes encoding cellulosomal proteins were found located outside of the cluster (*cel5A*, *cel5D*, *cel44O*)[46].

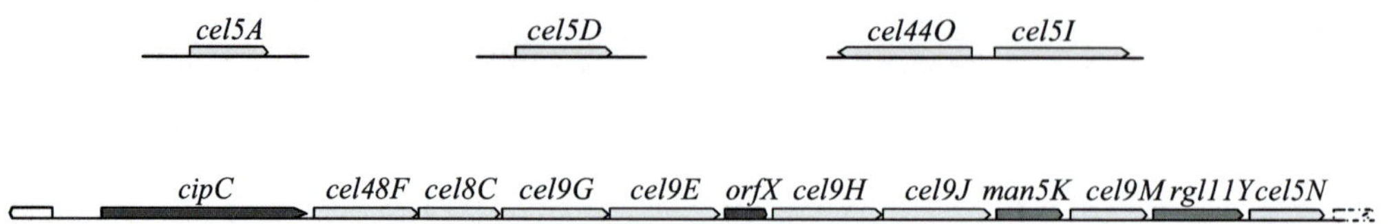

Fig. 1 : Genetic organization of the known cellulosomal genes. Cellulase genes are indicated by light gray arrows, and hemicellulase and pectinase genes by medium grey arrows. The genes encoding the scaffoldin and the other cohesin-containing protein are indicated in dark gray. White genes encode proteins not involved in cellulolysis. The number included into the gene name indicates the GH family of the encoded catalytic domain.

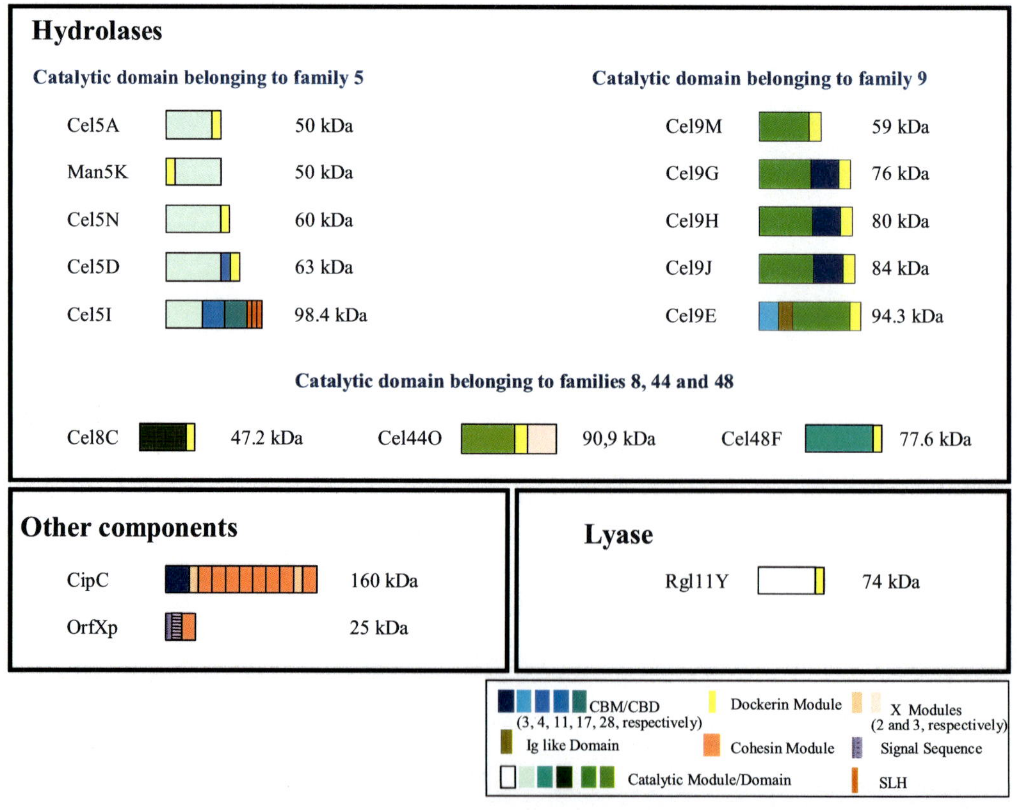

Fig. 2 : The modular structure of cellulosomal subunits and other components of the *C. cellulolyticum* cellulolytic/hemicellulolytic/pectinolytic system. The signal sequences were not shown, except for OrfXp.

As shown in Fig. 2, *C. cellulolyticum* cellulases and other enzymes associated to its cellulolytic system are multi-domain proteins. Among these domains, some are structurally and functionally independent modules, some are functionally linked to one or two other domains. Cellulosomal enzymes share a common module: the dockerin module, which is usually found at the C-terminus of the protein. Man5K, and Cel44O are the only two enzymes, in which the dockerin module has not been found at the C-terminus. In Cel48F, Cel5A, Cel8C, Man5K and Cel5N, only two domains were found: one catalytic module, composed of only one GH domain (in green), and one dockerin module (in yellow). Pagès *et al* [60] demonstrated that the dockerin module is involved in the formation of the cellulosomes by anchoring cellulases onto cohesin modules (in orange in the figure) of the CipC protein termed the "scaffolding protein". Several cellulosomal enzymes contain one or two additional domains linked to the GH domain. The majority of these domains are classified as members of Carbohydrate Binding Module (CBM) families. Some cellulases seem to be unique in terms of domain composition. It is the case of the major protein Cel48F, the only cellulase with a catalytic domain/module belonging to family 48 of glycosyl hydrolases, or of Cel9E, the second major cellulosomal protein in which the family 9 catalytic domain is associated to one Ig-like domain and one family 4 cellulose binding domain. Usually, all known bacterial cellulosomes contain only one type of GH48-containing enzyme, which always constitutes the major enzymatic component. Conversely, we can notice that *C. cellulolyticum* synthesizes three very homologous cellulases, which are composed of one family 9 catalytic domain, one cellulose binding domain belonging to family CBM_{3c} and one dockerin domain: the cellulases Cel9G, Cel9H and Cel9J. This redundancy may constitute another type of major component. As shown on Fig. 2, *C. cellulolyticum* also produces several enzymes containing a family 5 catalytic domain. Nevertheless, their substrate specificities can be quite different. The catalytic modules of Cel5A and Man5K only contain one family 5 catalytic domain; the first one is a cellulase and the second one a mannanase.

Some of these different cellulases: Cel5A, Cel5D, Cel8C, Cel48F, Cel9G (representative of the group Cel9G, H and J), Cel9E and Cel9M, have been over-produced, purified and characterized. Their activities on different substrates are given in Table 1. Neither Cel44O, nor Cel5N have been studied yet. However Cel44O is very similar to the C-terminal part of the *Clostridium thermocellum* Cel44J enzyme, which was characterized as a cellulase displaying some activity with xylan [2]. We can thus suppose that Cel44O is a cellulase.

The four endocellulases Cel5A, Cel5D, Cel8C and Cel9M have similar substrate hydrolysis spectra. They are very active on different kind of amorphous celluloses, CarboxyMethyl-Cellulose (CMC) and Phosphoric Acid-Swollen-cellulose (PAS-cellulose) [8,9,19,23,25,63,77], show little activity on xylan, except for Cel9M, and are unable to hydrolyze β-1,3 links. Nevertheless, they exhibit quite different catalytic parameters. Cel8C and Cel9M, as all family 8 and 9 enzymes, act by configuration inversion, whereas Cel5A, as all family 5 enzymes, acts by configuration retention. On the other hand, Cel5A is sensitive to cellobiose inhibition (K_i = 11 mM) whereas Cel8C is not [23,25]. The major product of crystalline cellulose hydrolysis by Cel9M is cellotetraose, and the enzyme was shown to be unable to hydrolyze cellotriose and cellotetraose, whereas the cellodextrinase activities of Cel5A and Cel8C were measurable even on cellotriose. Cel5D is composed of a GH5 catalytic domain, as Cel5A, but which is linked to a CBD belonging to the family CBM_{11}. This latter enzyme was not extensively characterized, and we do not know yet if any peculiar property is carried by the CBM_{11} domain [81]. From a structural point of view, these four endocellulases belong to three different GH families and thus display different 3-D topologies. The three dimensional structure of Cel5A and Cel9M have been determined by X-ray crystallography [19,63]. Cel5A has the architecture of the classic eightfold α/β barrel motif, whereas Cel9M diplays an (α/α)6-barrel topology (Fig. 3). The enzymatic site was found, in both cases, to be located into a cleft. The data, obtained from the cellobiose-Cel9M complex 3-D structure and from the analysis of the topology of the cleft, suggested that the enzymatic site displays 6 glucose binding sites corresponding to the subsites -4 to +2. Two flexible loops block both extremities of the cleft. They might be involved in the binding of a long cellulose chain during the endo attack.

Table 1: Activities of the *Clostridium cellulolyticum* cellulases on various substrates

Substrate	Activity (IU/µM) [(1)]						
	Cel5A[(2)]	Cel5D[(3)]	Cel8C[(2)]	Cel48F[(4)]	Cel9E[(2)]	Cel9G[(4)]	Cel9M[(4)]
CMC	2500	800-2800	2160	ND	13.5	1300	2500
Avicel	5.4	ND	0.8	13.4	5.8	5.2	2.5
PAS-cellulose	100	-	336	42.5	71.5*	38	57
BMCC	-	-	-	3.9	5.5	10.3	2.0
Xylan	280	+	6.2	ND	2.3	ND	-
Laminarin	ND	-	ND	ND	ND	ND	-
Lichenan	2090	-	528	ND	5.8	140[(2)]	-
pNP-cellobiose	1.5	-	ND	ND	34.8*	ND	ND

ND : not detected, - : not determined, + indicates that some activity has been detected but has not ben evaluated.
CMC : carboxymethyl-cellulose, PAS-cellulose : phosphoric acid swollen Avicel cellulose, BMCC : bacterial micro-crystalline cellulose,
pNP-cellobiose : paranitrophenyl-cellobiose. In laminarin, glucose monomers are linked by β-1,3 bonds, and in lichenan, by β-1,3 or β-1,4 bonds.

Cel48F and Cel9E look like Processive Cellulases, and Cel9G is an Unusual Endocellulase. These last three enzymes are significantly different and each of them presents several specific features, that were deduced from various studies: substrate specificities, sugar released during cellulose hydrolysis, structural studies, mutant protein characterization.

Cel48F was characterized as a processive endocellulase, which initially cuts the cellulose chain as an endocellulase and then go on using a processive mode of action [74,75]. This enzyme is active on the amorphous PAS-cellulose, but not on CMC (Table I), the carboxymethyl substituants fitting not very well in the active site which is partially closed by loops (Fig. 4). The active site is composed of a 25 Å long closed tunnel section, that is followed by an open cleft region, and carries ten sub-sites. The tunnel enables the enzyme to hold the substrate during the processive part of the mechanism [64,65,76]. Even though no crystal of the Cel48F catalytic module alone was obtained, and thus the structure of the catalytic site devoid of sugar was not observed, it is expected that the tunnel would initially be opened, like a cleft, to fix a cellulose chain issued from an amorphous region. After the initial binding of cellulose, the tunnel is closed and the enzyme proceeds to the first endo-cut and then processively produces cellobiose units from the reducing end of the chain. This enzyme hydrolyses rather well crystalline cellulose, such as Avicel and BMCC but is not active towards cellotriose G3 or pNP-cellobiose (Table 1). During the processive action, the cellulose substrate has to slide through the tunnel to continuously supply the leaving group site with sugar residues after the catalytic cleavage. The study of several complexes of the wild-type catalytic module and the inactive E55Q mutant with various natural oligosaccharides and thio-analogues showed two alternative substrate binding positions in the tunnel part of the active site (Fig. 4). These two positions, corresponding to a shift in the chain direction of about half a sugar unit, have different patterns of stacking interactions with aromatic residues present in the tunnel. These variable interactions might be of major importance to reduce the energy barrier during the sliding of the substrate chain through the tunnel in the processive action and to stabilize the substrate at the productive binding site. Multiple patterns are not observed in nonprocessive endocellulases, where only one sugar position is favored by aromatic stacking.

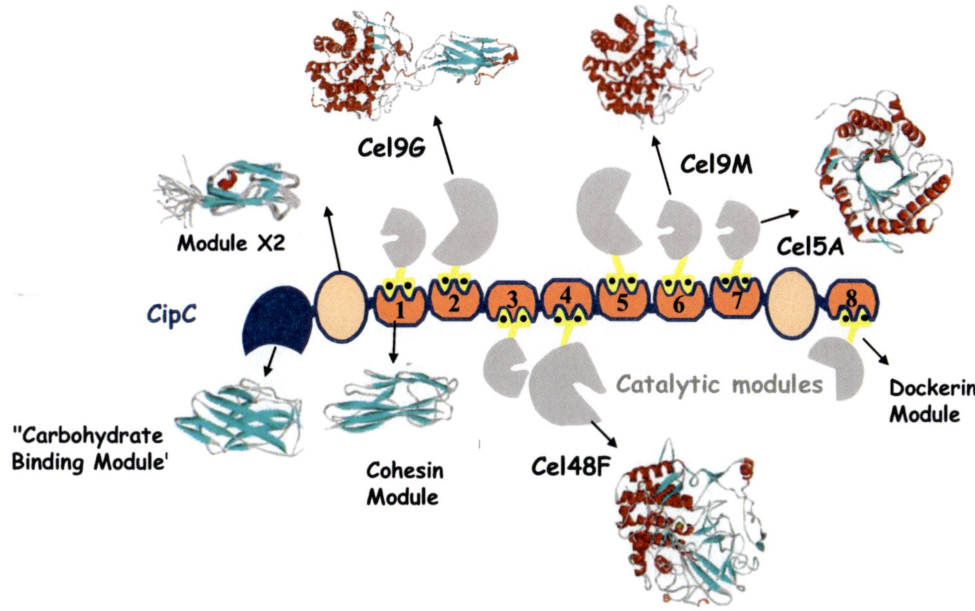

Fig. 3: The cellulosome scheme and structural units. Catalytic modules of the enzymes are in gray. Dockerin modules are in yellow; calcium ions are indicated by black points. The cohesin modules are numbered and colored in orange. Structures are from the PDB.

As Cel9M, Cel9G contains a catalytic domain belonging to family 9 of glycosyl hydrolases (GH9), but this domain is attached to a cellulose-binding domain that belongs to the family CBM_{3c} [5,28] having a nine-stranded antiparallel sandwich fold forming a jelly roll motif (Fig. 3). The two domains are connected by a 18-residue linker which forms many hydrogen bonds with these domains and helps to rigidify the position of one domain relative to the other. The GH9 domain alone is not active at all; the catalytic module is thus formed of two domains: the GH9 domain linked to the CBD. The two domains together form a extended flat platform onto which crystalline cellulose is suggested to bind and be fed into the active-site cleft for endolytic hydrolysis [51]. Cel9G hydrolyses efficiently CMC. Nevertheless, of the different types of cellulases characterized up to now in *C. cellulolyticum*, Cel9G displays the highest activity toward bacterial microcrystalline cellulose (Table I). It is thus tempting to propose that after cellulosomal attachment to crystalline cellulose, Cel9G would participate to the initial disruption of the tertiary structure of cellulose and subsequent endolytic attack on the liberated chains. Indeed, as also suggested by Tormo et al [92], the three-dimensional structure of family-3 CBM is ideal for disrupting the intrinsic H-bonding network of cellulose. This would then open the way for the synergistic attack by accompanying cellulases on the cellulosome. Nevertheless, when looking at the distribution of homologous residues among the different family-3 CBMs, including those found in the cellulosomal scaffoldin, the highest concentration of conserved residues in Cel9G is along the planar strip of residues carpeting the β-strands located in-line with the catalytic cleft of the GH9 domain. On the other hand, the binding capacity to Avicel displayed by the family 3c CBD of Cel9G is weaker than that of the CBM_{3a} of CipC [28]. It is thus more likely that the main role of CBD_{3c} would be to guide the chain into the catalytic cleft. Unfortunately, this hypothesis could not be assessed by any structural study. Long cello-oligosaccharides are not soluble and an oligosaccharide with a degree of polymerization (DP) of 15 would be necessary to form an enzyme substrate complex spanning the GH9 and CBD. A DP-10 synthetic thio-oligosaccahride was used but it did not lead to visualize any interaction on the CBD part of Cel9G. The fact that Cel9G may display processive mode of action was not ruled out.

The activities of Cel9E resemble those found for Cel48F. Cel9E however, among all the enzymes tested, shows a rather high activity on pNP-cellobiose (Table I) and the main product of its activity on Avicel is almost exclusively cellobiose. As a result, it has been characterized as a cellobiohydrolase [30]. As Cel9G, Cel9E contains more than 2 domains: a domain classified by sequence homology as one cellulose binding domain belonging to the family CBM_4 is associated to an Ig-like domain at the N-terminus of Cel9E (Fig. 2). As for Cel9G, the deletion of the CBD induces a total loss of activity, which shows that these CBDs are not independent modules but rather a part of the enzymatic site. This hypothesis was supported in the case of Cel9G by the results of structural studies.

Rgl11Y and Man5A, two non cellulase enzymes are encoded by genes which were found in the second part of the "*cel*" cluster. These enzymes are dockerin-borne proteins that were shown to be incorporated into the cellulosomes [62,69].

Rgl11Y is an endo-acting rhamnogalacturonan lyase which presents extensive sequence similarity with family 11 polysaccharide lyases of the carbohydrate-active enzyme database (http://afmb.cnrs-mrs.fr/CAZY). Calcium is required for its activity. Rgl11Y cleaves the α-L-Rha*p*-(1→4)-α-D-Gal*p*A glycosidic bond in the backbone of rhamnogalacturonan I (RGI) via a β-elimination mechanism. It requires galactan decoration of the RGI backbone.

Man5K is the only known *C. cellulolyticum* enzyme bearing a N-terminal dockerin domain. The enzyme is 20% more active with Locust Bean Gum (LBG) than with guar gum. Galactomannans are composed of a linear homopolymer of mannose linked by β-1,4 bonds, substituted with α-1,6-galactose side chains. In LBG and guar gum, the mannose/galactose ratios are 4:1 and 2:1, respectively. The study of the activity and the substrate preference profile of Man5K thus indicate that the enzyme is an endomannanase, active on main chains of galactomannans exhibiting a low degree of substitution with galactose side chains.

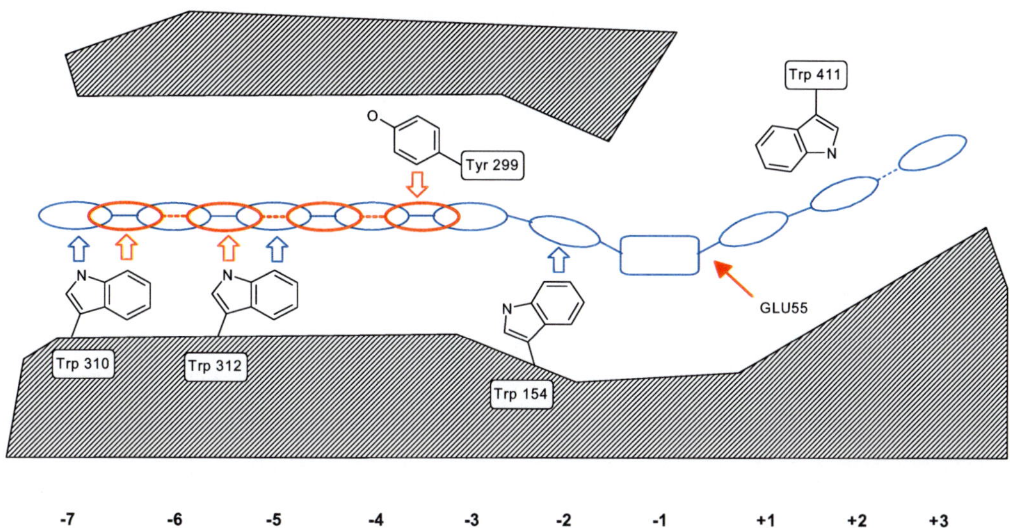

Fig. 4 : Scheme of the active site of the processive endocellulase Cel48F. The upper part of the tunnel (hatched) is formed by two pairs of loops. The position of the thio-oligosaccharide IG4 inhibitor (methyl 4-S-β-cellobiosyl-4-thio-β-cellobioside) was determined from the 3D structure analysis of complexes with the wild-type protein (red broken line). The natural oligosaccharides cellotetraose and cellohexaose co-crystallized with the catalytic mutant Cel48F E55Q show different interactions in the tunnel part of the active center (normal blue line). The possible aromatic stacking interactions between sugar rings and amino acid lateral chains are indicated by wide arrows. The position of the cleavage site is marked by a filled red arrow. Glu55 is the proton donor of the reaction ; it protonates the glycosidic oxygen, while an activated water molecule performs a nucleophilic attack on the anomeric carbon. Subsites are numbered (adapted from Parsiegla *et al* [63]).

When determined, the pH and temperature optima of all these enzymes were found to be around 6-7 and 50°C, respectively [23,25,81], except in the case of Rgl11Y (pH 8.5). At the optimum pH of most of the *C. cellulolyticum* cellulases (pH 6 to 7), Rgl11Y displays only 4 to 13% of its maximum activity. It has been suggested that the pH value of some plant tissues changes during microbial attack. This might indicate that the degradation of plant cell wall polysaccharides occurs sequentially according to the pH of the plant tissues.

Purified dockerin-containing proteins from *E. coli* overproducing strains are quite unstable and loose rapidly their dockerin module. The effect of this loss on the catalytic properties vary with the enzyme considered, but is never null. The most probable explanation is that the absence of the dockerin module might induce small conformational changes in the catalytic core which are responsible for the catalytic differences observed between the two forms of an enzyme.

The enzymatic and structural studies of some of the numerous cellulases encountered in the cellulosomes of *C. cellulolyticum* enlightened their diversity. The two processive Cel48F and Cel9E are the two major enzymatic components of the cellulosomes; these two enzymes exhibited quite different characteristics and might be involved in the degradation of cellulose in different ways. The various endocellulases studied also exhibited different characteristics, in particular when considering the degradation of cellodextrins. Among the endocellulases, Cel9G appeared as a special enzyme, regarding its efficiency in the crystalline cellulose degradation, probably due to the presence of the CBM_{3c} domain, which extends the active site cleft. Cel9G is not one of the major enzymes of the cellulosomes; however, two other proteins very similar to Cel9G are also synthesized by the bacterium: Cel9H (55% similar) is a hypothetical cellulosomal protein, and Cel9J (46% similar) has been identified as a cellulosomal protein [29]. It is thus reasonable to hypothesize that they act on crystalline cellulose as Cel9G does. Two of the corresponding genes are probably issued from duplication events. To date, 81 microbial cellulases containing a GH9 domain have been identified (http://afmb.cnrs-mrs.fr/CAZY). The major part of them includes additional domains, especially CBD which are required for enzymatic activity, as in the case of Cel9G and Cel9E. Though this study is far from being complete, the data collected to date show a large diversity of the cellulosomal enzymes. It is worth noting however that all these enzymes exhibit rather low activities on crystalline cellulosic substrates, suggesting that the efficiency of cellulosomes results from the activities of all the enzymes acting at the same time on the substrate. Nevertheless, the large number of different enzymes remains intriguing. Why are so many enzymatic proteins necessary to hydrolyze a so chemically simple substrate? It is possible, as suggested by the existence of the very similar Cel9G, Cel9J, Cel9H proteins, but also by the existence of many endocellulases, that several enzymes are redundant. It would thus be possible to construct simplified cellulosomes, as active as the wild-type ones, which would contain the "essential" enzymes of the system. The second major question raised by this multi-enzymatic system is the understanding of the cooperation of the different enzymes to achieve efficient cellulose degradation.

Synergism studies of couples of free enzymes provided some information. For example, Cel9E was found to exhibit strong synergism with several enzymes. Synergism appears to be more important between the cellobiohydrolase Cel9E and endocellulases such as Cel5A, Cel8C or Cel9M, especially on BMCC, the most crystalline substrate tested, but also appears to be very significant with the processive enzyme Cel48F and with Cel9G [9,30]. More interesting and realistic are the informations gathered from the study of different minicellulosomes designed to selectively incorporate specific enzymes into precise positions within the complexes. The cooperation of pairs of enzymes assembled on mini-scaffoldin containing or not a CBM was studied. Most ternary complexes were found to be more active than the corresponding free enzyme pairs on Avicel cellulose, even those including a mini-scaffoldin that do not contain any CBM. In most cases, bringing two different cellulases together in a close proximity is thus enough to trigger enhanced efficiency, especially on most crystalline and recalcitrant celluloses. Nevertheless, the presence of a CBM_{3a} module in the scaffoldin significantly increases the activity of the complexes, suggesting that the increase of activity observed when enzyme pairs are bound to a scaffoldin is due to both the proximity of the catalytic domains in the complex and the presence of the family 3a CBM. Among the enzymes tested, the pairs associating a processive enzyme to an endocellulase exhibited the highest specific activities on crystalline cellulose, especially Cel48F-Cel9G

[24]. These studies are largely described in a separate chapter entitled "Cellulosome chimeras, a powerful tool to investigate the functioning of cellulosomes with potential biotechnological applications".

3. THE CELLULOSOME SCAFFOLDIN

At the 5' end of the cluster is located the *cipC* gene, that encodes the cellulosome scaffolding protein (Fig. 1). The sequencing of approximately 3 kb upstream of the *cipC* gene revealed that it was the first gene of the "*cel*" cluster [46]. *C. cellulolyticum* secretes a glycosylated scaffoldin [29]. Its organization resembles that of *Clostridium cellulovorans*. Its N-terminal part is composed of a CBM_{3a} module followed by a X_2 module of unknown function and then of several hydrophobic modules, named cohesin modules (Fig. 2 and 3) [59,84]. Its C-terminal region contains a second X_2 module followed by the eighth cohesin domain, but does not contain any dockerin module, which might be involved in the cell surface anchorage, as in CipA, the scaffoldin of *C. thermocellum* cellulosomes [32]. The three types of modules encountered in CipC have been structurally studied (Fig. 3). The 160 amino-acid CipC CBM_{3a} is very similar to the CBM_{3a} from the CipA scaffoldin of *C. thermocellum* [83]. The structure reveals a nine-stranded jelly-roll topology with a calcium-binding site, a putative cellulose binding surface on a flat bottom face, and a shallow groove of unknown function on the top face, which also harbors the N and C-termini. The overall dimensions of the module are 48 Å X 28 Å X 23 Å. The NMR solution structure of the first X_2 module revealed an immunoglobulin-like fold with two β-sheets packed against each other and a $3_{.10}$ helix intercalated between them; its overall molecular dimensions are approximately 35 Å x 20 Å x 16 Å (Fig. 3). The structural study revealed that this 100 residue module has a surprising conformational stability. It is also highly soluble and might act as a conformational linker and solubility enhancer within the scaffoldin [57]. Indeed, the miniCipC_0 which contains the CBM_{3a} and the $X_{2\text{-}1}$ module does not precipitate after three hours at 50°C, while 50% of the CBM alone precipitate in the same conditions. The 144 residue cohesin 1 has an elongated shape with overall dimensions of 50 Å X 31 Å X 26 Å. Its topology is a β-barrel, with a jelly roll topology [87]. The 160,000 Da CipC scaffoldin is thus composed of a series of linked oblong modules as a pearl necklace. In order to measure the interactions between a cellulase and CipC, various mini-scaffoldins were constructed and over-produced in *E. coli*: miniCipC_0 which only contains the CBM_{3a} and the first X_2 module, and miniCipC1 which contains the first cohesin module in addition to the modules found in miniCipC_0. On the other hand, the Cel5A enzyme was used under three forms to specify the domain(s) involved in the binding with the scaffoldin. Indeed, proteolytic cleavage of the C-terminal end of Cel5A occurs during storage and leads to the mix of the entire protein, a derivative depleted of the C-terminal moiety of the dockerin domain, and a derivative depleted of the entire dockerin domain. Far Western blotting tests using the miniCipC proteins as probes and the various forms of Cel5A in the blot, and the reverse experiments showed that the entire dockerin domain of Cel5A is necessary to anchor the enzyme to the cohesin domain of the miniCipC protein [60]. Furthermore, it was shown that several other purified dockerin-bearing enzymes (Cel8C, Cel48F, Rgl11Y) can interact with the cohesin

C1 [60,62,75]. These high affinity interactions are influenced neither by the catalytic domain attached to the dockerin domain, nor by the noncohesin domains of CipC. The free cohesin1 from CipC and dockerin from Cel5A have the same K_D (2.5 x 10^{-10} M) as that of the entire Cel5A and a larger fragment of CipC (miniCipC_1) [27]. Dockerin domains are composed of two conserved segments of 22 residues: removal of the second segment abolishes the affinity for cohesin1, whereas modified dockerins harboring twice the first segment, twice the second, or both segments but in a reverse order have K_D values for cohesin1 in the same range as that observed for wild-type dockerin. These data indicate that if two segments are required for the complexation with the cohesin, segments 1 and 2 are similar enough to replace each other. The interaction was shown to be calcium dependent. Calcium overlay experiments revealed that the dockerin domain has one calcium binding site per conserved segment. Circular dichroism performed on wild-type and mutant dockerins indicates that this domain is well structured and that removal of calcium only weakly affects the secondary structure, which remains 40-45% helical [27].

Recombinant DNA manipulation offered the possibility of designing truncated proteins, and it was especially useful in case of the scaffolding protein organized as a sequence of eleven modules of three types. Cohesins 1 and 8 only share 60% of identity [59] There are the most divergent cohesins of the scaffoldin CipC. The chimeric MiniCipC_8 which contains the eighth cohesin in addition to miniCipC_0 has been constructed to evaluate the affinity of Cel5A for C8 in comparison with C1. Cel5A was found to bind to these two cohesins with similar affinity (Ka = 4.8 x 10^9 M^{-1})[59]. The complexes appeared to assemble via specific interactions between enzymic dockerin domains and scaffoldin cohesin domains, without any strict association of the different enzymes with one specific cohesin. The role of the cohesin domain as a receptor for the catalytic subunit has been clearly established.

The role of the two X_2 modules remains unclear. These hydrophilic modules might act as conformational linkers. Hydrophylic modules might also enhance the solubility of scaffoldin, that contain several hydrophobic modules (the cohesins). Doi *et al* suggested recently that the X_2 modules might play a role in binding the cellulosomes to the cell surface [18]. But this hypothesis was not supported by any experimental data, except the observed tight association of the cells and the cellulose substrate during growth, supposed to be mediated by the cellulosomes. The role of the CBM_{3a} of CipC in cellulose degradation was investigated by studying the synergistic action of this module with Cel5A [61]. The CBM_{3a} diplays high affinity towards insoluble cellulose; the K_D vary from 10^{-7} to 3 x 10^{-8} M, depending on the cellulose used. The activity of the dockerin-borne Cel5A associated to the miniCipC_1 was found to be slightly enhanced on insoluble cellulose, when compared to the activity of the enzyme alone. The synergism observed might result from the Cel5A concentration on the substrate due to its interaction with miniCipC_1. More interesting, is the two-fold and four-fold increase of activity obtained with the dockerin-lacking enzyme acting on Avicel and PAS-cellulose, respectively, when an excess of MiniCipC_1 (10 μM in a 1% cellulose suspension) was used (the enzymatic reaction was monitored by using 0.3 μM and 0.02 μM Cel5A for Avicel and PAS-cellulose degradation, respectively). This synergism can only result from a miniCipC_1 cellulose disrupting action, which may lead to the exposure of new sites that can be hydrolyzed by the enzyme. This hypothesis was confirmed by studying the effects of the

CBM_{3a} module of CipC grafting onto the catalytic module of Cel5A; the chimeric enzyme was found more active on all the insoluble substrates tested, especially PAS-cellulose.

A second gene encoding a cohesin-containing protein is located in the "*cel*" cluster (Fig. 1). Based on its sequence analysis, the 229 aa polypeptide of OrfXp can be divided into three distinct domains. The first domain, of about 25 aa, located at the N-terminus, resembles a signal sequence. It is followed by a Proline-Threonine-Serine rich region of 58 aa, and then by a cohesin domain (Fig. 2). This cohesin (named cohesin X) is only 35 and 28% identical to cohesin 1 and cohesin 8 from CipC, respectively. It was shown to interact with the cellulosomal dockerin-bearing enzymes. Nevertheless, the K_a was found, by using the BIAcore, 19-fold lower for cohesin X with Cel5A, compared to that for the miniCipC_1-Cel5A complex. Interestingly, the drop in K_a was almost exclusively due to the dissociation rate constant (k_{off}), the association rate constant (k_{on}) being almost unchanged. This indicates that the dissociation of the cohesin X-dockerin complex is much faster. OrfXp is a cell-associated glycosylated protein only found to be produced when cellulose is present in the medium. It might be involved in the cellulosome assembly, as a shuttle carrying the free cellulases to CipC, such as it has been proposed for the protein OlpA in *C. thermocellum* [78], but this role was never experimentally demonstrated for OrfXp.

4. Cellulosomes

Cellulosomes were easily purified from the residual cellulose of cultures by using water to eluate the adsorbed complexes. At least 14 bands were resolved on 1D SDS-PAGE of purified cellulosomes. The molecular masses of the corresponding polypeptides are ranging from 160 to 35 kDa [29]. The 160, 94 and 80.6 kDa fractions are the most abundant and correspond to the scaffolding protein CipC, and the cellulases Cel9E and Cel48F, respectively [29,75]. Complexes exhibit two major properties, the first one is their ability to hydrolyze different kind of celluloses: the totally amorphous and soluble CMC, partially amorphous cellulose PAS-cellulose, as well as highly crystalline celluloses like Avicel, but also other plant cell wall polymers, such as xylan or mannan [29]. The second property is the binding ability to cellulose which is due to the presence of the CBM_{3a}, located at the N-terminus of the largest and major subunit of the complexes: the scaffolding protein CipC. This latter protein assembles the complexes by the mean of the eight cohesins. The complete dissociation of the complexes into individual components requires harsh treatments, such as elevated temperatures and/or the presence of chaotropic agents. Many attempts to purify functional components from the complexes were unsuccessful [24,26,59,60,75].

Up to the years 2000, molecular analysis of the bacterial cellulosome has thus essentially been based on studies of recombinant proteins [7,17]. Functional studies "*in Clostridio*" began with the development of suitable laboratory protocols for effecting gene transfer. One first interesting genetic tool was transferred by conjugation: the broad host-range streptococcal transposon Tn*1545*. Conjugative transposons are indeed useful tools for insertional mutagenesis and mutational cloning [100]. Nevertheless, the yield of the *C. cellulolyticum* transconjugants obtained was low. Electro-transformation proved to be a more reliable method than conjugation for obtaining gene transfer, but only after the incoming

DNA was protected from restriction. *C. cellulolyticum* protects its own DNA from restriction by methylation, and donor DNA that has been methylated on the external C of the sequence 5'-CCGG-3'using either M.*BsuF*I *in vivo* or M.*Msp*I *in vitro* gives reproducible electro-transformation [42] with high apparent efficiency, when the transformants are selected in liquid media [90]. Several plasmids that are based on different replicons (pAMβ1, pIM13, pCB102, pIP404 and pWV01) exhibiting different copy numbers were introduced in *C. cellulolyticum* [42,90]. Plasmids that have been transferred to an organism in which they do not normally reside are frequently unstable [20,35,53]. In *C. cellulolyticum* however, both the pAMβ1 (tetha-replication) and the pIM13 (rolling circle, ssDNA plasmid) replicons were found moderately stable in the absence of selection (frequency of plasmid loss of about 5 X 10^{-3} per generation). The pCB102 replicon was shown to be rather less stable (4 X 10^{-2} per generation). Heterologous plasmids, as pJIR418 seem to be also structurally stable in this strain [90]. Gene transfer and vector disposal paved the way for functional analysis of the various cellulosome components *in vivo*. The *in vivo* modification of the cellulosome composition was obtained by genetic manipulation of the wild-type strain. Two different approaches were used. The first one consisted in mutant selection, and the second one in the construction of recombinant strains with an extra-gene carried by a plasmid.

The first cellulolysis mutants were by serendipity isolated. A *cipC* fragment interrupted by a 2.6 kb sequence was identified by screening a genomic DNA bank using a *cipC* probe. From the strain used to purify the genomic DNA, was isolated 2 series of *cipC* insertional mutants, only differing by the nature of the insertion sequence [47]. The foreign sequence was located in both cases in the same place, at the beginning of the sequence encoding cohesin number 7. Sequence analysis showed that the sequence inserted was either composed of one insertion sequence (named ISCce*1*) or the combination of two different endogenous insertion sequences (ISCce*2* inserted into ISCce*1*). In liquid culture, the growth of the mutant strains was not affected with cellobiose as substrate, but the strains were found to be unable to efficiently degrade crystalline cellulose and exhibited a poor growth on this substrate when compared to the wild-type strain [48]. Extensive molecular studies were performed on one mutant strain (*cipC*Mut1) [48]. The SDS-PAGE analysis of the proteins which were adsorbed on residual cellulose showed a very different pattern in comparison to the wild-type strain. Western blot analysis using specific antibodies revealed the presence of small amounts of a truncated form of CipC, but none of the enzymes encoded by the genes that follow *cipC* in the cluster were detected. The mutant enzymatic system bound to cellulose was found to be mainly deficient in micro-crystalline cellulose degradation, as it only retained 20% of the Avicel degradation activity of the wild-type enzyme system. The activities on amorphous cellulosic substrates and hemicellulose were less affected. The residual cellulolytic activity indicated that some of the proteins adsorbed on cellulose were cellulases, even if we did not detect any of the major cellulosomal components as Cel48F and Cel9E, etc... Molecular filtration analysis of the fractions revealed the absence of typical cellulosome complexes in mutant strain cultures (Fig. 5B). The major part of the proteins were of lower molecular weight than the complexes and might correspond to minor components encountered in the WT fraction. The poor cellulolytic activity of the mutant strain was not due to a low amount of truncated CipC, as the overproduction of CipC in a *trans*-complemented strain (*cipC*Mut1(pSOS*cipC*)) did not restore the wild-type cellulolytic phenotype (Fig. 5A). The

Cel⁻ phenotype was therefore mainly due to the absence of many cellulosomal enzymes, including the major processive cellulase Cel48F. The molecular filtration analysis of the complemented system suggested the formation of CipC-based complexes (Fig. 5B). This was confirmed by analysing the cellulolytic systems purified from the various strains (wild type, mutant, and complemented mutant strains) by native gel electrophoresis: complexes were assembled in the complemented strain. These complexes, of lower average molecular mass than the wild-type cellulosomes, were found to be also different in their composition. Western blotting experiment after SDS-PAGE using specific antibodies and a biotinylated mini-CipC showed that the complexes produced in the mutant and complemented strains contained at least a dozen of different dockerin-harboring proteins encoded by genes located outside of the "*cel*" cluster, such as Cel5A (Fig. 5C). Despite the presence of larger amount of cellulosome-like complexes compared to the mutant strain, their amount in the cellulose-adsorbed fraction remain lower in the *cipC*-complemented strain than in the wild-type. Low amounts of complexes and the absence of several enzymatic components thus appeared to greatly impair the degradation of crystalline cellulose. This study provided genetic evidence that the cellulases encoded by the genes located in the "*cel*" cluster are essential for the building of efficient cellulosomes. It revealed also that the genes of the "*cel*" cluster are transcriptionally linked. The absence of the proteins encoded by the genes located downstream of *cipC* was due to a strong polar effect. We do not know yet if very large mRNAs covering the major part of the cluster are synthesized. Such RNA are very unstable and difficult to visualize by Northern blotting; a 7.5 kb-long mRNA, large enough to carry the *cipC* ORF and the *cel48F* ORF, was detected with a *cipC* probe, supporting the hypothesis of the synthesis of large polycistronic mRNAs.

In such transcriptionally linked genes, the use of antisense technology is a suitable mean for reducing gene expression without any strong effect on adjacent genes. The effect of the over-produced antisense RNA (asRNA) directed to the translation initiation site of *cel48F* mRNA on the Cel48F synthesis and on the activities of the cellulosomes was studied [71]. The cellulolytic system secreted by the asRNA producing strain showed a three to four times decrease of Cel48F content compared to the control strain containing the empty plasmid. Conversely, the ratio Cel9E/CipC did not change. The decrease of the Cel48F amount was correlated with a 30% decrease of the activity of the cellulolytic system on Avicel cellulose. These results suggest that Cel48F plays an important role in crystalline cellulose degradation. This role has also been enlightened by *in vitro* enzymatic studies of minicellulosomes. The enzyme pairs including a processive cellulase (Cel48F or Cel9E) and Cel9G were shown to be the most efficient in the aggregated crystalline cellulose degradation.

In *C. cellulolyticum*, thirteen genes encoding cellulosomal catalytic subunits have been discovered so far (Fig. 1). Most of them code for cellulases. The composition of the cellulosomes is likely to vary with the amount of the various dockerin-containing enzymes available when the complexes are assembling. At least 22 dockerin-borne proteins can be produced by *C. cellulolyticum* [48], that all may be incorporated into cellulosomes, though CipC only contains eight cohesin domains. Thus, a whole population of complexes with different enzymatic compositions is concurrently secreted by the bacterium. Nevertheless, it has been shown that the cellulosomes can degrade, in addition to cellulose, other plant cell

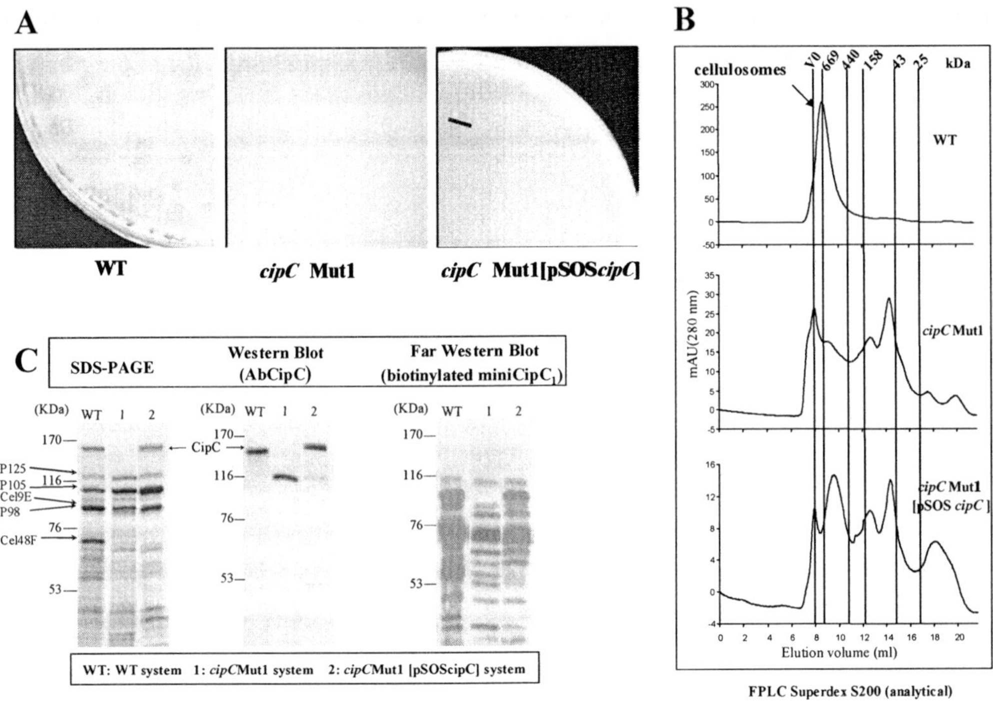

Fig. 5: The cellulolytic systems of the *cipC*Mut1 and *cipC*-complemented strains compared to the wild-type system. *A. Phenotype of C. cellulolyticum wild-type, cipCMut1 and cipCMut1(pSOScipC) on solid medium containing MN300 cellulose.* The solid bar corresponds to 6 mm. *B. Gel filtration FPLC analysis of the cellulolytic system of the various strains.* Vertical lines indicate the elution volume of molecular mass markers. AU: absorbance unit. *C. Composition of the cellulolytic systems.* Anti-CipC antibodies and biotinylated miniCipC$_1$ were used to reveal the scaffoldins in the Western blot and the dockerin-bearing components in the Far western blot experiments, respectively (adapted from Maamar *et al* [48])

wall polysaccharides [29], and that they contain xylanases [54], at least one genuine pectinase [62], and one mannanase Man5K [69]. The latter enzyme is present in low amounts in the cellulosomes which in turn display a low mannanase specific activity. The *man5K* gene was therefore an ideal target for over-expression. With such experiment, one can address the relationships between cellulosome composition and levels of expression of the genes encoding the enzymatic components. Man5K was successfully *trans*-produced using *C. acetobutylicum thiolase* derived promotors; the protein was surprisingly found almost exclusively in the cellulosomes. The major cellulases Cel48F and Cel9E were also found to be located almost exclusively in the complexes. These results highly support the hypothesis that the cohesin–dockerin interaction is not enzyme-specific in *C. cellulolyticum*, as already suggested by the identical affinity of Cel5A for the two most divergent cohesins 1 and 8. The enzymatic composition of cellulosomes appears to be controlled by the relative amounts of the enzymes available to be incorporated into the complexes. In the Man5K over-producing strains, the mannanase units synthesized in high amounts might replace some natural cellulosomal enzymes or interact with unoccupied cohesins. The Man5K enriched cellulosome fraction displayed 20-fold-higher specific activities than control fractions on galactomannan substrates, whereas the specific activity on crystalline cellulose was reduced by 20%. *C. cellulolyticum* is naturally able to adapt the enzyme composition of its cellulosomes in response to growth conditions. Cellulosomes produced by cells growing in xylan-containing medium exhibited a different polypeptide composition when compared to cellulosomes produced by cells growing in cellulose-containing medium. Indeed, the proportions of the high molecular weight enzyme fractions, which contain Cel9E and Cel48F, decrease, whereas the low molecular weight enzyme fractions, which contain the xylanases (54 and 31 kDa), simultaneously increase [54].

5. Non Cellulosomal Proteins

Cel5I is encoded by a gene located outside of the "*cel*" cluster. The protein is composed of a N-terminal GH5 catalytic domain associated to two different domains classified as members of the CBM 17 and 28 families, respectively and three C-terminal S-layer homology (SLH) domains (GenBank access code: AY077754). Furthermore Cel5I lacks a dockerin domain, thus indicating that this protein is probably associated to the cell wall but not to the CipC protein. The assignment of the ^{1}H, ^{13}C and ^{15}N resonances of the 22.5 kDa CBM_{28} module (198 amino acid residues) indicates that as many CBMs which belong to various families, CBM_{28} is mainly organized in β-strands [58].

In addition to this first non cellulosomal component, two unknown proteins, termed P105 and P98 (with apparent molecular mass of 105 and 98 kDa) were found to be major components of the "cellulolytic system" of the *cel48F* asRNA producing strain [71]. These proteins were eluted with cellulosomes from the washed residual cellulose of cultures. They also were found to be major components of the "system" produced by the *cipC* insertional mutant. P105 and P98 were not co-immunoprecipitated with the cellulosomes and failed to interact with biotinylated $MiniCipC_1$ in far Western blot; they could be non cellulosomal components of the "cellulolytic system". They are also found, but in lower amounts in the wild-type cellulolytic system (Fig. 5). The N-terminal sequence of the P105 protein did not match with any known cellulase, whereas the N-terminus of P98 shares 44% identity with the N-terminal region of family six catalytic domains of *Humicola insolens* Cel6A and Avicelase 2 [71,97]. They are expected to participate, in addition to Cel5I, to the non cellulosomal cellulolytic system.

6. Cellulolysis

The primary event in the degradation of cellulose is the tight adhesion of *C. cellulolyticum* to the substrate. Growth on Avicel cellulose occurs when bacteria are stuck to insoluble cellulose; 80% of the biomass is detected on the insoluble substrate [31]. This adhesion is probably mediated by cell-associated cellulosomes and/or cell-attached cellulases. Although the mechanism of the cellulosome-cell association has not been investigated yet, indirect evidence that it is mediated by specific molecular species has been furnished by the study of cellulosomes produced by cells cultured on xylan-containing medium. These cellulosomes were found to be not associated with cells. This suggests that *C. cellulolyticum* is able to regulate the cell association of its cellulosomal complexes, depending on the insoluble substrate used [54].

Carbon metabolism has not been widely investigated in cellulolytic bacteria. Most of the physiological studies were performed on ruminal bacteria [52,79,91]. However, among the cellulolytic clostridia, *C. cellulolyticum* is the only strain, in which the carbohydrate metabolic pathways have been investigated. Carbon flow was investigated in batch and chemostat cultures, (i) by measuring the intracellular metabolic intermediates and the final exocellular fermentation products, (ii) by evaluating the levels of oxidized and reduced nucleotides and of adenylate pool (ATP, ADP and AMP) and, (iii) by assaying some clue enzymes of the metabolic pathways. The strain was shown to better control its catabolic pathway in culture on mineral salt-based medium, than in culture on rich medium [38].

Extensive studies of the fermentation metabolism were performed using cellobiose, which is the major end product of the cellulolysis. In rich medium, *C. cellulolyticum* metabolizes only a small quantity of soluble carbohydrates [33]. Continuous culture studies revealed that, in all tested dilution rates, acetate was the main product of cellobiose metabolism, and lactate and ethanol were minor products. Production of reduced compounds such as ethanol and lactate is not sufficient to re-oxidize all the NADH produced through the Embden-Meyerhof-Parnas pathway. NADH is also regenerated at a slow rate into NAD+ by the NADH-ferredoxin-reductase and hydrogenase pathway (Fig. 6), leading to H_2/CO_2 ratios greater than one. The electron flow from NADH (E'_0 = -320 mV) via ferredoxin to protons (E'_0 = -420 mV) is thermodynamically unfavorable [43] under standard conditions and needs interspecies H_2 transfer to be efficient [56]. *C. cellulolyticum* is not able to cope with carbon overflow by inducing metabolic shift to produce reduced compounds, such as *C. acetobutylicum* [34]. Carbon metabolism stops and thus growth is limited at high carbon flow; this was explained by a saturation of the NADH reoxidation path. NADH-NAD ratios as high as 57 to 1 were evaluated [67]. A NADH-NAD+ ratio higher than 1 was found to significantly inhibit the *in vitro* activity of glyceraldehyde-3-phosphate dehydrogenase (GAPDH), suggesting a possible limitation of the carbon flow as previously described in several other clostridia [34].

In synthetic medium, *C. cellulolyticum* shows a better control of the carbon and electron flow. Major differences in the regulation of the carbon flow between carbon-limited and carbon-sufficient continuous cultures have been reported [38,39]. As the dilution rate increases in cellobiose-limited chemostats, the metabolic pathways towards ethanol and lactate contribute to balance the reducing equivalents supplied by acetate formation [38].

Under cellobiose-saturated conditions, the redox balance is essentially maintained by NADH-ferredoxin-reductase-hydrogenase and ethanol dehydrogenase activities [39], and the carbon flow is equilibrated by overflows, i.e., exopolysaccharide, extracellular protein, amino acid, and pyruvate excretions. In nitrogen-limited conditions, glucose 6-phosphate (G6P) and glucose 1-phosphate (G1P), connected by phosphoglucomutase, constitute important branch points for distributing the carbon fluxes inside and outside the cells (Fig. 6) [37]. Thirty percent of the original cellobiose consumed is converted to cellotriose, glycogen and polysaccharides rather being used for generation of metabolic energy and biosynthetic precursors. In contrast to many bacterial species, *C. cellulolyticum* do not accumulates glycogen only when growth is limited, e.g. by phosphorus, sulfur or nitrogen in the presence of an excess of carbon, but it synthesizes similar quantities of glycogen during growth phases in carbon- and nitrogen-sufficient media, in carbon excess and nitrogen limitation, as well as in carbon limitation and nitrogen excess. Glycogen appears to be simultaneously synthesized and degraded during rapid growth; It might be a mean for limiting the flux of carbon through glycolysis, avoiding the accumulation of intracellular metabolites, such as pyruvate. All these biosynthesis (cellotriose, glycogen and polysaccharides) also dissipate the excess of energy. It appears that pyruvate and acetyl coenzyme A are important branch points for regulating the redox balance of the carbon metabolism, whereas the "G1P-G6P" branch points control the excretion of carbon surplus and the excess of energy dissipation. These last regulations are needed since *C. cellulolyticum* seems to be unable to regulate the rate of its cellobiose consumption.

Bacterial growth on cellulose differs from that on cellobiose by the necessity for bacteria, first to adhere to the substrate, and second, to hydrolyze it into soluble sugars. It is thus expected that cells grow under soluble substrate limitation. In fact, soluble sugars liberated from cellulose are the substrates for growth and are incorporated by cells almost as rapidly as they are formed [13]. This hypothesis was confirmed by similar characteristics observed for the metabolisms in carbon-limited chemostat when cellobiose was used as the substrate and in cellulose-limited continuous culture. Carbon-limited chemostat studies using cellulose as the substrate showed that *C. cellulolyticum* is well adapted to this culture condition; neither a shift from an acetate-ethanol fermentation to a lactate-ethanol fermentation, nor pyruvate overflow at high catabolic rates were observed [13,16]. The true growth yield obtained on cellulose was higher (50.5 g of cells per mol of hexose equivalent) than on cellobiose (36.2 g mol^{-1}), suggesting that the energy-efficient transport system of soluble cellodextrins allows to reach higher biomass on cellulose than on cellobiose, for the same hexose equivalent quantity consumed [88]. In continuous culture, the rate constant (k) of MN301 cellulose degradation was determined using the model equation $S_R/S_0 = k\ t_R + x$ [66], where S_R and S_0 are the concentrations of cellulose in the feed reservoir and in the culture vessel, respectively, and t_R the retention time (1/dilution rate). Cellulose digestion was found to follow first-order kinetics, with a rate constant of cellulose removal of 0.05 h^{-1}. This was similar to what was found for *Ruminococcus albus* (with Avicel cellulose), but 3.4 times lower than that was found for *C. thermocellum* (0.17 h^{-1} with Avicel cellulose), which is the highest one so far determined [13]. Nevertheless, these values have to be considered with precaution, because they might be greatly influenced by the characteristics of the cellulose used and the temperature of the culture. This slow rate of cellodextrin release regulates the entering carbon flow, leading to a lower μ, and so metabolism is not as distorted as observed with soluble substrates. The shift from cellulose-limited to cellulose-sufficient conditions is accompanied by an increase of both the acetate/ethanol ratio and lactate biosynthesis. In cellulose-sufficient continuous culture, the ATP/ADP and NADH/NAD ratios were higher. A part of the catabolized sugar and hence ATP are no longer associated with biomass production. Lactate and pyruvate leakage rise. Nevertheless, the strain seems to be able to regulate and optimize carbon metabolism under cellulose-saturated conditions [15]. The overflows are more limited under cellulose-excess conditions, when compared to cellobiose-excess conditions. Such differences were related to the carbon consumption rate, which is lower on cellulose than on cellobiose.

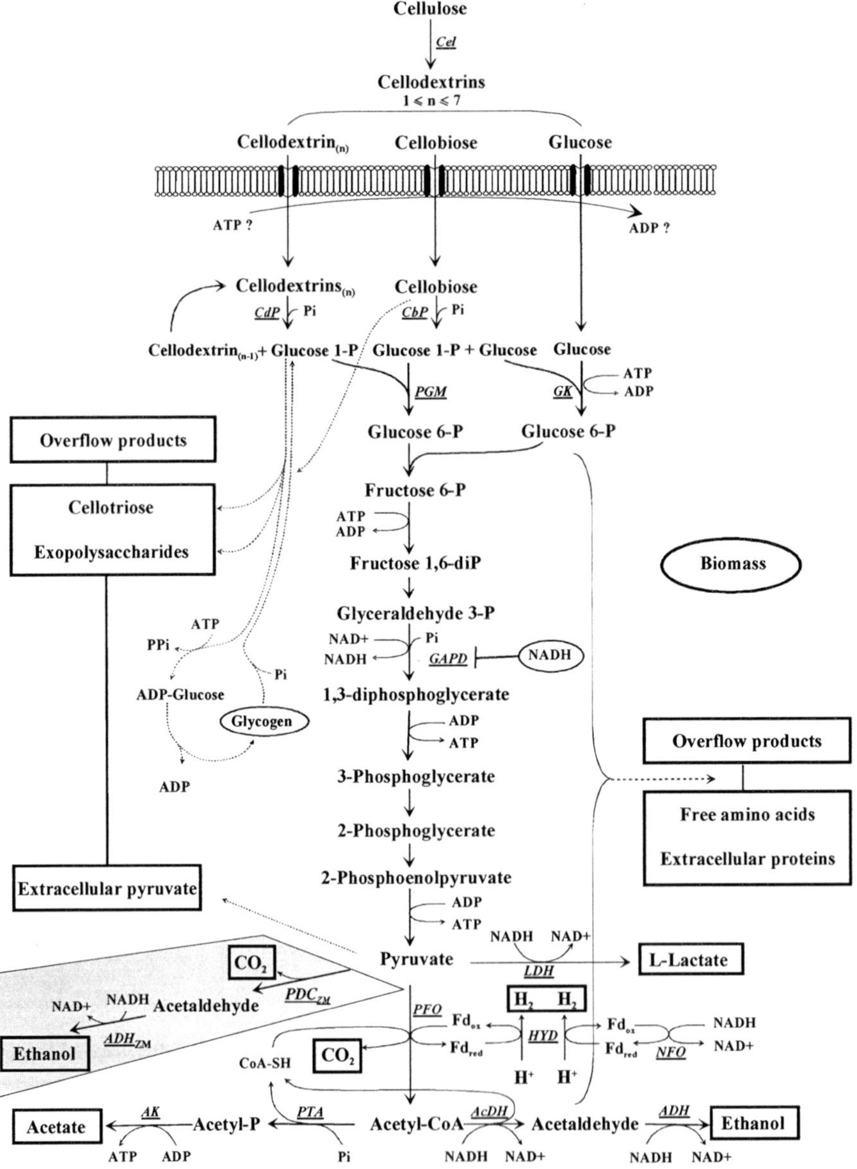

Fig. 6: Carbon and electron flows within the metabolic pathways of *C. cellulolyticum* grown on cellobiose or cellulose. Pathways leading to overflow products are indicated by dashed lines. The pathway which was cloned in the engineered strain CC-pMG8 is shadowed in gray (adapted from Payot et al [67], Desvaux *et al* [13-15] and Guedon *et al* [36,38]). *AcDH*: acetaldehyde dehydrogenase; *ADH*: alcohol dehydrogenase; ADH_{ZM}: alcohol dehydrogenase from *Zymomonas mobilis*; *AK*: acetate kinase; *CbP*: cellobiose phosphorylase; *CdP*: cellodextrin phosphorylase; *Cel*: cellulases; *GAPD*: glyceraldehyde-3-phosphate dehydrogenase; *GK*: glucokinase; *HYD*: hydrogenase; *LDH*: Lactate dehydrogenase, *NFO*: NADH-ferredoxin-oxidoreductase; PDC_{ZM}: pyruvate decarboxylase from *Zymomonas mobilis*; *PFO*: pyruvate-ferredoxin-oxidoreductase; *PGM*: phosphoglmucomutase; *PTA*: phosphotransacetylase; CoA-SH: coenzyme A; ox: oxidized, red: reduced; n: number of hexose residues.

Taken together these data indicated that *C. cellulolyticum* is restricted to a cellulolytic lifestyle and not adapted to the use of a carbon source in excess. Biotechnological exploitation of such a micro-organism requires genetic improvements. The heterologous expression in *C. cellulolyticum* of an artificial pyruvate decarboxylase-alcohol dehydrogenase operon, formed by genes from *Zymomonas mobilis,* allowed to reduce the accumulation of pyruvate, previously identified as one of the inhibitory compounds of bacterial growth. The recombinant strain (CC-pMG8) grows at the same specific rate, compared to the reference strain, but shows a 180% increase in cell dry weight, and a 150% increase in cellulose consumption [36].

7. Conclusion

Up to now, thirteen genes encoding cellulosomal catalytic subunits have been discovered in *C. cellulolyticum*. The molecular study of the first cellulolysis minus mutant strain *cipC*Mut1 revealed that at least 22 dockerin-containing components are potentially synthesized by *C. cellulolyticum*. Because CipC contains only eight cohesins, only eight enzymes can bind to a single scaffoldin. Thus, the wild-type complexes may widely vary in their composition. It was shown by overproducing a mannanase subunit in *C. cellulolyticum*, that the enzymatic composition of cellulosomes is regulated by the relative amounts of the enzyme available to be incorporated into the complexes. Some of the enzymatic subunits encoded by the genes located in the *cel* cluster must be essential for the degradation of crystalline cellulose, as shown by the non cellulolytic phenotype of the *cipC*Mut1(pSOS*cipC*). Many of these components have been extensively studied. Nevertheless, the role of each subunit and the relationships with the numerous other components remain quite mysterious. The large number of different cellulases belonging to GH family 5 and family 9 present in the same organism is consistent with the ability of these families to vary their active site architectures. Presuming a cooperative digestion of cellulose, one can expect these enzymes act on different cellulose conformations or for different attack modes. In order to gain new insights on the individual roles of the essential enzymes, we tested an asRNA strategy to selectively inhibit the synthesis of the major processive cellulase Cel48F. Although successful, this strategy was shown not to be well adapted to the study of major components synthesized from the *cel* cluster. The synthesis inhibition was found to be (i) uncomplete and, (ii) indirectly affect at least the synthesis of CipC, if not other proteins synthesized from the *cel* cluster of genes. Polycistronic mRNAs carrying *cel48F* were supposed to be affected in their stability by the hybridized asRNAs.

Nevertheless, the development of some genetic tools adapted to this *Clostridium* strain opened up the possibility of *in vivo* studies. Regularory gene or structural gene cloning are feasible and will offer the possibility of new investigations. The challenge is now to succeed in the creation and study of knock-out *cel* mutants and random mutants, which would give us important information on the respective role of the different cellulosome components and the identification of critical genes involved in the regulation of the *cel* and *cip* gene expression. Two new insertion sequences, ISCce*1* and ISCce*2*, were identified in *C. cellulolyticum*. ISCce*2*, which belongs to the IS256 family, was found to be mostly inserted into ISCce*1* [47] and thus does not appear to be a good candidate for the construction of a random mutagenetic tool. On the other hand, ISCce*1* is copied at least twenty times in the genome, and the study of several of its target sequences didn't reveal any consensus sequence. This IS might be modified by adding a resistance marker for developing a transposon-based mutagenesis system.

Finally, the composition of the cellulosomes produced by *Clostridium cellulolyticum* can now be engineered by structural gene or regulatory gene expression. It is thus possible to design, by genetic engineering, functionally tailor-made cellulosomes displaying preferential and/or new substrate specificity. New enzymatic activities might be added in order to tentatively increase the rate of hydrolysis of complex polysaccharides, as those found in the plant cell wall. Cellulolytic microorganisms are regarded as having one of the most promising biotechnological potentials. Nevertheless, cellulolytic clostridia use small quantities of carbohydrates due to early inhibition of the metabolism and growth. *C. cellulolyticum* metabolism was extensively studied. These detailed studies of carbon and electron flux revealed the catabolic behavior of the strain with cellulosic substrates and suggested potential improvements by a metabolic engineering strategy. The *trans*-production of two heterologous enzymes transforming pyruvate to ethanol led to the elimination of intracellular pyruvate in excess allowing an increase of the cellulose degradation and biomass production.

The development of industrial processes to convert cellulose directly to feedstocks might consist in the modification of non cellulolytic strain already possessing valuable metabolic capabilities, e.g., yeast for the ethanol production, *Clostridium acetobutylicum* for solvent production, etc..., into cellulolytic strain by *trans*-expression of a cellulolytic system. Preliminary studies on the production of minicellulosomes by *C. acetobutylicum* are promising. The hybrid miniscaffoldin Scaf3, in which the cohesin 3 module from the scaffoldin CipA of *C. thermocellum* (see the chapter on minicellulosomes) was fused with the C-terminal region of miniCipC$_1$, has been successfully produced by the solventogenic *Clostridium* [70]. Attempts to produce enzymes are underway. This would constitute the starting point for the development of an industrial process to convert directly cellulosic materials into solvents.

References

Ahn, H. J., and L. R. Lynd. 1996. Cellulose degradation and ethanol production by thermophilic bacteria using mineral growth medium. *Appl Biochem Biotechnol 57-58*:599-604.

Ahsan, M., M. Matsumoto, S. Karita, T. Kimura, K. Sakka, and K. Ohmiya. 1997. Purification and characterization of the family J catalytic domain derived from the Clostridium thermocellum endoglucanase CelJ. *Biosci Biotechnol Biochem 61*:427-31.

Akila, G., and T. S. Chandra. 2003. A novel cold-tolerant Clostridium strain PXYL1 isolated from a psychrophilic cattle manure digester that secretes thermolabile xylanase and cellulase. *FEMS Microbiol Lett 219*:63-7.

Andrianarisoa, B. 1986. Synergie d'action entre deux bactéries cellulolytiques Clostridium cellulolyticum et Cellulomonas uda-Etudes physiologique et biochimique. State thesis. University of Aix-Marseille I, France, Marseille.

Arai, T., H. Ohara, S. Karita, T. Kimura, K. Sakka, and K. Ohmiya. 2001. Sequence of celQ and properties of CelQ, a component of the Clostridium thermocellum cellulosome. *Appl Microbiol Biotechnol 57*:660-6.

Bagnara-Tardif, C., C. Gaudin, A. Belaich, P. Hoest, T. Citard, and J. P. Belaich. 1992. Sequence analysis of a gene cluster encoding cellulases from Clostridium cellulolyticum. *Gene 119*:17-28.

Bayer, E. A., Y. Shoham, and R. Lamed 2000, posting date. *The prokaryotes: an evolving electronic resource for the microbiological community, 3rd ed.* Cellulose-decomposing bacteria and their enzyme systems. Springer-Verlag, New York, N.Y M. Dworkin, S. Falkow, E. Rosenberg, K.-H. Schleifer, and E. Stackebrandt (ed.). [Online.]

Belaich, A., H. P. Fierobe, D. Baty, B. Busetta, C. Bagnara-Tardif, C. Gaudin, and J. P. Belaich. 1992. The catalytic domain of endoglucanase A from Clostridium cellulolyticum: effects of arginine 79 and histidine 122 mutations on catalysis. *J Bacteriol 174*:4677-82.

Belaich, A., G. Parsiegla, L. Gal, C. Villard, R. Haser, and J. P. Belaich. 2002. Cel9M, a new family 9 cellulase of the Clostridium cellulolyticum cellulosome. *J Bacteriol 184*:1378-84.

Blanco, A., A. Juarez, and F. I. Pastor. 1991. Overproduction of a Clostridium cellulolyticum endoglucanase by mutant strains of Escherichia coli. *FEMS Microbiol Lett 65*:221-6.

Burrell, P. C., C. O'Sullivan, H. Song, W. P. Clarke, and L. L. Blackall. 2004. Identification, detection, and spatial resolution of Clostridium populations responsible for cellulose degradation in a methanogenic landfill leachate bioreactor. *Appl Environ Microbiol 70*:2414-9.

Cailliez, C., L. Benoit, J.-P. Thirion and H. Petitdemange. 1992. Characterization of 10 mesophilic cellulolytic clostridia isolated from municipal solid waste digestor. *Current Microbiology 25*:105-112.

Desvaux, M., E. Guedon, and H. Petitdemange. 2001. Carbon flux distribution and kinetics of cellulose fermentation in steady-state continuous cultures of Clostridium cellulolyticum on a chemically defined medium. *J Bacteriol 183*:119-30.

Desvaux, M., E. Guedon, and H. Petitdemange. 2000. Cellulose catabolism by Clostridium cellulolyticum growing in batch culture on defined medium. *Appl Environ Microbiol 66*:2461-70.

Desvaux, M., E. Guedon, and H. Petitdemange. 2001. Kinetics and metabolism of cellulose degradation at high substrate concentrations in steady-state continuous cultures of Clostridium cellulolyticum on a chemically defined medium. *Appl Environ Microbiol 67*:3837-45.

Desvaux, M., and H. Petitdemange. 2001. Flux analysis of the metabolism of Clostridium cellulolyticum grown in cellulose-fed continuous culture on a chemically defined medium under ammonium-limited conditions. *Appl Environ Microbiol 67*:3846-51.

Doi, R. H., and A. Kosugi. 2004. Cellulosomes: plant-cell-wall-degrading enzyme complexes. *Nat Rev Microbiol 2*:541-51.

Doi, R. H., A. Kosugi, K. Murashima, Y. Tamaru, and S. O. Han. 2003. Cellulosomes from mesophilic bacteria. *J Bacteriol 185*:5907-14.

Ducros, V., M. Czjzek, A. Belaich, C. Gaudin, H. P. Fierobe, J. P. Belaich, G. J. Davies, and R. Haser. 1995. Crystal structure of the catalytic domain of a bacterial cellulase belonging to family 5. *Structure 3*:939-49.

Ehrlich, S. D., C. Bruand, S. Sozhamannan, P. Dabert, M.F. Gros, L. Jannière, and A. Gruss. 1991. Plasmid replication and structural stability in Bacillus substilis. *Res Microbiol 142*:869-873.

Faure, E., C. Bagnara, A. Belaich, and J. P. Belaich. 1988. Cloning and expression of two cellulase genes of Clostridium cellulolyticum in Escherichia coli. *Gene 65*:51-8.

Faure, E., A. Belaich, C. Bagnara, C. Gaudin, and J. P. Belaich. 1989. Sequence analysis of the Clostridium cellulolyticum endoglucanase-A-encoding gene, celCCA. *Gene 84*:39-46.

Fierobe, H. P., C. Bagnara-Tardif, C. Gaudin, F. Guerlesquin, P. Sauve, A. Belaich, and J. P. Belaich. 1993. Purification and characterization of endoglucanase C from Clostridium cellulolyticum. Catalytic comparison with endoglucanase A. *Eur J Biochem 217*:557-65.

Fierobe, H. P., E. A. Bayer, C. Tardif, M. Czjzek, A. Mechaly, A. Belaich, R. Lamed, Y. Shoham, and J. P. Belaich. 2002. Degradation of cellulose substrates by cellulosome chimeras. Substrate targeting versus proximity of enzyme components. *J Biol Chem 277*:49621-30.

Fierobe, H. P., C. Gaudin, A. Belaich, M. Loutfi, E. Faure, C. Bagnara, D. Baty, and J. P. Belaich. 1991. Characterization of endoglucanase A from Clostridium cellulolyticum. *J Bacteriol 173*:7956-62.

Fierobe, H. P., A. Mechaly, C. Tardif, A. Belaich, R. Lamed, Y. Shoham, J. P. Belaich, and E. A. Bayer. 2001. Design and production of active cellulosome chimeras. Selective incorporation of dockerin-containing enzymes into defined functional complexes. *J Biol Chem 276*:21257-61.

Fierobe, H. P., S. Pages, A. Belaich, S. Champ, D. Lexa, and J. P. Belaich. 1999. Cellulosome from Clostridium cellulolyticum: molecular study of the Dockerin/Cohesin interaction. *Biochemistry 38*:12822-32.

Gal, L., C. Gaudin, A. Belaich, S. Pages, C. Tardif, and J. P. Belaich. 1997. CelG from Clostridium cellulolyticum: a multidomain endoglucanase acting efficiently on crystalline cellulose. *J Bacteriol 179*:6595-601.

Gal, L., S. Pages, C. Gaudin, A. Belaich, C. Reverbel-Leroy, C. Tardif, and J. P. Belaich. 1997. Characterization of the cellulolytic complex (cellulosome) produced by Clostridium cellulolyticum. *Appl Environ Microbiol 63*:903-9.

Gaudin, C., A. Belaich, S. Champ, and J. P. Belaich. 2000. CelE, a multidomain cellulase from Clostridium cellulolyticum: a key enzyme in the cellulosome? *J Bacteriol 182*:1910-5.

Gelhaye, E., H. Petitdemange, and R. Gay. 1993. Adhesion and growth rate of Clostridium cellulolyticum ATCC 35319 on crystalline cellulose. *J Bacteriol 175*:3452-8.

Gerngross, U. T., M. P. Romaniec, T. Kobayashi, N. S. Huskisson, and A. L. Demain. 1993. Sequencing of a Clostridium thermocellum gene (cipA) encoding the cellulosomal SL-protein reveals an unusual degree of internal homology. *Mol Microbiol 8*:325-34.

Giallo, J., C. Gaudin, J. P. Belaich, E. Petitdemange, and F. Caillet-Mangin. 1983. Metabolism of glucose and cellobiose by cellulolytic mesophilic Clostridium sp. strain H10. *Appl Environ Microbiol 45*:843-9.

Girbal, L., Croux, C., Vasconcelos, I., Soucaille, P. 1995. Regulation of metabolic shifts in Clostridium acetobutylicum ATCC 824. *FEMS Microbiol Rev 17*:287-297.

Gruss, A., and S. D. Ehrlich. 1988. Insertion of foreign DNA into plasmids from gram-positive bacteria induces formation of high-molecular-weight plasmid multimers. *J Bacteriol 170*:1183-90.

Guedon, E., M. Desvaux, and H. Petitdemange. 2002. Improvement of cellulolytic properties of Clostridium cellulolyticum by metabolic engineering. *Appl Environ Microbiol 68*:53-8.

Guedon, E., M. Desvaux, and H. Petitdemange. 2000. Kinetic analysis of Clostridium cellulolyticum carbohydrate metabolism: importance of glucose 1-phosphate and glucose 6-phosphate branch points for distribution of carbon fluxes inside and outside cells as revealed by steady-state continuous culture. *J Bacteriol 182*:2010-7.

Guedon, E., S. Payot, M. Desvaux, and H. Petitdemange. 1999. Carbon and electron flow in Clostridium cellulolyticum grown in chemostat culture on synthetic medium. *J Bacteriol 181*:3262-9.

Guedon, E., S. Payot, M. Desvaux, and H. Petitdemange. 2000. Relationships between cellobiose catabolism, enzyme levels, and metabolic intermediates in Clostridium cellulolyticum grown in a synthetic medium. *Biotechnol Bioeng 67*:327-35.

He, Y. L., Y. F. Ding, and Y. Q. Long. 1991. Two cellulolytic Clostridium species: Clostridium cellulosi sp. nov. and Clostridium cellulofermentans sp. nov. *Int J Syst Bacteriol 41*:306-9.

Hungate, R. E. 1944. Studies on cellulose fermentation. I. The culture and physiology of an anaerobic cellulose-digesting bacterium. *J Bacteriol 48*:499-513.

Jennert, K. C., C. Tardif, D. I. Young, and M. Young. 2000. Gene transfer to Clostridium cellulolyticum ATCC 35319. *Microbiology 146 Pt 12*:3071-80.

Jungermann, K., R. K. Thauer, G. Leimenstoll, and K. Decker. 1973. Function of reduced pyridine nucleotide-ferredoxin oxidoreductases in saccharolytic Clostridia. *Biochim Biophys Acta 305*:268-80.

Koch, A. L. 1997. Microbial physiology and ecology of slow growth. *Microbiol Mol Biol Rev 61*:305-18.

Koch, A. L. 2001. Oligotrophs versus copiotrophs. *Bioessays 23*:657-61.

Maamar, H. 2003. Etude in vivo du système cellulolytique de Clostridium cellulolyticum: caractérisation du premier mutant d'insertion cipC. PhD. University of Aix-Marseille I, Marseille.

Maamar, H., P. de Philip, J. P. Belaich, and C. Tardif. 2003. ISCce1 and ISCce2, two novel insertion sequences in Clostridium cellulolyticum. *J Bacteriol 185*:714-25.

Maamar, H., O. Valette, H. P. Fierobe, A. Belaich, J. P. Belaich, and C. Tardif. 2004. Cellulolysis is severely affected in Clostridium cellulolyticum strain cipCMut1. *Mol Microbiol 51*:589-98.

Madarro, A., J. L. Peña, J. L. Lequerica, S. Vallés, R. Gay, and A Flors. 1991. Partial Purification and characterization of the cellulases from Clostridium cellulolyticum H10. *J Chem Tech Biotechnol 52*:393-406.

Madden, R. H., Bryder, M.J., and Poole N.J. 1982. Isolation and characterization of an anaerobic cellulolytic bacterium Clostridium papyrosolvens sp. nov. *Int J Syst Bacteriol 32*:87-91.

Mandelman, D., A. Belaich, J. P. Belaich, N. Aghajari, H. Driguez, and R. Haser. 2003. X-Ray crystal structure of the multidomain endoglucanase Cel9G from Clostridium cellulolyticum complexed with natural and synthetic cello-oligosaccharides. *J Bacteriol 185*:4127-35.

Martin, S. A. 1994. Nutrient transport by ruminal bacteria: a review. *J Anim Sci 72*:3019-31.

Minton, N. P., J.K. Brehm, T.J. Swinfield, S.M. Mauchline, N. Bodsworth, and J.D. Oultram. 1993. Clostridial cloning vectors, p. 119-150. In D. R. Woods (ed.), *The clostridia and biotechnology*. Butterworth-Heinemann, Boston.

Mohand-Oussaid, O., S. Payot, E. Guedon, E. Gelhaye, A. Youyou, and H. Petitdemange. 1999. The extracellular xylan degradative system in Clostridium cellulolyticum cultivated on xylan: evidence for cell-free cellulosome production. *J Bacteriol 181*:4035-40.

Monserrate, E., S. B. Leschine, and E. Canale-Parola. 2001. Clostridium hungatei sp. nov., a mesophilic, N2-fixing cellulolytic bacterium isolated from soil. *Int J Syst Evol Microbiol 51*:123-32.

Morvan, B., Rieu-Lesme, F., Fonty, G., Gouet, P. 1996. In vitro interactions between rumen H2-producing cellulolytic micro-organisms and H2-utilizing acetogenic and sulfate reducing bacteria. *Anaerobe 2*:175-180.

Mosbah, A., A. Belaich, O. Bornet, J. P. Belaich, B. Henrissat, and H. Darbon. 2000. Solution structure of the module X2 1 of unknown function of the cellulosomal scaffolding protein CipC of Clostridium cellulolyticum. *J Mol Biol 304*:201-17.

Mosbah, A., C. Tardif, O. Bornet, O. Valette, B. Henrissat, and H. Darbon. 2002. Assignment of the 1H, 13C, and 15N resonances of the 22,5 kDa CBM28 module of the cellulase Cel5I of Clostridium cellulolyticum. *J Biomol NMR 23*:157-8.

Pages, S., A. Belaich, H. P. Fierobe, C. Tardif, C. Gaudin, and J. P. Belaich. 1999. Sequence analysis of scaffolding protein CipC and ORFXp, a new cohesin-containing protein in Clostridium cellulolyticum: comparison of various cohesin domains and subcellular localization of ORFXp. *J Bacteriol 181*:1801-10.

Pages, S., A. Belaich, C. Tardif, C. Reverbel-Leroy, C. Gaudin, and J. P. Belaich. 1996. Interaction between the endoglucanase CelA and the scaffolding protein CipC of the Clostridium cellulolyticum cellulosome. *J Bacteriol 178*:2279-86.

Pages, S., L. Gal, A. Belaich, C. Gaudin, C. Tardif, and J. P. Belaich. 1997. Role of scaffolding protein CipC of Clostridium cellulolyticum in cellulose degradation. *J Bacteriol 179*:2810-6.

Pages, S., O. Valette, L. Abdou, A. Belaich, and J. P. Belaich. 2003. A rhamnogalacturonan lyase in the Clostridium cellulolyticum cellulosome. *J Bacteriol 185*:4727-33.

Parsiegla, G., A. Belaich, J. P. Belaich, and R. Haser. 2002. Crystal structure of the cellulase Cel9M enlightens structure/function relationships of the variable catalytic modules in glycoside hydrolases. *Biochemistry 41*:11134-42.

Parsiegla, G., M. Juy, C. Reverbel-Leroy, C. Tardif, J. P. Belaich, H. Driguez, and R. Haser. 1998. The crystal structure of the processive endocellulase CelF of Clostridium cellulolyticum in complex with a thiooligosaccharide inhibitor at 2.0 A resolution. *Embo J* 17:5551-62.

Parsiegla, G., C. Reverbel-Leroy, C. Tardif, J. P. Belaich, H. Driguez, and R. Haser. 2000. Crystal structures of the cellulase Cel48F in complex with inhibitors and substrates give insights into its processive action. *Biochemistry 39*:11238-46.

Pavlostathis, S. G., Miller, T.L., Wolin, M.J. 1988. kinetics of insoluble cellulose fermentation by continuous cultures of Ruminococcus albus. *Appl Environ Microbiol 54*.

Payot, S., Guedon, E., Cailliez, C., Gelhaye, E., Petitdemange, H. 1998. Metabolism of cellobiose by Clostridium cellulolyticum growing in continuous culture: evidence for decreased NADH reoxidation as a factor limiting growth. *Microbiology 144*:375-384.

Pérez-Martinez, G., L. Gonzales-Candelas, J. Polaina, and A. Flors. 1988. Expression of an endoglucanase gene from Clostridium cellulolyticum in Escherichia coli. *J Ind Microbiol 3*:365-371.

Perret, S., A. Belaich, H. P. Fierobe, J. P. Belaich, and C. Tardif. 2004. Towards designer cellulosomes in Clostridia: mannanase enrichment of the cellulosomes produced by Clostridium cellulolyticum. *J Bacteriol 186*:6544-52.

Perret, S., L. Casalot, H. P. Fierobe, C. Tardif, F. Sabathe, J. P. Belaich, and A. Belaich. 2004. Production of heterologous and chimeric scaffoldins by Clostridium acetobutylicum ATCC 824. *J Bacteriol 186*:253-7.

Perret, S., H. Maamar, J. P. Belaich, and C. Tardif. 2004. Use of antisense RNA to modify the composition of cellulosomes produced by Clostridium cellulolyticum. *Mol Microbiol 51*:599-607.

Petitdemange, E., F. Caillet, J. Giallo, and C. Gaudin. 1984. Clostridium cellulolyticum sp. nov., a cellulolytic mesophilic species from decayed grass. *Int J. Syst. Bacteriol. 34*:155-159.

Ravinder, T., M. V. Swamy, G. Seenayya, and G. Reddy. 2001. Clostridium lentocellum SG6--a potential organism for fermentation of cellulose to acetic acid. *Bioresour Technol 80*:171-7.

Reverbel-Leroy, C., A. Belaich, A. Bernadac, C. Gaudin, J. P. Belaich, and C. Tardif. 1996. Molecular study and overexpression of the Clostridium cellulolyticum celF cellulase gene in Escherichia coli. *Microbiology 142 (Pt 4)*:1013-23.

Reverbel-Leroy, C., S. Pages, A. Belaich, J. P. Belaich, and C. Tardif. 1997. The processive endocellulase CelF, a major component of the Clostridium cellulolyticum cellulosome: purification and characterization of the recombinant form. *J Bacteriol 179*:46-52.

Reverbel-Leroy, C., G. Parsiegla, V. Moreau, M. Juy, C. Tardif, H. Driguez, J. P. Belaich, and R. Haser. 1998. Crystallization of the catalytic domain of Clostridium cellulolyticum CeLF cellulase in the presence of a newly synthesized cellulase inhibitor. *Acta Crystallogr D Biol Crystallogr 54 (Pt 1)*:114-8.

Roig, V., H. P. Fierobe, V. Ducros, M. Czjzek, A. Belaich, C. Gaudin, J. P. Belaich, and R. Haser. 1993. Crystallization and preliminary X-ray analysis of the catalytic domain of endoglucanase from Clostridium cellulolyticum. *J Mol Biol 233*:325-7.

Salamitou, S., M. Lemaire, T. Fujino, H. Ohayon, P. Gounon, P. Beguin, and J. P. Aubert. 1994. Subcellular localization of Clostridium thermocellum ORF3p, a protein carrying a receptor for the docking sequence borne by the catalytic components of the cellulosome. *J Bacteriol 176*:2828-34.

Shi, Y., P. J. Weimer, and J. Ralph. 1997. Formation of formate and hydrogen, and flux of reducing equivalents and carbon in Ruminococcus flavefaciens FD-1. *Antonie Van Leeuwenhoek 72*:101-9.

Shima, S., Y. Igarashi, and T. Kodama. 1991. Nucleotide sequence analysis of the endoglucanase-encoding gene, celCCD, of Clostridium cellulolyticum. *Gene 104*:33-8.

Shima, S., Y. Igarashi, and T. Kodama. 1993. Purification and properties of two truncated endoglucanases produced in Escherichia coli harbouring Clostridium cellulolyticum endoglucanase gene celCCD. *Appl Microbiol Biotechnol 38*:750-4.

Shima, S., Y. Igarashi, and T. Kodama. 1991. Molecular cloning of a new endoglucanase gene from Clostridium cellulolyticum and its expression in Escherichia coli. *Appl Biochem Biotechnol 35*:233-236.

Shimon, L. J., S. Pages, A. Belaich, J. P. Belaich, E. A. Bayer, R. Lamed, Y. Shoham, and F. Frolow. 2000. Structure of a family IIIa scaffoldin CBD from the cellulosome of Clostridium cellulolyticum at 2.2 A resolution. *Acta Crystallogr D Biol Crystallogr 56 Pt 12*:1560-8.

Shoseyov, O., M. Takagi, M. A. Goldstein, and R. H. Doi. 1992. Primary sequence analysis of Clostridium cellulovorans cellulose binding protein A. *Proc Natl Acad Sci U S A 89*:3483-7.

Sleat, R., and R. A. Mah. 1985. Clostridium populeti sp. nov., a cellulolytic species from a woody-biomass digestor. *Int J Syst Bacteriol 35*:160-163.

Sleat, R., R.A. Mah, and R. Robinson. 1984. Isolation and characterization of an anaerobic, cellulolytic bacterium, Clostridium cellulovorans sp. nov. *Appl Environ Microbiol 48*:88-93.

Spinelli, S., H. P. Fierobe, A. Belaich, J. P. Belaich, B. Henrissat, and C. Cambillau. 2000. Crystal structure of a cohesin module from Clostridium cellulolyticum: implications for dockerin recognition. *J Mol Biol 304*:189-200.

Strobel, H. J., Caldwell, F.C., Dawson, K.A. 1995. Carbohydrate transport by the anaerobic thermophile Clostridium thermocellum LQRI. *Appl Environ Microbiol 61*:4012-4015.

Sukhumavasi, J., O.K. Shimisu, and K. Lieno. 1988. Clostridium josui sp. nov., a cellulolytic moderate thermophilic species from Thaï compost. *Int J Bacteriol 38*:179-182.

Tardif, C., H. Maamar, M. Balfin, and J. P. Belaich. 2001. Electrotransformation studies in Clostridium cellulolyticum. *J Ind Microbiol Biotechnol 27*:271-4.

Thomas, S., and J. B. Russell. 2004. The effect of cellobiose, glucose, and cellulose on the survival of Fibrobacter succinogenes A3C cultures grown under ammonia limitation. *Curr Microbiol 48*:219-23.

Tormo, J., R. Lamed, A. J. Chirino, E. Morag, E. A. Bayer, Y. Shoham, and T. A. Steitz. 1996. Crystal structure of a bacterial family-III cellulose-binding domain: a general mechanism for attachment to cellulose. *Embo J 15*:5739-51.

Van Dyke, M. I., and A. J. McCarthy. 2002. Molecular biological detection and characterization of Clostridium populations in municipal landfill sites. *Appl Environ Microbiol 68*:2049-53.

Varel, V. H. 1989. Reisolation and characterization of Clostridium longisporum, a ruminal sporeforming cellulolytic anaerobe. *Arch Microbiol 152*:209-14.

Varel, V. H., and W. G. Pond. 1992. Characteristics of a new cellulolytic Clostridium sp. isolated from pig intestinal tract. *Appl Environ Microbiol 58*:1645-9.

Varel, V. H., R. S. Tanner, and C. R. Woese. 1995. Clostridium herbivorans sp. nov., a cellulolytic anaerobe from the pig intestine. *Int J Syst Bacteriol 45*:490-4.

Varrot, A., S. Hastrup, M. Schulein, and G. J. Davies. 1999. Crystal structure of the catalytic core domain of the family 6 cellobiohydrolase II, Cel6A, from Humicola insolens, at 1.92 A resolution. *Biochem J 337 (Pt 2)*:297-304.

Warnick, T. A., B. A. Methe, and S. B. Leschine. 2002. Clostridium phytofermentans sp. nov., a cellulolytic mesophile from forest soil. *Int J Syst Evol Microbiol 52*:1155-60.

Yang, J. C., D. P. Chynoweth, D. S. Williams, and A. Li. 1990. Clostridium aldrichii sp. nov., a cellulolytic mesophile inhabiting a wood-fermenting anaerobic digester. *Int J Syst Bacteriol 40*:268-72.

Young, D. I., V.J. Evans,J.R. Jefferies, K.C.B. Jennert, Z.E.V. Phillips, A. Ravagnani, and M. Young. 1999. Genetic methods in clostridia. *Methods Microbiol 29*:191-207.

In: Cellulosome
Eds: V. Uversky and I. A. Kataeva, pp. 261-270
ISBN 1-59454-950-8

Chapter XII

REGULATION OF CELLULASE BIOSYNTHESIS IN *CLOSTRIDIUM THERMOCELLUM*

Michael Newcomb and J.H. David Wu*
Department of Chemical Engineering, University of Rochester, Rochester, NY, USA.

ABSTRACT

C. thermocellum produces a cellulase system consisting of a huge protein complex called the cellulosome. The cellulosome contains a large number of subunits, most of them glycosyl hydrolases. The core of the cellulosome is a 250-kDa non-catalytic subunit, CipA, which serves as a scaffold for the catalytic subunits. CipA contains a series of nine highly homologous domains, termed cohesin domains, which serve as receptors for the catalytic subunits. Binding to the cohesin domains is mediated by a highly conserved duplicated sequence of 22 amino acid residues called the dockerin domain, which is found mostly at the C-terminus of each cellulosomal catalytic subunit. CipA further contains a cellulose-binding domain (CBD), which anchors the array of the catalytic components on the cellulose surface. In addition to the cellulosome, *C. thermocellum* produces many free, non-cellulosomal glycosyl hydrolases.

Most studies on the *C. thermocellum* cellulase system have so far focused on molecular cloning and characterization of glycosyl hydrolases as well as the structure-function relationship of the cellulosomal and noncellulosomal components. Little is known about how the biosynthesis of these proteins is regulated. The task is obviously hampered by the large number of the genes and proteins involved. The cellulosome components can be classified into three categories: 1) the scaffolding protein (CipA); 2) the dockerin-containing catalytic subunits; and 3) the scaffoldin-anchorage proteins, which anchor the cellulosome or individual enzymes to the cell surface (such as OlpA, OlpB, SdbA, and Orf2p). Analysis of the *C. thermocellum* draft genome sequence indicates that category 2

* Correspondence concerning this article should be addressed to J.H. David Wu, PhD University of Rochester Department of Chemical Engineering Room 206 Gavett Hall Rochester, NY 14627-0166. Tel: 585-275-8499; Fax: 585-273-1348; Email: davidwu@che.rochester.edu.

alone comprises at least 72 different genes. The study is further complicated by many non-cellulosomal enzyme components produced by the bacterium. The question of how the bacterium coordinates the biosynthesis of such a large number of biomass-degrading enzymes is intriguing. Elucidation of the regulatory mechanisms is important not only for understanding the cellulase biology, but also for engineering deregulated, hypercellulolytic strains. The knowledge obtained will also serve as a paradigm for the cellulosomes produced by other microorganisms since the *C. thermocellum* complex is the most extensively studied cellulosome so far.
We have submitted the *C. thermocellum* genomic DNA to the DOE Joint Genome Institute (JGI) for sequencing and a draft genome sequence is now available [http://genome.jgi-psf.org/draft_microbes/cloth/cloth.home.html]. This provides an unprecedented opportunity for studying the regulatory mechanism at the genomic level. In this article, we briefly review the recent progress, highlighting the results of some studies facilitated by the genome sequence.

Keywords: Cellulosome, Biomass Energy, Cellulose Degradation, Negative Control, Operon, Repressor, Anaerobic Bacterium, Thermophiles

Carbon Source Regulation

It has been well documented that the carbon source regulates true cellulase biosynthesis in *C. thermocellum* [1,6,15,20,23]. True cellulase production is much less when grown on cellobiose than cellulose [1,15,26,30]. The observation has been recently confirmed by Dror et al. [3,4]. On the other hand, total endoglucanase activity is not significantly affected by the carbon source [9,13,24,26], although expression of individual endoglucanases can be regulated [21,25].

Cellulase activity is detected when the bacterium is grown on the following substrates: cellulose, cellobiose, glucose, fructose and sorbitol [15]. Some of the substrates require a long (> 100 h) lag phase. Growth on fructose or sorbitol leads to a high rate of cellulase production despite the long lag. On the contrary, the long lag on glucose did not lead to a high rate of cellulase synthesis, suggesting the repressive role of glucose. High enzyme production on fructose is presumably due to the lack of such repressive effect [26, J.H.Wang and J.H.D.Wu, Abstr. Ann. Mtg. Soc. Ind. Microbiol. p. 25, 1989]. The high yield of cellulase production on fructose is repressed by adding cellobiose to the medium [15].

Regulation of Individual Cellulase Components

As mentioned above, culture broth from the cellulose-grown culture has higher cellulase activities against amorphous or crystalline cellulose than that from the cellobiose-medium, indicating that cellulose induces the biosynthesis of the cellulase component(s) important for cellulose degradation. However, the enzymatic activity against the soluble substrate carboxymethyl cellulose (CMC) is not significantly influenced by the carbon source. Therefore, not all the cellulase components are induced by cellulose. Due to the large number of the cellulases and hemicellulases produced by this bacterium, identification of specific components that are regulated by the carbon source has been a challenging task. The hurdle will soon be overcome by our ability to take genomic and proteomic approaches to catalog extracellular proteins whose biosynthesis depends on the carbon source, employing techniques such as two-dimensional protein gel electrophoresis or DNA microarray.

Although experimental results from genome-wide study are yet to be seen in the literature, conventional immunoassays have revealed that both CipA and CelS, the two most abundant components of the cellulosome, are regulated by the carbon source. Using enzyme-linked immunosorbent assay (ELISA) and an antibody against a segment of the cellulosome scaffoldin (CipA), Zhang and Lynd [31] reported that in batch cultures the CipA production is nine-fold higher in Avicel (cellulose) cultures than in cellobiose cultures. Furthermore, in continuous cultures, the CipA production is 1.3- to 2.4-fold greater on Avicel at the same dilution rates as on cellobiose. Dror et al. [3] reported that the production of CelS (Cel48A), the most abundant cellulosomal catalytic component and the only well-characterized cellobiohydrolase in the cellulosome [16,17,27,28,29], is induced by cellulose as revealed by SDS-PAGE and western blot analysis using an anti-CelS antiserum. CelS is almost completely lacking from the cellulosome produced in the cellobiose medium. On the other hand, several proteins in the region corresponding to 67 kDa were only observed when the cells were grown on cellobiose. N-terminal amino acid sequence analysis indicated that these proteins are XynC (M_r = 66,147), CelB (M_r = 61,063), and CelG (M_r = 59,769). Later study [5] revealed that *xynC* was upregulated in the cellobiose medium as determined by RNase protection assays. These results suggest that cellulose and cellobiose each induces different cellulase genes. The underlying mechanisms, however, remain to be investigated.

Using Real-Time PCR to quantify gene expression, Stevenson and Weimer reported [25] that *cipA*, *celS* and *manA* are significantly upregulated by growth on cellulose when compared to cellobiose.

EFFECT OF GROWTH RATE

Dror et al. [3] combined the data from the batch and continuous cultures to show that the *celS* transcription, quantified by RNase protection assays using a ^{32}P-labeled antisense RNA probe, is influenced by growth rate. The transcription level decreases as the growth rate increases. Growth rate-dependent transcription was observed in chemostats limited by carbon (cellobiose) or nitrogen (ammonia sulphate), although the *celS* transcription level was much lower in the nitrogen-limited chemostat. The results indicate that the *celS* expression is influenced by growth rate. It would be tempting to speculate that sigma factors of the RNA polymerase are involved in the observed growth dependence. However, despite the fact that two potential promoters, similar to σ^A- and σ^B-dependent promoters of *Bacillus subtilis*, respectively, were identified in the upstream region, primer extension analysis indicated that the relative activity of the two promoters remained constant under the conditions studied. Thus, the molecular mechanism underlying the observed growth rate-dependence of the *celS* transcription remains unclear.

The same type of methodology was used to study the expression of the scaffoldin-related genes. The results showed that the expression of *cipA*, *olpB*, and *orf2p*, which are contiguous on the chromosome, is affected by growth rate under carbon (cellobiose) limitation, a higher growth rate leading to a lower transcription level [4]. On the other hand, the transcription of *sbdA* is not significantly affected by growth rate. Primer extension analysis identified two transcription start sites at -81 and –50 bp upstream of the translation start site of *cipA*, respectively. The promoters showed homology to sigma factors σ^A- and σ^L-dependent promoters of *B. subtilis*, respectively. Only low levels of transcription were detected from the σ^A-like promoter; it was active only under carbon limitation. Transcription from the σ^L-like promoter was found under all conditions.

Growth dependence of the *cipA* transcription was corroborated by the results from immunoassays at the protein level. Zhang and Lynd reported [31] that the CipA level decreases with an increasing growth rate in cellobiose- or Avicel-limited continuous cultures. Expression in the Avicel medium is 1.3 to 2.4-fold higher as mentioned above. However, analysis of the chemostat data revealed that the specific production of CipA dramatically declines when cellobiose concentration increases from near zero to 0.2 g/liter, suggesting that cellobiose exerts a repressive effect on the CipA biosynthesis.

Chemostat was also used to study the effect of growth rate on the transcription of a number of endoglucanases and a xylanase. It was found that the expression of *celB*, *celG*, and *celD* is growth rate dependent [5]. However, there is no clear relationship between growth rate and the expression of *xynC*.

Recently, using Real-Time PCR to quantify gene expression, Stevenson and Weimer [25] confirmed the negative correlation between the growth rate and the transcript levels of *celS* and *cipA* in cellobiose-limited chemostats. In addition, they reported the same correlation for *manA*. However, in contrast to the report mentioned above [31], they did not find such negative correlation in cellulose-limited chemostats.

Cellulase Gene Clusters

Most of the cloned cellulase and hemicellulase genes of *C. thermocellum* are not contiguous in the genome [12]. However, many gene clusters have recently been found (Fig. 1). The genes in a cluster may be regulated coordinately and expressed as a polycistronic messenger. In fact, several of the scaffoldin-related genes are arranged in a cluster consisting of two tandem operons [8]. The genes in the cluster encode CipA, the main scaffolding protein, OlpB, Orf2p, and OlpA. OlpB, Orf2p and OlpA each contains three C-terminal domains termed surface layer homology (SLH) repeats and are presumed to anchor the cellulosome or individual enzymes to the cell surface.

In addition to the scaffoldin gene cluster, gene clusters encoding catalytic subunits have also been found. These clusters may consist of genes encoding cellulosomal and non-cellulosomal enzymes. For example, CelI, a non-cellulosomal endoglucanase, is clustered with CelN, a cellulosomal cellulase, and a putative cellulosome structural component, CseP [34]. CelI and CelN both hydrolyze phosphoric acid-swollen cellulose, but produce different products. While CelI is a progressive endoglucanase producing mostly cellotetraose that slowly degrades to smaller products, CelN is a non-progressive endoglucanase producing long oligosaccharides, which continuously break down into smaller sugars. *celI* and *celN* are on the same strand of the chromosome, while the third gene of the cluster, *cseP*, is on the opposite strand. Therefore, *cseP* is likely transcribed independently from *celI* and *celN*. CseP was examined for enzymatic activities and none was found; however, western blot analyses revealed that a large amount of the protein exists in the cellulosome. CseP is homologous to a spore-coat assembly protein, CotH, of *B. subtilis*. It has been suggested to play a structural role in the cellulosome.

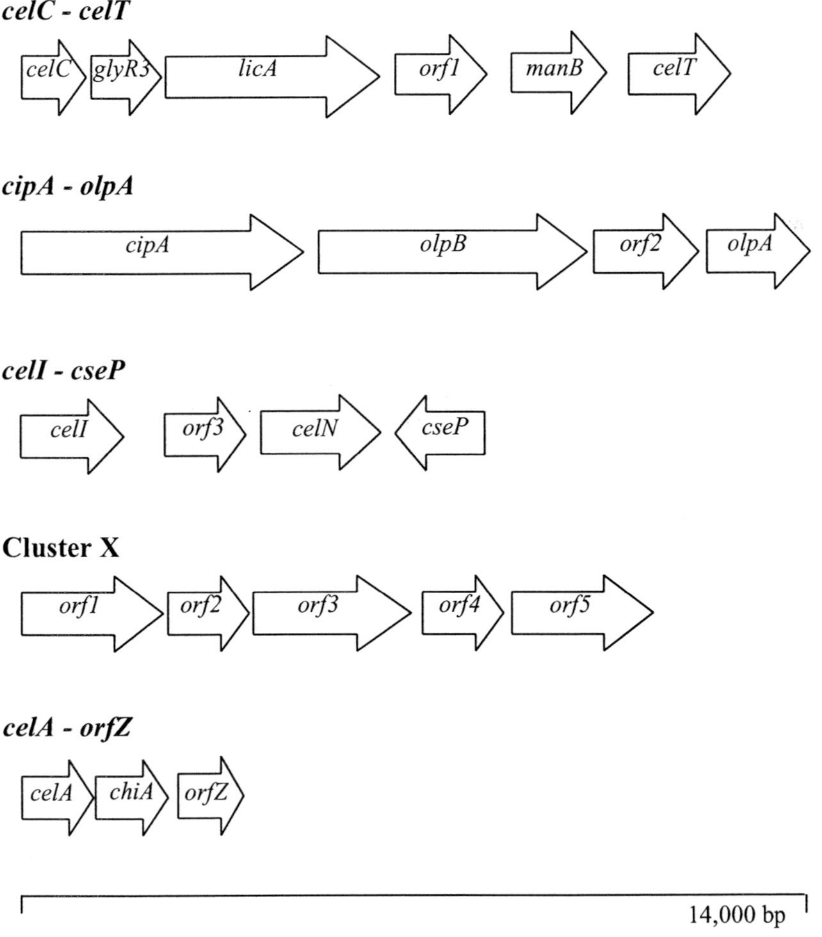

Figure 1. Major cellulase gene clusters found in *C. thermocellum*. Cluster X was recently discovered [M. Newcomb and J.H.D. Wu, unpublished results] and none of its components has been experimentally characterized so far.

Another cluster that mixes cellulosomal and non-cellulosomal genes consists of six members, including the genes encoding CelC, a LacI-homologous protein GlyR3, LicA, and a potential membrane protein with eight putative transmembrane regions [7]. Immediately downstream of this cluster lie the genes encoding ManB [18] and CelT [19]. CelC is a non-cellulosomal endo-β-1,3-1,4-glucanase [7]. LicA is a non-cellulosomal endo-β-1,3-glucanase.

ManB is cellulosomal mannanase [18]. Finally, CelT is a cellulosomal family 9 glycosyl hydrolase [19]. All of the genes are on the same strand of the chromosome. We will describe this gene cluster again in the next section entitled Negative Regulation.

A third gene cluster in *C. thermocellum* contains *celA*, *chiA*, and *orfZ* [33]. CelA is a non-cellulosomal endoglucanase. ChiA is the only chitinase found in the cellulosome. OrfZ has a dockerin domain and a domain of unknown function. A transcriptional terminator sequence is found downstream from each of the genes of the cluster. The terminator immediately downstream from *celA* is active; co-transcription of *celA* and *chiA* does not seem to occur [2].

The final major gene cluster known so far was recently found by analyzing the *C. thermocellum* genome sequence [M. Newcomb and J.H.D. Wu, unpublished data]. This new cluster (Cluster X) consists of five genes that encode different dockerin-containing proteins. To date, this is the largest cluster of dockerin sequence-containing genes found in *C. thermocellum*. So far none of the genes in this cluster have been cloned or characterized. Interestingly, distances between the open reading frames are very short, some as short as 16 bp. The close proximity between the five genes strongly suggests transcription regulation in the form of an operon. The first gene of this cluster encodes a putative family 2 glycosyl hydrolase with a type 4 CBD (cellulose-binding domain) and a dockerin domain. No other family 2 glycosyl hydrolases have been found in the draft genome. The encoded protein is homologous to β-galactosidases found in many other organisms. The second gene of the cluster encodes a putative family 43 glycosyl hydrolase. Like the first gene of the cluster, gene 2 contains a type 4 CBD and a dockerin domain. The gene product is homologous to many bacterial xylanases. A domain of unknown function followed by a type 4 CBD and a dockerin domain are encoded by gene 3 of the cluster. The fourth gene of the cluster encodes a protein consisting of a putative esterase domain, a type 4 CBD and a dockerin domain. BLAST search showed significant homology between its putative esterase domain and the feruloyl esterase domain of XynZ [10,11] of *C. thermocellum*. The final gene of this cluster encodes a putative family 5 glycosyl hydrolase with a type 4 CBD, a fibronectin-like domain and a dockerin domain.

Other smaller gene clusters found in the genome may have resulted from gene duplication (Fig. 2). The gene pairs of *celK-cbhA* [32] and *xynB-xynA* [14] are the examples. More cellulase and hemi-cellulase gene clusters may be discovered after the complete genome sequence of *C. thermocellum* becomes available.

NEGATIVE REGULATION

The mechanisms through which *C. thermocellum* regulates the expression of its cellulase and hemicellulase genes are largely unknown. Existence of gene clusters described above suggests that some cellulase/hemicellulase genes are regulated in the form of an operon. In an operon, a transcription regulator binds to the promoter region, activating or repressing the operon. Searching the genome for genes encoding possible transcription factors yielded three genes that are homologous to *lacI*, which encodes the repressor of the *lac* operon of *Escherichia coli* [22]. The proteins were named GlyR1, GlyR2, and GlyR3 (formerly CelR1,

CelR2, and CelR3, respectively). Each of these proteins has two putative domains: a helix-turn-helix (HTH) DNA-binding domain that is N-terminal to a sugar-binding domain (Fig. 3). Structure similarity between these proteins and LacI suggests that the GlyR proteins are negative transcription regulators. GlyR3 is particularly interesting because it is part of the *celC* gene cluster. As mentioned above, the cluster encodes CelC, GlyR3, LicA, a putative membrane protein, ManB, and CelT.

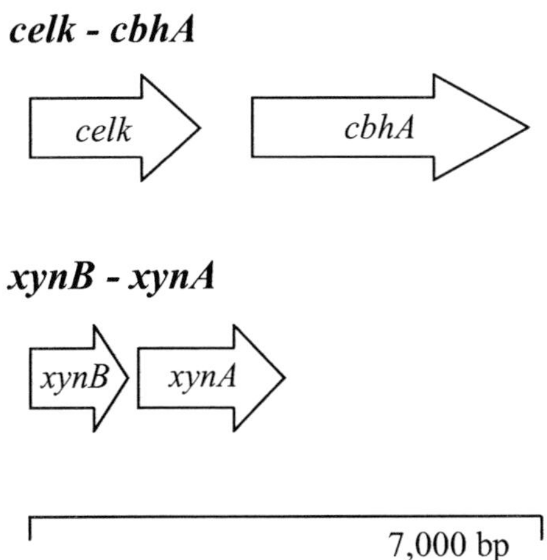

Figure 2. Gene sets found in *C. thermocellum* that most likely resulted from gene duplication.

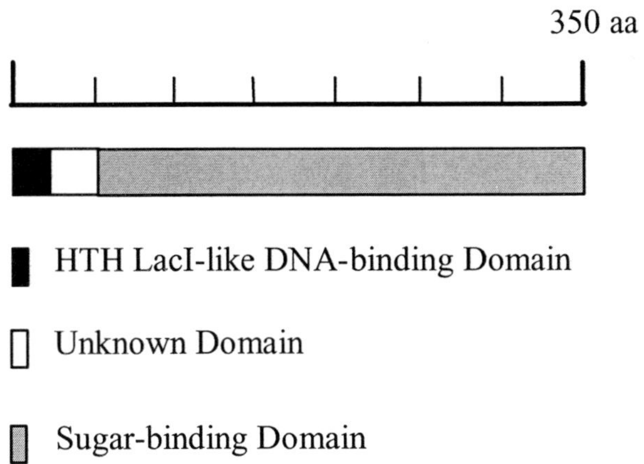

Figure 3. GlyR1, GlyR2 and GlyR3 are LacI-homologous, putative transcription regulators and consist of two distinct domains. As described in the text, the HTH (helix-turn-helix) domain of GlyR3 shows similar structure to its corresponding domain in LacI. The sugar-binding domain, on the other hand, is different.

Homologous structure modeling showed that the folding of the DNA-binding domain of GlyR3 is very similar to that of LacI; however, the simulated structure of the sugar-binding domain is very different [22]. These results suggest that GlyR3's DNA-binding activity is regulated by a sugar other than lactose. Our experimental data verified this postulation and demonstrated that GlyR3 indeed binds to a specific sequence at the *celC* promoter region and the binding is inhibited by laminaribiose, a β-1,3 linked glucose dimer [22, M. Newcomb and J.H.D. Wu, unpublished data]. These results indicate that the *celC* gene cluster forms an operon and laminaribiose induces the operon by inactivating the binding of GlyR3 to the *celC* promoter. Since GlyR3 is encoded as part of the gene cluster, the protein should serve as an auto-regulator. GlyR1 and GlyR2 do not bind to the *celC* promoter [M. Newcomb and J.H.D. Wu, unpublished data]. Their functions remain to be elucidated.

Conclusion

How the bacterium commands more than a hundred genes to degrade cellulosic biomass, how their gene products are assembled into the cellulosome, and how the cellulosomal and non-cellulosomal enzymes coordinate their actions against the substrate are intellectually interesting and challenging questions. Understanding the complicated regulatory pathways is vital to determining how *C. thermocellum* so efficiently breaks down cellulose. Knowing how many enzymes are expressed under specific growth conditions will provide hints as to what proteins are needed to best create a designer cellulosome.

The study on how the cellulase and hemicellulase biosynthesis is regulated in *C. thermocellum* is uniquely challenging because of the complexity of the enzyme system and the substrate. As it stands now, our understanding is very primitive and a lot of work remains to be done. The future study will have to address the following aspects: 1) identification of enzyme components whose expression is influenced by a particular carbon source in the medium; 2) the molecular mechanism underlying each regulatory pathway; 3) relationship between the cellulase biosynthesis and the cellular metabolic activities; and 4) relationship between the cellulase biosynthesis and the cellulase activities. The availability of the genome sequence will expedite the study and rapid progress is expected in the near future.

Acknowledgements

This work was supported by a grant from DOE (DE-FG02-94ER20155).

References

Bayer, E. A.; Setter, E.; Lamed, R. *J. Bacteriol.* 1985, 163, 552-559.

Beguin, P.; Rocancourt, M.; Chebrou, M. C.; Aubert, J. P. *Mol. Gen. Genet.* 1986, 202, 251-254.

Dror, T. W.; Morag, E.; Rolider, A.; Bayer, E. A.; Lamed, R.; Shoham, Y. *J. Bacteriol.* 2003, 185, 3042-8.

Dror, T. W.; Rolider, A.; Bayer, E. A.; Lamed, R.; Shoham, Y. *J. Bacteriol.* 2003, 185, 5109-5116.

Dror, T. W.; Rolider, A.; Bayer, E. A.; Lamed, R.; Shoham, Y. *J. Bacteriol.* 2005, 187, 2261-2266.

Freier, D.; Mothershed, C. F.; Wiegel, *J. Appl. Environ. Microbiol.* 1988, 54, 204-211.

Fuchs, K.-P.; Zverlov, V. V.; Velikodvorskaya, G. A.; Lottspeich, F.; Schwarz, W. H. Microbiol. 2003, 149, 1021-1031.

Fujino, T.; Beguin, P.; Aubert, J.-P. *J. Bacteriol.* 1993, 175, 1891-1899.

Garcia-Martinez, D. V.; Shinmyo, A.; Madia, A.; Demain, A. L. *Eur. J. Appl. Microbiol. Biotechnol.* 1980, 9, 189-197.

Grepinet, O.; Chebrou, M.-C.; Beguin, P. *J. Bacteriol.* 1988, 170, 4582-4588.

Grepinet, O.; Chebrou, M. C.; Beguin, P. *J. Bacteriol.* 1988, 170, 4576-4581.

Guglielmi, G.; Beguin, P. *FEMS Microbiol. Lett.* 1998, 161, 209-215.

Hammerstrom, R. A.; Claus, K. D.; Cughlan, J. W.; McBee, R. H. *Arch. Biochem. Biophys.* 1955, 56, 123-129.

Hayashi, H.; Takehara, M.; Hattori, T.; Kimura, T.;Karita, S.; Sakka, K.; Ohmiya, K. *Appl. Microbiol. Biotechnol.* 1999, 51, 348-357.

Johnson, E. A.; Bouchot, F.; Demain, A. L. J. *Gen. Microbiol.* 1985, 131, 2303-2308.

Kruus, K.; Andreacchi, A.; Wang, W. K.; Wu, J. H. *Appl. Microbiol. Biotechnol.* 1995, 44, 399-404.

Kruus, K.; Wang, W. K.; Ching, J.; Wu, J. H. *J. Bacteriol.* 1995, 177, 1641-1644.

Kurokawa, J.; Hemjinda, E.; Arai, T.; Karita, S.; Kimura, T.; Sakka, K.; Ohmiya, K. *Biosci. Biotechnol. Biochem.* 2001, 65, 548-554.

Kurokawa, J.; Hemjinda, E.; Arai, T.; Kimura, T.; Sakka, K.; Ohmiya, K. *Appl. Microbiol. Biotechnol.* 2002, 59, 455-461.

Lamed, R.; Kenig, R.; Setter, E.; Bayer, E. A. *Enzyme Microb. Technol.* 1985, 7, 37-41.

Mishra, S.; Beguin, P.; Aubert, J. P. *J. Bacteriol.* 1991, 173, 80-85.

Newcomb, M.; Wu, J. H. D. Transcription regulators of the *Clostridium thermocellum* cellulase system; K. Ohmiya, K. Sakka, S. Karita, T. Kimura, M. Sakka and Y. Onishi. Uni Publishers: Tokyo, 2004; 148-154.

Nochur, S. V.; Roberts, M. F.; Demain, A. L. *FEMS Microbiol. Lett.* 1990, 71, 199-204.

Shinmyo, A.; Garcia-Martinez, D. V.; Demain, A. L. *J. Appl. Biochem.* 1979, 1, 202-209.

Stevenson, D. M.; Weimer, P. *J. Appl. Environ. Microbiol.* 2005, 71, 4672-4678.

Wang, J. H. Sugar utilization and cellulase formation by *Clostridium thermocellum*; MS Thesis, 1989, University of Rochester.

Wang, W. K.; Kruus, K.; Wu, J. H. *J. Bacteriol.* 1993, 175, 1293-1302.

Wang, W. K.; Wu, J. H. *Appl. Biochem. Biotechnol.* 1993, 39-40, 149-158.

Wang, W. K.; Kruus, K.; Wu, J. H. *Appl. Microbiol. Biotechnol.* 1994, 42, 346-352.

Wu, J. H. D. Cooperative degradation of crystalline cellulose by the *Clostridium thermocellum* cellulase components; PhD Thesis, 1987, Massachusetts Institute of Technology.

Zhang, Y. H.; Lynd, L. R. *J. Bacteriol.* 2005, 187, 99-106.

Zverlov, V. V.; Velikodvorskaya, G. A.; Schwarz, W. H.; Kellermann, J.; Staudenbauer, W. L. *Appl. Microbiol. Biotechnol.* 1999, 51, 852-859.

Zverlov, V. V.; Fuchs, K.-P.; Schwarz, W. H. *Appl. Environ. Microbiol.* 2002, 68, 3176-3179.

Zverlov, V. V.; Velikodvorskaya, G. A.; Schwarz, W. H. *Microbiol.* 2003, 149, 515-524.

In: Cellulosome
Eds: V. Uversky and I. A. Kataeva, pp. 271-303
ISBN 1-59454-950-8

Chapter XIII

CELLULOSOMES OF ANAEROBIC FUNGI

***Lars G. Ljungdahl*[1,*]*, Huub J.M. Op den Camp*[2]*, Harry J. Gilbert*[3]*, H. R. Harhangi*[2]*, P.J.M. Steenbakkers*[2] *and Xin-Liang Li*[4]**

[1]Department of Biochemistry and Molecular Biology, the University of Georgia, Athens, Georgia, USA;
[2]Department of Microbiology, IWWR, Faculty of Science, Radboud University, Nijmegen, The Netherlands;
[3]Department of Biological and Nutritional Sciences, University of Newcastle upon Tyne, UK;
[4]Fermentation of Biotechnology and Research Unit, National Center for Agricultural Utilization Research, USDA/ARS, Peoria, IL, USA.

ABSTRACT

Anaerobic fungi discovered by Orpin (J. Gen. Microbiol. 1975, *91*, 270) live in close contact with bacteria and other microorganisms in the rumen and caecum of herbivorous animals, where they digest ingested plant food. There are seventeen morphologically distinct anaerobic fungal species described belonging to five genera. They have been isolated from over fifty different herbivorous animals. Anaerobic fungi have efficient system to solubilize lignocellulose and hydrolyze plant carbohydrates. They produce all necessary enzymes for degradation of cellulose and hemicelluloses. The enzymes are exocellular and found free or bound to large cellulosomal and polycellulosomal complexes. Enzymes recognized in the cellulosomal complexes include endo- and exocellulases, β-glucosidases, xylanases, β-glucanases, lichenases, mannanases, α-methylglucuronidases, acetyl-xylan esterases, and feruloyl esterases. All enzymes associated with cellulosomes from anaerobic fungi are modular containing in addition to

* Correspondence concerning this article should be addressed to: L. G. Ljungdahl, Department of Biochemistry and Molecular Biology, The University of Georgia, Athens, GA 30602-7229, USA. Tel.: 706 542 1846; Fax: 706 542 2222; E-mail: larsljd@bmb.uga.edu.

the catalytic modules duplicated docking modules in similarity with enzymes of the cellulosomes of anaerobic bacteria. The presence of scaffolding proteins are also recognized but a scaffoldin from an anaerobic fungus has still to be isolated. Serpins, which are inhibitors of trypsin-like proteinases, are associated with the cellulosomes. The non-catalytic docking domains (NCDD) although having the same functions as dockerins of cellulosomes of bacteria, do not have any sequence homology to them. Judging sequences of the cellulosomal enzymes it is apparent that they originate from different sources including anaerobic and aerobic bacteria, and aerobic fungi. Gene transfer between the anaerobic fungi seems prevalent as is gene duplication.

Keywords: anaerobic fungi, cellulosomes, (hemi)cellulolytic enzymes, gene transfer, serpins, dockerins.

Introduction

It has been estimated that about 150 x 10^9 metric tons of plant biomass is produced per year by photosynthetic fixation of carbon dioxide. The biomass consisting mostly of cellulose (50%), hemicelluloses (25%), and lignin (25%) is ultimately in nature degraded by fungal and bacterial consortia to CO_2, which is returned to the atmosphere [84]. Microorganisms degrading plant biomass are abundant in nature. They include aerobic and anaerobic fungi and bacteria, many of which grow at extreme conditions of temperature and pH. One important habitat for microorganisms degrading plant tissue is the intestine of herbivorous animals, which use plant material for food. These animals do not have glycoside hydrolases needed for the hydrolysis of cellulose and hemicelluloses and thus they are completely dependent on the microorganisms they harbor in their intestines. In this chapter we will discuss the glycoside hydrolytic enzyme system of anaerobic fungi. These fungi were discovered by Orpin [91, 92].

Microorganisms degrading cellulose and plant polysaccharides generally have in common that they produce multiple cellulases and other glycoside hydrolases. These enzymes are usually modular. Thus, they may have additional catalytic modules, which may have similar activity or they may have different catalytic activities making the enzymes bifunctional. The most common modules besides the catalytic are carbohydrate-binding modules (CBMs), which bind the hydrolytic enzymes to the substrates. Other modules of glycoside hydrolases are dockerins. These modules are found in enzymes, which are attached to large exocellular organelles called cellulosomes [10,29,105]. The dockerins bind to cohesin modules of scaffolding proteins thus forming the cellulosome complex. Additional modules, but less common, found in glycoside hydrolases are immunoglobulin(Ig)-like modules and modules designated X. These modules contribute to the thermostability and catalytic activity of glycoside hydrolases [68,69].

It is now apparent that the cellulolytic enzyme systems of aerobic fungi and bacteria differ from those of anaerobic bacteria [114]. Studies of the aerobic fungus *Trichoderma reesei*, the aerobic bacterium *Cellulomonas fimi*, and the anaerobic bacterium *Clostridium thermocellum*; the most investigated cellulolytic microorganisms, exemplify this. *T. reesei* and *C. fimi* produce several well separated cellulolytic enzymes, which are relatively easy to purify and when combined effectively hydrolyze cellulose. In contrast *C. thermocellum* and several other anaerobic bacteria produce large extracellular complexes, named cellulosomes, first discovered by Lamed et al. [70]. The cellulosome from *C. thermocellum* has a molecular mass of about 2 x 10^6 Dalton or larger and it forms polycellulosomes of around 100 million Dalton, which are attached to the bacterial surface and the cellulose substrate [39,89]. It consists of at least 26 different polypeptides, most of which are cellulases or other glycoside hydrolases. As already mentioned these enzymes are modular and they have dockerins, which bind the enzymes to internally repeated domains called cohesins of a scaffolding polypeptide [9]. Details of cellulosomes from *C.thermocellum, Acetivibrio cellulolyticus, Bacteroides cellulosolvens, Clostridium cellulovorans* and *Ruminococous albus* can be found in chapters of this book and in several excellent reviews [9-12,47,48,114].

Anaerobic Fungi: Their Habitats and Properties

Bacteria, protozoa and anaerobic fungi are the three major groups of microorganisms that inhabit the gastrointestinal tract of herbivorous animals. Of these groups, the anaerobic fungi were discovered most recently [92]. The discovery of them was at first met with some skepticism, since they earlier had been considered flagellated zoospores of protozoa. Also the existence of strictly anaerobic fungi was questioned. Now it is acknowledged, that anaerobic fungi are present in many if not most herbivorous animals and that they play a significant role in the digestion of plant material in these animals [111,118]. The classification of the anaerobic fungi has been somewhat controversial. They are of the kingdom Eumycota in the subdivision Mastigomycotina, class Chytridiomycetes, order Neocallimasticales and family Neocallimasticaceae [65]. At present there are 17 morphologically distinct anaerobic fungal species described belonging to 5 genera. Fungi belonging to the genera *Caecomyces, Neocallimastix* and *Piromyces* are monocentric, whereas those belonging to *Orpinomyces* and *Anaeromyces* are polycentric [7,111,118].

Analyses of 18S ribosomal RNA clearly indicate that anaerobic fungi are related to each other [72]. Sequences of their genes encoding enzymes show high homology. Furthermore DNA analyses reveal that the anaerobic fungi have very low GC content ranging from 13% for a *Neocallimastix* sp. [25] to 22% for *Caecomyces communis* [14]. These observations indicate that the fungi although morphologically different, constitute a very homogeneous group [72,111]. Life cycles of the mono- and polycentric fungi are described by several authors and discussed in the review by Trinci et al. [111].

Anaerobic fungi have been isolated from over 50 different herbivorous animals. They have the ability to utilize different carbohydrates of plants including cell walls. The fermentation products are formate, acetate, ethanol, lactate, carbon dioxide, hydrogen and some times a small amount of succinate [8,20,86]. Analyses of culture supernatants and extracts from anaerobic fungi have revealed that they have very active and different glycohydrolytic enzymes including those related to the hydrolysis of hemicelluloses (Fig. 1; 104).

One of the main features of anaerobic fungi is that they do not possess mitochondria. Instead they have hydrogenosomes, which produce H_2 and CO_2 from pyruvate and malate obtained by fermentation of carbohydrates. Anaerobic fungi are very sensitive to oxygen. [124,125]. They are capable of oxygen respiration when the oxygen concentration in liquids is less than 10 μM, but a liquid oxygen content of 1.4 μM causes 50% inhibition of H_2 production. It has been proposed, that hydrogenosomes are of mitochondrial origin. Mitochondria have, but hydrogenosomes at least those from anaerobic fungi seem to lack DNA. However, sequencing of anaerobic fungal genes encoding enzymes found in hydrogenosomes showed that these enzymes are homologous to similar enzymes from mitochondria [1]. Like mitochondria hydrogenosomes are involved in electron transport and energy generation [87,88]. A scheme connecting glucose metabolism with that of the hydrogenosome has been presented by Akhmanova et al. [2]. Discussion of the hydrogenosome, its relation to mitochondria, and its importance for eukaryotes in anaerobic environments are given by Hackstein et al. [59] and Voncken [112].

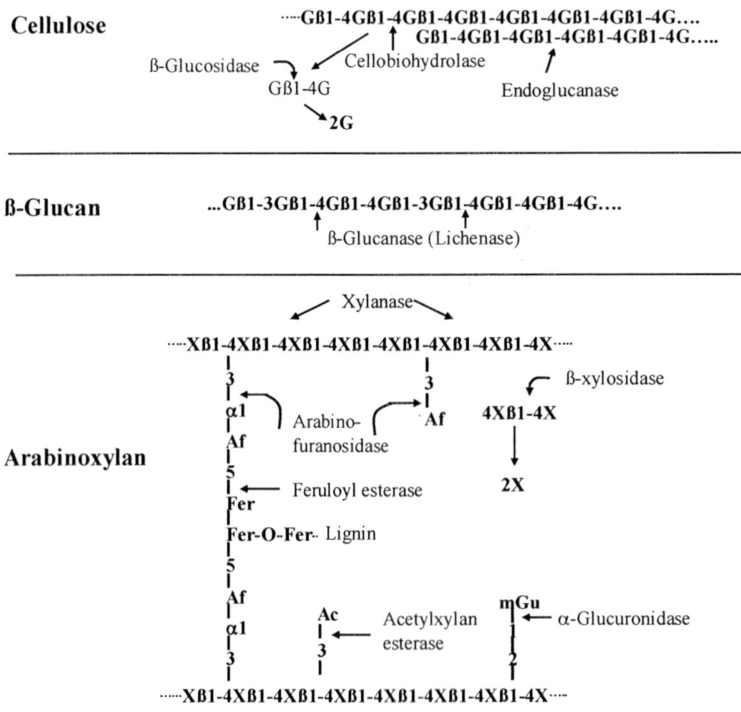

Figure 1. Structures of cellulose, β-glucan, and arabinoxylan and the enzymes needed for their hydrolysis according to Selinger et al., [104].

Techniques for collecting samples containing anaerobic fungi from saliva, feces and intestine are discussed by Wubah [118]. Borneman et al. [20] isolated five morphologically different anaerobic fungi from digestures obtained from a rumen-fistulated cow essentially according to the following procedure. Before the ruminal samples were collected the animal was fed a high fiber diet to stimulate growth of the fungi. The digesture was collected under carbon dioxide to ensure anaerobic conditions in thermos bottles previously heated with warm water (over 40°C). The isolation of the fungi was done using the Hungate technique as modified by Caldwell and Bryant [28] or by using plastic petri dishes kept in anaerobic glove boxes with an atmosphere of 90% CO_2, 5% nitrogen, and 5% hydrogen. The medium used was that of Caldwell and Bryant [28] containing cellobiose, xylan or cellulose and fortified with 4,000 units of penicillin per ml, 260 units per ml of streptomycin, and chloroamphenicol 30 μg to inhibit bacterial growth. Individual colonies developed were subcultured to ensure purity. To maintain the fungal cultures, they were transferred to fresh media every five to seven days maintaining anaerobic conditions and a temperature of 40°C. Our experience is that the fungi are very sensitive to temperature and oxygen. The five different isolates obtained included two monocentric (MC-1 and MC-2) and three polycentric (PC-1, PC-2 and PC-3) species as shown in Fig.2. The MC-1 and MC-2 isolates resembled members of *Piromyces* and *Neocallimastix*, respectively. PC-2 and PC-3 were later considered to be *Orpinomyces* species [119].

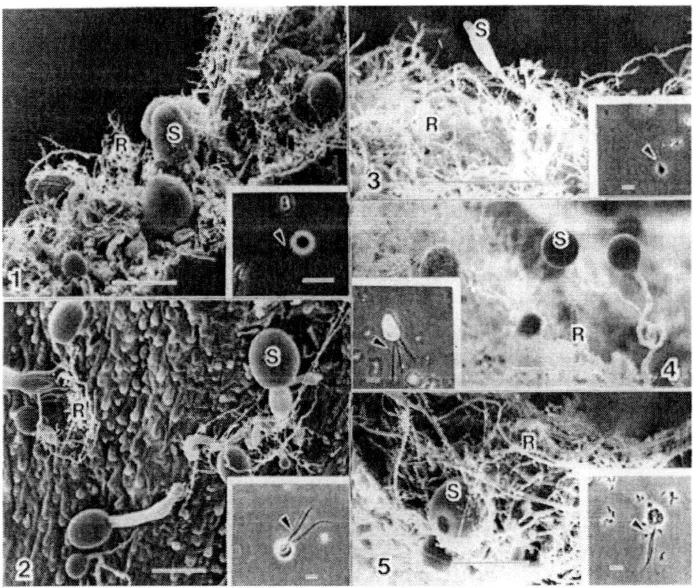

Figure 2. Colonization of anaerobic fungi on CBG leaf blades showing development of sporangia (S), rhizoidal structures (R), and uni- or multiflagellate zoospores (inset, arrowheads) for each of five isolates, Bars, 50 m; inset bars, 10 m. A. Monocentric isolate MC-1 with uniflagellate zoospore. B, Monocentric isolate MC-2 with multiflaggellate zoospore. C. Polycentric isolate PC-1 with uniflagellate zoospore. D. Polycentric isolate PC-2 with multiflagellate zoospore. E. Polycentric isolate PC-3 with multiflagellate zoospore [From Borneman et al., 20].

Cellulosomes of Anaerobic Fungi

Realizing that anaerobic fungi effectively hydrolyze cellulose, hemicelluloses, and other plant materials, the question must be asked whether they produce separate enzymes like the aerobic cellulolytic fungi and bacteria or do they have a cellulosomal-type cellulolytic enzymes system similar to that of *C. thermocellum* and other anaerobic bacteria. The answer is that they produce cellulosomal-type complexes and in addition some free separate glycolytic enzymes. Wilson and Wood [116] were first to report about a cellulosomal type of complex produced by anaerobic fungi. They isolated from the culture fluid of 9 days cultures of *Neocallimastix frontalis* grown on Whatman cellulose filter paper a complex with a molecular mass of 670 kDa. The complex, which had very high activity toward cotton fiber, consisted of at least 6 different proteins with masses from 56 to 126 kDa. Cellulosomal type complexes are also produced by strains of *Piromyces* [4, 45-46, 53, 110] and *Orpinomyces* sp. PC-2 [76].

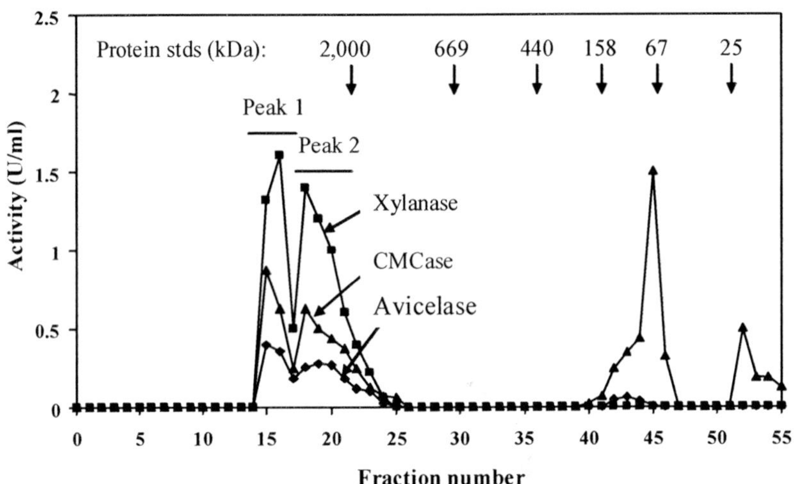

Figure 3. Gel-filtration profile using a Sepharose B6 column of proteins extracted by distilled water from residual Avicel obtained from a four-days old culture of *Orpinomyces* sp. PC-2.

In studies of an endoxylanase from *Piromyces* sp. E2, Teunissen et al. [110] noticed that xylanase active proteins emerged from Sephacryl S-300 gel filtration columns as a high molecular mass protein complex of about 1,200 kDa and as a low molecular mass protein of about 20 kDa. The complex was very stable but it was dissociated with zwitterionic detergent CHAPS. Xylanase was purified from the dissociated complex and this xylanase was found to have properties identical to the xylanase purified from the low molecular fraction. Dijkerman et al. [45,46] studying the complex from *Piromyces* sp. E2 found it to have a mass between 1,200 to 1,400 kDa and with avicelase, endoglucanase, and β-glucosidase activities. It contained at least 10 polypeptides. Also observed was that the major part of the cellulase activity produced by the fungus when it was grown on cellulose resided in the complex. Having cellulose in the growth medium induced the formation of the complex. Ali et al. [4] isolated a complex from a *Piromyces* sp. (*P. equi*) having a molecular mass of 670 kDa, which binds to cellulose and is eluted from it with water. The complex contained at least 10 polypeptides with masses from 50 to 190 kDa several of which had endoglucanase, xylanase and mannanase activities [63]. Fanutti et al. [53] obtained the primary structure of a xylanase A and a mannanase A from *Piromyces*. The enzymes in addition to the catalytic modules contained almost identical reiterated noncatalytic 40-residues sequences, which were found to bind specifically to 97 and 116 kDa polypeptides, which are components of the complexes isolated from *Piromyces* and *Neocallimastix patriciarum*, respectively. These properties indicate that the reiterated sequences are functioning as a noncatalytic docking domain

(NCDD) similarly to dockerins present in enzymes bound to cellulosomes from anaerobic bacteria. That NCDDs bind specifically to the 97 and 116 kDa proteins identifies them as scaffolding polypeptides of the complexes. The data presented by Fanutti et al. [53] suggest that anaerobic fungi, indeed, have cellulosomes. Although NCDDs function as dockerins, throughout this article we will use the designation NCDD for them, because their sequences are completely different from those of the bacterial dockerins [98,107].

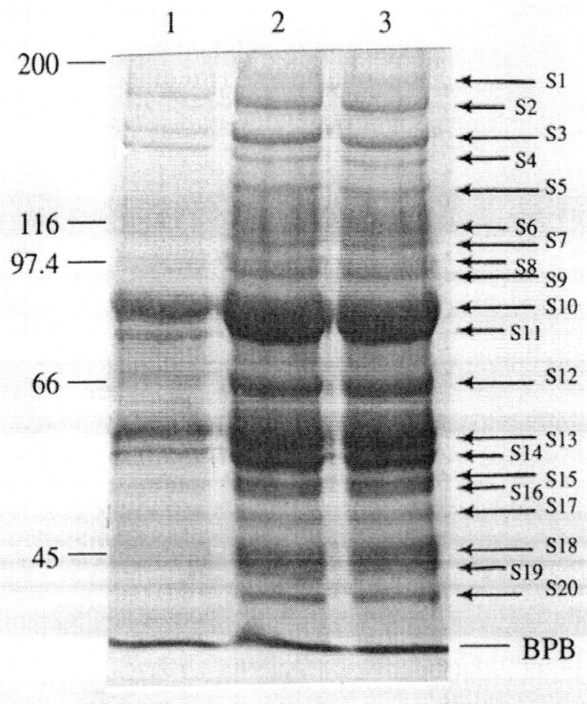

Figure 4. SDS-PAGE of proteins of peaks 1 and 2 from Fig. 3. Numbers on the left constitutemolecular mass of protein standards in kDa; numbers on the right are protein bands S1 to S20. Lane 1, 50 μg of protein from peak 1; lane 2, 200 μg of protein from peak 1; lane 3, 200 μg of protein from peak 2.

The cellulosomes from *Orpinomyces* sp. PC-2 were isolated from 3 - 4 days cultures grown on Avicel or Coastal Bermuda grass. Although cellulosomes were found in the supernatant of the growth culture, most of them were found tightly bound to residual substrate particles from which they were easily isolated using methods similar to those used for isolation of cellulosomes from *C. thermocellum* [83]. The insoluble substrate particles recovered from the growth culture by centrifugation were washed with 0.1 M buffer, pH 7, or salt solutions to remove proteins with little or no glycoside hydrolase activity. Subsequent washing with distilled or deionized water eluted a protein fraction with high activity against carboxymethylcellulose, Avicel and oat spelt xylan. The protein in the distilled water fraction was applied to a Sepharose 6B gel-filtration column 6B. Most of the protein eluted in two fractions (Fig. 3, peaks 1 and 2) in front of Blue Dextran used to determine the void volume of the column. This indicated that the molecular masses were over 2,000 kDa. Some additional protein with cellulase activity was eluted at places corresponding to molecular masses of about 70 and 24 kDa. Proteins of the two high molecular mass peaks had activities of xylanase and cellulase (Avicelase and CMCase, Fig 3); the lower molecular mass fractions possessed CMCase and some Avicelase activity. Fractions of peaks 1 and 2 were analyzed by SDS-PAGE (Fig. 4). Their protein profiles were identical and included at least 20 proteins with molecular masses from about 30 to 190 kDa. More detailed evaluation of the enzymatic activities of the proteins of peaks 1 and 2 revealed, that in addition to those listed above they hydrolyzed *p*-nitrophenyl(pNP)-glucoside, pNP-cellobioside, pNP-arabinoside, pNP-xyloside, and FAXX (Table 1), which showed that the peaks contained a range of glycohydrolases, and feruloyl esterase. The rate of hydrolysis of oat spelt xylan was very high indicating the presence of enzymes active against hemicelluloses.

Table 1. Specific activities of the cellulose/hemicellulase complex isolated from *Orpinomyces* sp. PC-2 culture.

S u	S p
C	87
Avicel	10.2
pNP-glucoside	23.0
pNP-cellobioside	17.4
O	11
pNP-xyloside	12.6
F	4.
pNP-arabinoside	0.005

*FAXX means *O*-{5-*O*-[(*E*)-feruloyl]-α-L-arabinofuranosyl}-(1→3)-*O*-β-D-xylopyranosyl-(1→4)-D-xylopyranose.

Proteins of peaks 1 and 2 were examined by electron microscopy using negative staining as described by Mayer et al. [89] for EM studies of the *C. thermocellum* cellulosome. This revealed that peaks 1 and 2 consisted of distinct particles of sizes 80 (peak 1) and 25 (peak 2) nanometers. Assuming the particles were spherical they would have molecular masses of about 3,000 and 80,000 kDa, respectively which are sizes corresponding to the cellulosome and polycellulosome of *C. thermocellum* [39]. The complexes were found to bind to Avicel as observed with electron microscopy (Fig. 5).

The cellulosomal complexes of *Orpinomyces* sp. PC-2 are located along the fungal mycelium and particularly at the surface of the growing tips (Fig.6) This was demonstrated using antibodies against XynA, a component of the *Orpinomyces* cellulosome [77], and gold labeling technique. The demonstration that the cellulosomes bind to both cellulose and to the surface of the fungal mycelium indicates that they mediate the attachment of the mycelium to the plant tissue. That the cellulosomes are mostly clustered at the growing tips of the mycelium indicate that the mycelium has available enzymes, which should allow it to penetrate through the hemicellulose/lignin layer of the plant tissue into the cellulose. This has been observed with the use of scanning electron microscopy [3].

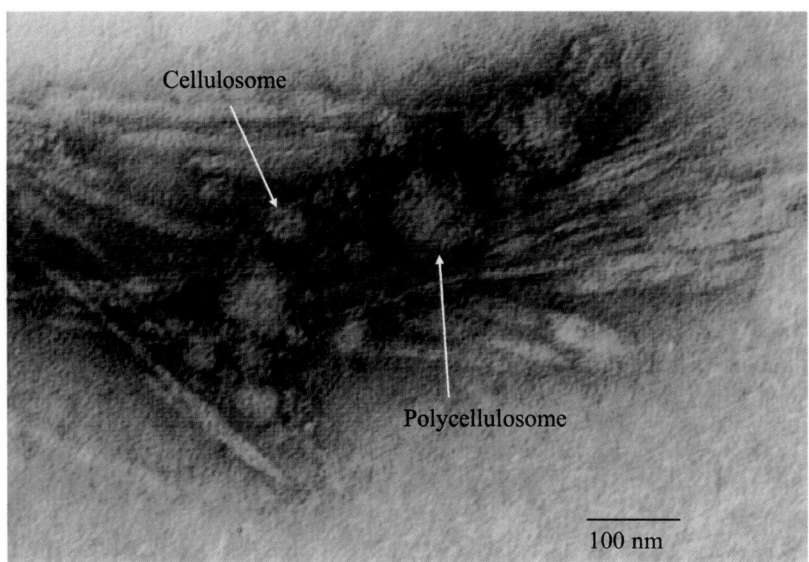

Figure 5. Electron microscopy using negative staining as described by Mayer et al., [89] showing cellulosomes of 25 nm in diameter, and polycellulosomes of 80 nm from *Orpinomyces* PC-2. Calculation of molecular mass indicates the cellulosome to have a mass of 3,000 kDa and the polycellulosome of 80,000 kDa.

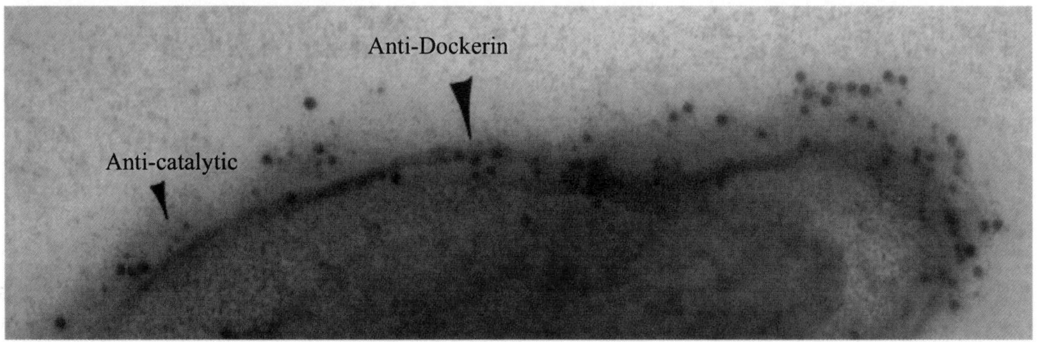

Figure 6. Localization of the *Orpinomyces* sp. PC-2 cellulosomes on the filamentous rhizoid as proven with gold-labeled antibodies against the catalytic and the dockerin domains of cellulosomal XynA.

Enzymes of Cellulosomes from Anaerobic Fungi

Anaerobic fungi effectively hydrolyze a wide range of plant carbohydrates including cellulose and hemicelluloses. Thus, one may expect that they have a complete set of enzymes shown in Fig.1. This has been demonstrated for *Neocallimastix, Piromyces*, and *Orpinomyces* species. Most of the enzymes are bound and found in cellulosomal complexes, but free enzymes have also been isolated. The enzymes with few exceptions are modular, and those containing in addition to the catalytic modules NCDDs are considered being associated with cellulosomes. Glycoside hydrolases or carbohydrate-active enzymes, of which several hundreds have been sequenced, are classified according to the sequences and modular structures of the catalytic modules. This classification also includes carbohydrate binding modules [37, 38].

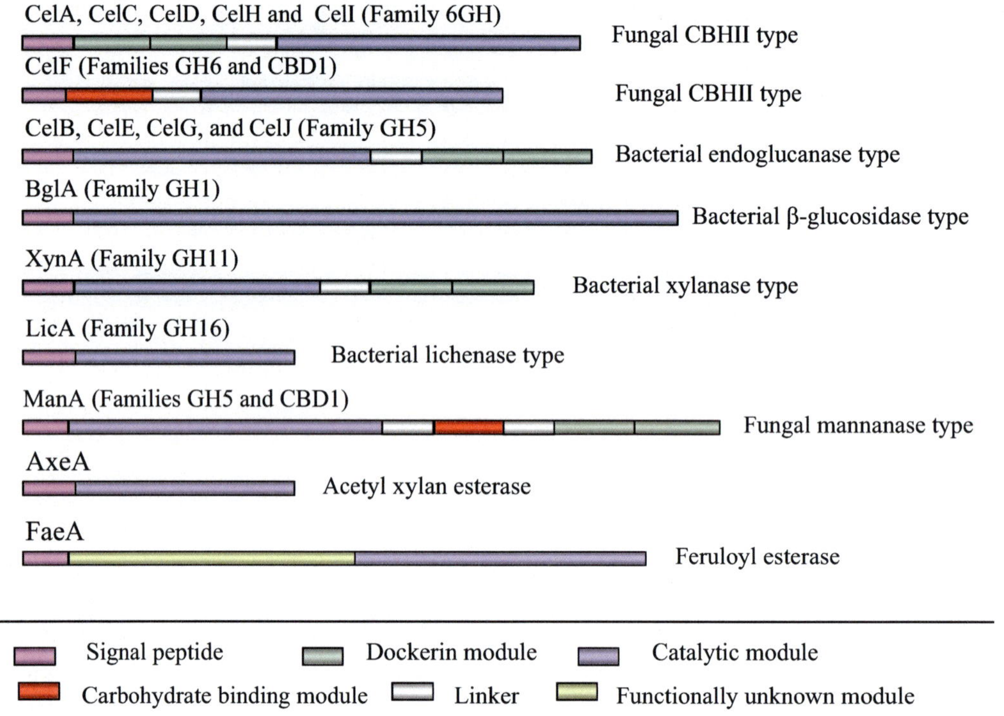

Figure 7. Module structures of glycoside hydrolases from *Orpinomyces* sp. PC-2 based on sequencing of cDNAs.

Figure 7 shows enzymes from *Orpinomyces* sp. PC-2, which are typically found in anaerobic fungi. Genes encoding these enzymes were all isolated from an *Orpinomyces* sp. PC-2 cDNA library in phage λZAPII [33] using mRNA from cells grown on Avicel and xylan. To isolate if possible a full spectrum of genes encoding the carbohydrate-active enzymes of the fungus from the library three different approaches were employed. Cellulase A (CelA), CelB, CelC, CelE, CelF, xylanase A (XynA), acetyl xylan esterase (AxeA) and lichenase A (LicA) were isolated by screening for enzyme-producing plaques using substrates for the enzymes. Genes for feruloyl esterase (FaeA) and β-glucosidase (BglA) were obtained by first determining the amino terminal and an internal peptide sequence of the enzymes. Based on these sequences oligonucleotide probes were constructed, which were used to isolate cDNAs by reverse genetic procedures. Genes for CelH, CelI, CelJ and mannanase (ManA) were obtained by a method specifically selecting genes encoding enzymes, which in addition to the catalytic modules have NCDDs binding the enzymes to cellulosomes of anaerobic fungi [107]. Using this method, which will be discussed later in connection with properties of NCDDs and scaffolding proteins from anaerobic fungi, it was found that *Piromyces* sp. E2 and *Orpinomyces* sp. PC-2 contain 19 and 16 distinctive non-catalytic dockerins, respectively. This indicates that the two fungi have at least 19 and 16 different catalytic polypeptides bound to their respective cellulosomes. As shown in Fig. 7, eleven of the enzymes listed have NCDDs. An additional cellulase CelK has been identified having NCDD (Xin-Liang Li, personal com.). These enzymes may be considered cellulosome associated. This means for the *Orpinomyces* sp. PC-2 cellulosome, that there are at least six additional polypeptides with NCDDs yet to be recognized bound to the cellulosome. CelF, BglA, LicA, AxeA and FaeA do not have dockerins and they must be considered to be free enzymes.

Cellulases

There are ten cellulases listed in Fig.7. The catalytic modules of six of these enzymes belong to glycoside hydrolases family 6, the other five belong to family 5 [37-38]. CelF of family 6 differs from all the other cellulases in that it does not have a NCDD; instead it has a carbohydrate-binding module (CBM) belonging to family 1. In addition the gene encoding CelF has an 111-bp intron located in the CBM region [31]. The catalytic module of CelF has identity levels between 56 to 76% with CelA, CelC, and CelD, but only between 34 to 40% with CelH and CelJ [80]. It has also high homology with CELA from *N. patriciarum* and cellobiohydrolase from the aerobic fungus *Trichoderma reesei* [43]. Introns disrupting coding sequences are considered a criterion for eukaryotic origin of genes, since introns in genes of prokaryotic organisms are rare. Thus, the presence of an intron in *CelF* and its homology with cellobiohydrolases from aerobic fungi indicate that family 6 cellulases in both aerobic and anaerobic fungi have the same origin. In contrast to family 6 glycoside hydrolases, family 5 glycoside hydrolases CelB, CelE, CelG, and CelJ appear to have an origin similar to bacterial endoglucanases. CelB has 83.1% identity with CelB from *N. patriciarum* including both the catalytic module and the NCDD [126] and also with CelA from *N. frontalis* [55]. These cellulases have high homology with several endoglucanases from both anaerobic and aerobic bacteria [77]. *N. pariciarum* like *Orpinomyces* sp. PC-2 have multiple cellulases. Thus in addition to CelB Xue et al. [123] isolated CelA and CelC. CelA was considered to be a cellobiohydrolase, whereas CelC like CelB is an endoglucanase. Three family 5 endoglucanases CelA, CelB29, and CelB2 have been isolated from *Orpinomyces joyonii* [81, 96]. CelA has a N-terminal catalytic module followed by two repeated sequences of 163 and 165 amino acid residues, respectively. The repeated sequences not needed for activity are homologous with a noncatalytic module of unknown function of an endoglucanase (EG3) from *Fibrobacter succinogenes*. However, the catalytic modules of CelA and EG3 were not homologous indicating a shuffling of enzyme domains between bacteria and fungi occur in the rumen. CelB29 and CelB2 have 57% homology, but CelB29 consists of 477 amino acids whereas CelB2 has 388. The missing residues are from the N- and C- terminal regions. Both enzymes seem to have NCDDs and be cellulosomal bound. A remarkable cellulase, CelD, has been isolated from *N. patriciarum*. It has three multi-functional catalytic modules with high endoglucanase, cellobiohydrolase and xylanase activities [122]. The three modules separated by proline, threonine and glutamic acid rich linkers have high degree of identity and appear to be functionally very similar, which was shown by expressing the catalytic modules separately and determining their activities [5]. The C-terminal end of CelD contains a NCDD, which suggest the enzyme being a part of the *N. patriciarum* cellulosome. *Piromyces equi* has two unusual endoglucanases Cel5A and Cel45A [52]. Cel5A is large with a molecular mass of 194,847 Da. It contains three repeats of the 40-residues sequences of a NCDD; one is at the N-terminal, the other two at the C-terminal. Between these NCDD sequences it has four modules belonging to family 5 glycoside hydrolases. It was suggested, that Cel5A may have arisen by the fusion of two endoglucanases genes encoding tandem catalytic sites with NCDDs. Cel45A is the only family 45 glycoside hydrolase so far found in anerobic fungi and it suggests it is originated from an aerobic fungal gene. Cel45A is unusual also in that it has three of the NCDD repeats, all three located at the N-terminal.

The cellulosome of *Piromyces* sp. E2 has three dominant proteins of 55, 80, and 90 kDa. A cDNA of 2.8 kb containing an open-reading frame of 2,334 nucleotides encodes a 778-residue protein, Cel9A with a molecular mass of 83,811 Da that has the N-terminus of the 90 kDa cellulosomal protein [108]. Cel9A has a 21-residue N-terminal signal peptide, a 445-residue glycoside hydrolase family 9 catalytic module which is followed by a 160 amino acids module of unknown function, a threonine-rich linker and three copies of the NCDD sequence. Of the three copies the first is classified as dockerin type 2 with four cysteine residues, and the following two copies as type 3 with six cysteine residues [108]. Modelling of the NCDD sequences indicates that the cysteine residues form disulfide bridges as found in the structure of NCDD from *P. equi* [98]. The family 9 glycoside hydrolase includes several clostridial cellulosomal proteins but Cel9A seems to have the highest homology to enzyme E4 from *Thermobifida fusca* (45% identity and 60% similarity). The structure of the *T. fusca* E4 has been determined [101]. Homology modelling of Cel9A reveals several features similar with E4 enzyme although the E4 contains a CBM apparently involved in orienting the substrate toward the active site. The Cel9A does not have the CBM.

β-Glucosidases

Cellobiose is the main product of the hydrolysis of cellulose by endoglucanases and cellobiohydrolases. However, cellobiose is a major inhibitor of cellulases. Thus, to have an efficient cellulolytic enzyme system it requires the synergistic action of endoglucanases, cellobiohydrolases, and β-glucosidase. The latter hydrolyzes cellobiose to glucose (Fig. 1). β-Glucosidases have been isolated from the culture fluid from *N. frontalis* Eb188 [74,75,64,115] *Piromyces* sp. E2 [109] and *Orpinomyces* sp. PC-2 [32]. These enzymes have in common high molecular masses from about 80,000 to 160,000, are found both in the cultural fluid and in the cytoplasm, and contain varying amount of carbohydrates. More recently two β-glucosidase genes (*cel1A and cel3A*, accession numbers AJ276438, and AY172977) have been sequenced from *Piromyces* sp. E2 [61,106] and one gene (*bglA* AF016864) from *Orpinomyces* sp. PC-2 [79]. BglA and Cel1A are very similar. They belong to glycoside hydrolase family 1, have molecular masses of 75,227 and 75,800 Da with an homology of 72%, respectively, lack NCDDs and are not part of the cellulosomes, are glycosylated to varying degrees, and did not produce active proteins when expressed in *E. coli*. Active enzymes were obtained for BglA in the yeast *Saccharomyces cerevisiae* and for Cel1A in *Pichia pastoris*. The enzymes have significant homology with β-glucosidases from bacterial and plant sources but not to those of aerobic fungi. In fact a phylogentic tree with known sequences of family 1 glycoside hydrolases shows that Cel1A and BglA form a distinct separate cluster [62]. Cel1A was highly expressed in *Piromyces* grown on fructose. Cel3A β-glucosidase belongs to glycoside hydrolases family 3 and its catalytic site associates more with fungal β-glucosidases than with those of bacteria. It has a molecular mass of 91,457 Da and modular structure with the catalytic site at the N-terminal followed by an auxiliary module and three copies of NCDDs typical for cellulosomal proteins. Cel3A is the first β-glucosidase known to be associated with a cellulosome, of which it is a minor component. However, it is responsible for the cellulosomes activity to produce glucose.

Xylanases

Orpin and Letcher [95] first observed that *N. frontalis* grows on polysaccharides including cellulose, starch and xylan. Later Lowe et al. [86] found that culture supernatants of a *Neocallimastix* sp. R1 growing on larchwood xylan had high levels of xylanase activity, and that arabinose and xylose rapidly accumulated in the medium. The five anaerobic isolates from a rumen of cow MC-1, MC-2, PC-1, PC-2 and PC-3 representing *Piromyces, Neocallimastix,* and *Orpinomyces* species when grown on Coastal Bermuda grass had in the culture supernatant from five to seven times more xylanase than cellulase activity [20]. In addition β-glucosidase, β-xylosidase, acetyl xylan esterase, and esterases releasing ferulic and *p*-coumaric acids from plant walls were present in the culture media [19]. This indicates that the anaerobic fungi have the enzymes needed for the complete and efficient degradation of most hemicelluloses.

Three xylanases, XYLA, [57], XYLB [15], and XynC [82] have been described for *N. patriciarum.* XYLA and XynC belong to glycoside hydrolases of family 11, which contains several bacterial xylanases. The molecular mass of XYLA is 66,192 Da. After an N-terminal signal peptide it has two 225-amino acid repeated sequences. Following them are a threonine/proline linker, and a tandem 40-residues C-terminal, which clearly constitutes a NCDD. The N-terminal repeated sequences exhibit 82.9% identity and they constitute two distinct catalytic modules displaying similar substrate specificities, which suggest gene duplication. XynC has a mass of 50,192 Da. It has one catalytic module, which has high identity with those of XYLA. It is connected to a C-terminal CBM of Family 1, and a NCDD. Activity of the enzyme was not changed by deletion of the NCDD, but the CBM was required for optimal activity of the enzyme. That homology observed between catalytic modules of family 11 xylanases from anaerobic fungi and rumen anaerobic bacteria suggests a horizontal transfer of genes between the bacteria and the fungi [57]. XYLB of *N. patriciarum* has a mass of 88,066 Da [15]. It has an unusual module structure. It starts from the N-terminals with a 21-residue signal peptide, followed by a 304-amino acid sequence constituting a family 10 catalytic module. It is followed by two regions designated B and C. The B region has 455 amino acid residues. It contains an octapeptide with the sequence XSKTLPGG where the X being either S, N or K. The sequence TLPG is repeated 57 times. The role for the B region is not known, but it resembles much shorter linker regions of from 10 to 50 residues found in many glycoside hydrolases with S, T, and P as prominent amino acids. The C region constitutes the C-terminus of XYLB. It has 70 amino acid residues and may contain a CBM of family one, but its sequence differs to the extent, that it may have some other function. A gene encoding an endoxylanase from *N. patriciarum* 27 was cloned and expressed in *E. coli* HB101 [71]. The enzyme located in the periplasmic space of the *E. coli* host. In addition to hydrolyzing oat spelt xylan it hydrolyzed carboxymethyl cellulose at a rate about 28% of that of xylan. It bound to crystalline cellulose but lacked activity against it.

A number of xylanases have been isolated from *N. frontalis*. Two 1,4-β-xylan endohydrolases, xylanase I and xylanase II with molecular masses 45 and 70 kDa, respectively, belonging to family 10 were shown to be active toward both xylan and carboxymethyl cellulose [58]. Xylanase I did not have any glycosylation. It adsorbed on crystalline cellulose indicating the presence of a CBM. In contrast, xylanase II was a glycoprotein and it did not bind to cellulose. Different cDNAs of 2.3 and 0.95 kb designated *xyn3* and *xyn4*, respectively, have been isolated from an expression library of *N. frontalis* [49]. The two genes were not homologous although they encoded endoxylanases. The product of *xyn3*, XYN3, was characterized. It has a molecular mass of 66,000 Da and consists of 607 amino acids. After a signal peptide of 31 residues it has two homologous (88.3 % identity) modules of 223 amino acid residues each, which when cloned separately had endoxylanase activity and belonging to glycoside hydrolases family 11. The C-terminal end has the non-catalytic reiterated 40-residues sequence synonymous with NCDDs indicating that XYN3 is cellulosome associated. The recombinant enzyme was expressed in *E. coli*, but it formed inclusion bodies and was not secreted. The N-terminal catalytic module of XYN3 designated XYN3A was however, successfully expressed in and secreted by *Kluyveromyces lactis* and *Penicillium roqueforti* [50]. The secreted enzyme from *P. roqueforti* was heavily glycosylated having a mass of 58 kDa compared to the non-glycosylated form of XYN3A with a mass of 35 kDa.

A cellulosomal complex and a free protein fraction obtained from *Piromyces* sp. strain E2 were both found to have xylanase activity. The two xylanses were purified to homogeneity and biochemically characterized [110] The enzymes seem to be identical having the same molecular mass of 12.5 kDa, the same K_m for oat spelt xylan (3 mg · ml^{-1}), and V_{max} (2,600 IU · mg^{-1} of protein). The products were mostly xylo-oligosaccharides indicting the enzymes to be endoxylanases. The enzymes were not active against Avicel and carboxymethyl cellulose. A cDNA library constructed from a *Piromyces* sp. (*P. equi)* isolated from the caecum of a horse [4, 94] contains the gene encoding a xylanase (XYLA) with a molecular mass of 68,049 Da containing 625 amino acid residues [53]. The enzyme expressed in *E. coli* hydrolyzed oat spelt xylan and wheat and rye arabinoxylan, but had no activity against mannan, arabinan, galactan or cellulosic polysaccharides. The N-terminal started with a typical signal peptide, which was followed by two modules of 223 amino acids that had 57% identity to each other and sequence identity with glycoside hydrolases of Family 11. These modules were separated by a 40-residue repeated sequence, which was identified as a NCDD. The position of a NCDD between two apparent catalytic sites is rather unusual. NCDDs are mostly located at C-termini and somewhat less common at N-termini. Cloning of the C-terminally located catalytic module led to the expression of an active xylanase. In contrast the expressed N-terminal module was inactive.

A gene, *xynA*, encoding for XylA has been isolated from a cDNA library of the polycentric fungus *Orpinomyces* sp. strain PC-2 [78]. The open reading frame (ORF) of *xynA* encodes a xylanase (XynA) of 362 amino acids with a calculated molecular mass of 39,542 Da. This size is about half of the sizes for XYLA from *Piromyces* [53] XYN3 from *N. frontalis* [49], and XYLA from *N. patriciarum* [57]. The size differences are due to the presence of one catalytic module in the *Orpinomyces* XynA whereas there are two in the XYLAs from *Piromyces* and *Neocallimastix.* The *Orpinomyces* XynA has from its N-terminus a signal peptide, a glycoside hydrolase family 11 catalytic module, a linker, and a NCDD (Fig. 7). The catalytic module has about 90% identity with those of the enzymes from the *Neocallimastix* species, but only 34% with XYLA from the *Piromyces*. The very high homology between the catalytic sites of xylanases and also cellulases from *Neocallimastix* and *Orpinomyces* suggest horizontal transfer of genes between these fungi and that they are related on the phylogenetic level although they differ morphologically. Antibodies were produced separately against synthetic peptides based on sequences of the catalytic (OPX1) and the NCDD (OPX2) modules of *Orpinomyces* XynA. Proteins from supernatants of *Orpinomyces* sp. PC-2 and *N. frontalis* EB188 [6] separated by SDS-PAGE were analyzed by Western blots with the OPX1 and OPX2 antibodies. OPX1 reacted with two and one proteins from the *Orpinomyces* and one from *N. frontalis*, respectively. The masses of the proteins were 39.5 and 29 kDa for *Orpinomyces* and 65 kDa for *N. frontalis.* The latter corresponds with XYLA from *N patriciarum* [57] and XYN3 from *N. frontalis* [49]. The 39.5 kDa protein corresponds to XylA of *Orpinomyces*. The 29 kDa protein seems to correspond to a xylanase of *Orpinomyces* believed to be a truncated (loss of the NCDD) form of XylA [35]. In contrast OPX2 reacted with about 10 different extracellular proteins from both the *Orpinomyces* and *Neocallimastix* fungi indicating that many proteins from them have NCDDs and thus may be bound to the cellulosomes of the two fungi. The catalytic module of XylA from *Orpinomyces* PC-2 cloned into *E. coli* was expressed at a high level, but not secreted. About four g of xylanase with a specific activity of 3,500 μmol / min /mg using birchwood xylan as substrate was obtained from 100 g of *E. coli* cells. This activity is among the highest reported for a xylanase [42]. The high activity of the xylanase is just an example of generally very active enzymes from anaerobic fungi.

β-Glucanases / Lichenases

Mixed linked 1,3-1,4-β-D-glucans are major parts of cell walls of cereals, such as barley and oat [26]. 1,3-1,4-β-D-Glucan 4-glucanohydrolase (β-glucanase; lichenase) (EC 3.2.1.73) cleaves β-1,4 linkages adjacent to β-1,3 bonds in glucans. β-Glucanases are present in plants and bacteria and include a LicB of *C. thermocellum,* which is associated with the cellulosome [102]. A 971-bp cDNA has been isolated from the gene library [33] of *Orpinomyces* sp. PC-2 [34]. It encodes LicA (1,3-1,4-β-D-glucanase / lichenase), which has 245 amino acids and a molecular mass of 27,929 Da. It has a 29 amino acid N-terminal signal peptide, which is followed by a catalytic module belonging to glycoside hydrolase family 16, and it lacked a NCDD indicating that LicA is a free enzyme and is not associated with the cellulosome. The enzyme has high homology (>50%) to several sequences of bacterial β-glucanases, which suggest a bacterial origin of the gene. The *licA* gene inserted into *E. coli* was expressed and 95% of the enzyme was found in the culture fluid. The high secretion of LicA may be relate to a finding that β-glucanases expressed in *E. coli* alters the permeability of the outer membrane and allows leakage of periplasmic enzymes to the culture medium [27]. With barley β-glucan as substrate LicA had a specific activity of 5,320 μmol/min/mg. It did not hydrolyze laminarin, carboxymethylcellulose or xylan. Attempts to find the enzyme in *N. frontalis* EB188 were unsuccessful.

Mannanases

Ali et al. [4] recognized several mannanases (EC 3.2.1.78) all binding to a cellulosomal complex from *P. equi*. Three of them have been sequenced; MANA by Fanutti et al. [53] and MANB and MANC by Millward-Sadler et al. [90]. In addition, the sequence of one mannanase (ManA) also cellulosomal from *Orpinomyces* is known [120]. The catalytic modules of the three mannanases from *P. equi* belong to glycoside hydrolases family 26 and they displayed extensive sequence identity with each other. MANA has 606 amino acid residues and a molecular mass of 68,055 Da. Corresponding data for MANB and MANC are 571 and 569 amino acid residues and masses of 64,403 and 64,156, respectively. MANA is somewhat larger than MANB and MANC, which seems to depend on that MANA has three copies of the NCDD sequence whereas MANB and MANC have the commonly present reiterated NCDD sequence. The module organization is identical for the three mannanases. Starting with the N-termini they have typical signal peptides, which are followed by the catalytic modules, linkers, and at the C-termini the NCDDs. That the three mannanases have the docking modules indicate that they are components of the *P. equi* cellulosome. The very high similarity between the three mannanases, which seem to have bacterial origin, suggests that they arose through gene duplication. The *P. equi* mannanases hydrolyze Carob, galactomannan, and ivory nut mannan, but had no activity against barley β-glucan, carboxy methyl cellulose, Avicel, xylan, galactan or arabinan.

ManA from *Orpinomyces* sp. PC-2 is very different from the mannanases from *P. equi*. The mass of the enzyme calculated from the cDNA is 64,425 Da consisting of 579 amino acid residues. Starting from the N-terminus it has a signal peptide of 20 amino acid residues, which is followed by a 350 amino acid residues catalytic module belonging to glycoside hydrolases family 5. A linker, a CBM of family 1, a linker, and a NCDD follow the catalytic module (Fig. 7). The catalytic module and the CBM have homology with similar modules of mannanases and other glycoside hydrolases of family 5 from aerobic fungi and both seem to have fungal origin. In contrast the NCDD is typical only for anaerobic fungi. Attempts to express the catalytic module of ManA with the signal peptide and the CBM in *E. coli* were unsuccessful. It was, however, expressed in *S. cerevisiae,* and mannanase activity was found cell associated and secreted to the culture medium. The V_{max} and the K_m for the purified enzyme with locust bean mannan was 179 U/mg and 3.4 mg/ml, respectively. It did not hydrolyze barley β-glucan, carboxymethylcellulose or xylan. The CBM of ManA expressed separately binds preferably to insoluble polysaccharides that include cellulose and hemicelluloses. It actually binds better to crystalline cellulose and insoluble xylan than to mannan, which suggests, that the CBM facilitates the binding of the *Orpinomyces* cellulosomal complex to lignocellulolytic structures.

Acetylxylan Esterases

From 20 – 50% of xylose moities in xylan may be acetylated at O-2 or O-3 positions [36]. Acetylxylan esterases (EC 3.1.1.72) remove these O-acetyl groups, which hinder the hydrolysis of xylan by xylanases [13]. Acetylxylan esterases have been found in monocentric and polycentric anaerobic fungi [20], and they are present in several bacteria including those of the rumen and in aerobic fungi. Dalrymple et al. [41] screened a cDNA expression library of *N. pariciarum* for esterase genes using as substrates β-naphthyl acetate and a model cinnamoyl ester compound. Four genes designated *bnaA, bnaB, bnaC, and bnaD* encoding proteins BnaA-D were found with acetyl esterase activity but lacking activity against cinnamoyl esters. BnaA with 393 amino acids contain a signal sequence, catalytic module, linker, and NCDD. A BLAST search did not show any sequence match with the catalytic module of BnaA. BnaB with 389 amino acids has a module composition similar to BnaC. Its catalytic module did not have sequence homology with BnaA, but homology with modules of unknown function of XynB a xylanase from *Ruminococcus flavefaciens* and CelE and CelX of *C. thermocellum* [41]. The unknown region of *C. thermocellum* CelE has been now shown to have esterase activity. BnaB, not fully sequenced, has a catalytic module with some homology with the one from BnaC. These findings indicate the presence of two different types of acetylxylan esterases in *N. patriciarum*. BnaD is an intracellular enzyme belonging to an acetylcholin-esterase-cotaining family of esterases and lipases [41]. Purified recombinant BnaA cloned and expressed in *E. coli* had a specific activity of 76 U/mg with soluble acetylated birchwood xylan as substrate, but it had very little activity against spear grass (*Heteropogon contortus*) or delignified spear grass. In combination with the recombinant xylanase, XynA, also from *N. patriciarum,* a substantial increase of hydrolysis of the three substrate were obtained as determined by measuring the release of both xylose and acetate demonstrating a synergetic interaction between the acetylxylan esterase and the xylanase [40].

A cDNA obtained from the *Orpinomyces* sp. PC-2 library codes from an acetylxylan esterase, AceA, which has 313 amino acids and a calculated mass of 34,845 Da [17]. It has 56% amino acid identity with the acetylxylan esterase, BnaA, from *N patriciarum* described above, but AceA is devoid of NCDD, which is present in BnaA. Thus the *Orpinmyces* enzyme is considered a free enzyme in contrast to the *Neocallimastix* enzyme, which is cellulosome associated. Like BnaA the AceA works synergetically with a xylanase, XynA, from *Orpinmyces* toward acetylated xylan. The rate of hydrolysis being about tripled.

Cinnamoyl Esterases (Feruloyl and *p*-Coumaroyl Esterases)

One of the most important cross-linkages in plant cell walls is the linkages between two ferulic acid side chains in xylan molecules (Fig. 1) [113]. They drammatically increase plant cell walls` mechanical properties and have been likened to spot welding of steel mesh frames. [66]. Evidence has been presented that ferulates couple with lignin monomers in grasses and that this is directing cell-wall cross-linking during plant growth and development [99]. This generates stable and recalcitrant structures not easily degraded by bacteria. Anaerobic fungi colonize lignified tissue and their fungal rhizoids penetrate into and through cell walls resulting in weakening of lignified structures [3]. It is likely that cinnamoyl esterases are responsible for these results. Borneman et al. [21] studying five different anaerobic fungi isolated from rumen of a cow Borneman et al. [20] found that ferulic acid and *p*-coumaric acid was released from Coastal Bermuda grass upon treatment with culture fluid from the fungi. Subsequently, one *p*-coumaroyl (CAE) and two feruloyl esterases (FAE-I and FAE-II) were purified from *Neocallimastix* sp. MC-2 culture supernatants combined with distilled water extracts of residual cell wall fragments [22,23]. Substrates for the enzymes were *O*-[5-*O*-((*E*)-*p*-coumaroyl)-α-L-arabinofuranosyl]-(1→3)-*O*-β-D-xylopyranosyl-(1→4)-D-xylopyranose (PAXX) and *O*-{5-*O*-[(*E*)-feruloyl]- α-L-arabinofuranosyl}-(1→3)-*O*-β-D-xylopyranosyl-(1→4)-D-xylopyranose (FAXX) isolated from Coastal Bermuda grass. The native coumaroyl esterase had an apparent mass of 11 kDa, but under denaturating condition the mass was 5.8 kDa indicating that the enzyme may exist as a dimer. The specific activity of the coumaroyl esterase with PAXX was 4.91 U/mg, with FAXX 0.05 U/mg and it had no acetylxylan esterase activity indicating it is rather specific for *p*-coumaroyl esters. FAE-I and FAE-II are monomeric and have molecular masses of 69 and 24 kDa, respectively. FAE-II had a specific activity with FAXX of 11.3 U/mg, PAAX 0.02 U/mg and no activity with acetylxylan. FAE-I was less specific. Its specific activity with FAXX was 2.89, PAXX O.94, and acetylxylan 0.009 U/mg. Synergism was observed between the esterases and xylanases using Coastal Bermuda grass as substrate [24]. The release of *p*-coumaroyl, feruloyl, and reducing sugars when the enzymes were combined were about four times for coumaroyl groups, seven times for feruloyl groups, and ten times for reducing sugars over that with the enzymes alone.

A feruloyl esterase, FaeA, has been purified from *Orpinomyces* sp. PC-2 [18]. A gene from the cDNA library of *Orpinomyces* sp. PC-2 encodes the mature enzyme together with four residues of a signal peptide (Accession no. AF1643510). The mature enzyme has 530 amino acids and a calculated molecular mass of 59 kDa. After the signal peptide at the N-terminus there is a 200 amino acid module of unknown function, which does not have homology with any sequence in a BLAST search. It is followed by a C-terminal module, which contains a 230 residues sequence homologous with modules of XynY and XynZ having feruloyl esterase activity. The two xylanases are bifunctional cellulosomal enzymes from *Clostridium thermocellum* with separate modules with xylanase and feruloyl esterase activities [16]. The structures have been determined for the feruloyl esterase modules of XynY [97] and XynZ [103]. There is no evidence for that FaeA is connected to the *Orpinomyces* cellulosome.

A modular cinnamoyl esterase (EstA) is produced by *P. equi* [54]. The *estA* gene encodes the enzyme, which has 536 amino acids and a mass of 55,540 Da. Thus it is of a size similar to FAE-I purified from *Neocallimastix* strain MC-2, and FaeA from *Orpinomyces* sp. PC-2. EstA has a N-terminal signal sequence followed by a docking module NCDD, which in turn is followed by an unusually large linker with several 13-residues repeats (NQGGGMPWGDFGG). The C-terminal 270- residues module of EstA contains the esterase catalytic site. It was expressed in *E. coli* and purified. It hydrolyzed methyl caffeate, methyl *p*-coumarate, methyl ferulate, 4-nitrophenyl acetate, 1-naphthyl acetate, FAXX, and Ara_2F (*O*-[2-*O*(E)-(1-feruloyl)- α-L-arbino-furanosyl]-(1→5)-L-arabinose). Thus it has rather broad specificity as was shown also for FAE-I, but unlike FAE-I, EstA containing a NCDD appears be associated with the cellulosome of *P. equi.*

(4-O-Methyl)-D-glucuronidase (α-glucuronidase)

A substituent in arabinoxylan is 4-*O*-methyl-α-D-glucopyranosyluronic acid, which is normally linked to the 2 position of xylopyranosyl residues (Fig.1). Wood and Wilson [117] reported the presence in *P. communis* of the enzyme α-(4-*O*-methyl)-D-glucuronidase, which hydolyzes the bond between the glucuronic acid and the xylose residue. *P. communis* (formerly *Piromonas*) is an anaerobic fungus isolated from sheep digesta [93]. They did not find the enzyme in *N. frontalis* and *N. patriciarum.* Gel filtration of a concentrated culture filtrate from *P. communis* showed the β-glucuronidase to be associated with protein fractions ranging from 100 to 300 kDa and the possibility that it is a part of the cellulosome is not excluded. An interesting approach to introduce the α-glucuronidase into *N. frontalis* has been presented by Durand et al. [51]. They biolistically introduced a plasmid containing a bacterial β-glucuronidase gene, which was fused to the promoter sequence of the enolase gene from *N. frontalis.* The enolase promoter sequence was responsible for expression of the α-glucuronidase gene. The expression was detected using a histochemical assay for α-glucuronidase.

Serpins in Cellulosomes

Serpins are a number of homologous proteins, many of which function as serine proteinase inhibitors [56]. Steenbakkers et al. [106] have identified a gene in *Piromyces* sp. E2, that encodes a protein having a module homologous with vertebrate neuroserpins and also having an NCDD indicating the protein is cellulosome-associated. This suggests that the cellulosome of *Piromyces* has a trypsin-like proteinase inhibitor.

The sequence of the *C. thermocellum* genome is now available and a search for genes encoding proteins homologous with serpins resulted in the identification of three serpins; CHTE 0485 containing 437 amino acids, CHTE 1412 with 604 amino acids, and CHTE 1413 with 601 amino acids [67,106]. The sequence identity between the two largest of these serpins is 68% and they share a similar module structure with a putative signal peptide at the N-terminus, a 100 amino acid fibronectin type III module, a bacterial dockerin, and the C-terminal serpin module. The presence of dockerins in the two largest serpins of *C. thermocellum* suggests, that they are cellulosomes components. The smaller serpin does not have a dockerin and it apparently is a free agent in the cultural fluid. At present serpins have been found only in cellulosomes from *C. thermocellum* and *Piromyces* sp. E2. It is likely, that they are present in cellulosomes from other bacteria and anaerobic fungi. One may at present only speculate about the function of serpins in the cellulosomes. The cellulosomes are exocellular and exposed to the evironment outside the cell where they may be attacked by trypsin-like proteinases from different sources. The serpins may protect the cellulosomes from such attacks.

NonCatalytic Docking Domains (NCDDs) and Scaffolding Proteins

The catalytic and other protein components associated with the cellulosomes of anaerobic fungi contain typically two copies of a conserved 40-amino acid cysteine–rich noncatalytic docking domain NCDD. The two copies being interspaced by a short linker specific for each cellulosomal protein. Degenerate primers were designed based on the conserved sequences of the noncatalytic modules of *Piromyces* sp. E2 and *Orpinomyces* sp. PC-2. Using PCR with the primers and cDNA from *Orpinomyces* and genomic DNA from *Piromyces* as templetes 69 clones were obtained from *Piromyces* and 45 clones from *Orpinomyces* representing 19 and 16 distinctive clones of NCDDs, respectively. Each distinctive clone representing a protein with a NCDD that binds to the cellulosome [107]. Several of the sequences of these distinctive NCDDs corresponded to previously identified NCDD sequences in catalytic proteins from *Piromyces* and *Orpinomyces*. The recognition of previously identified sequences showed that the cloning strategy was successful, and as discussed earlier in this article it was used to isolate NCDD-containing proteins not previously recognized. The sequences of known NCDDs from different anaerobic fungi have been aligned based on the number, position, and context of cysteine residues (Fig. 8). It showed that the NCDD sequences could be divided into three different types or subfamilies. Types 1 and 3 contain 6 cysteine residues, whereas Type 2 has only four. All sequences except one had vicinal cysteines. Type 1 NCDDs seem to be confined to hemicellulolytic enzymes.

The three-dimensional structure of the N-terminal NCDD (dockerin) from Cel45A from *P. equi* has been determined [98]. This NCDD belongs to Type 2 having four cysteine residues [107]. The structure contains a novel fold of 42 residues and includes from N-terminal end of the NCDD a helical stretch, a short β-strand, and a helical turn followed by three short β-strands, which form an antiparallel sheet structure. The N- and C- termini are close to each other. The structure resembles a cradle or bench. It is hold together by two disulfide bridges by pairing the N-terminal cysteine with the N-terminal vicinal cysteine and the C-terminal vicinal cysteine with the most C-terminal cysteine. Reduction of the disulfides results in loss of structure indicating that the functional NCDD is oxidized. Three amino acids tyrosine (Y8), tryptophan (W35), and aspartic acid (D23) located within the cradle are essential for the docking. Mutation of either Y8 or W35 with alanine completely abolishes docking of the NCDD to the *P. equi* cellulosome. Mutation of D23 reduces the binding by two orders of magnitude. Y8, W35, and D23 are conserved in all NCDDs until now identified except in two cases in which Y8 is replaced by a tryptophan or a phenylalanine (Fig.8). Docking of NCDDs to the anaerobic fungal cellulosome does not require calcium, which is required for the dockerin/cohesin interaction in the cellulosome of *C. thermocellum.* The anaerobic fungal NCDD binds to cellulosomes from several fungal species, which contrasts with the bacterial cellulosome systems where the dockerins bind specifically to cohesins of the same species.

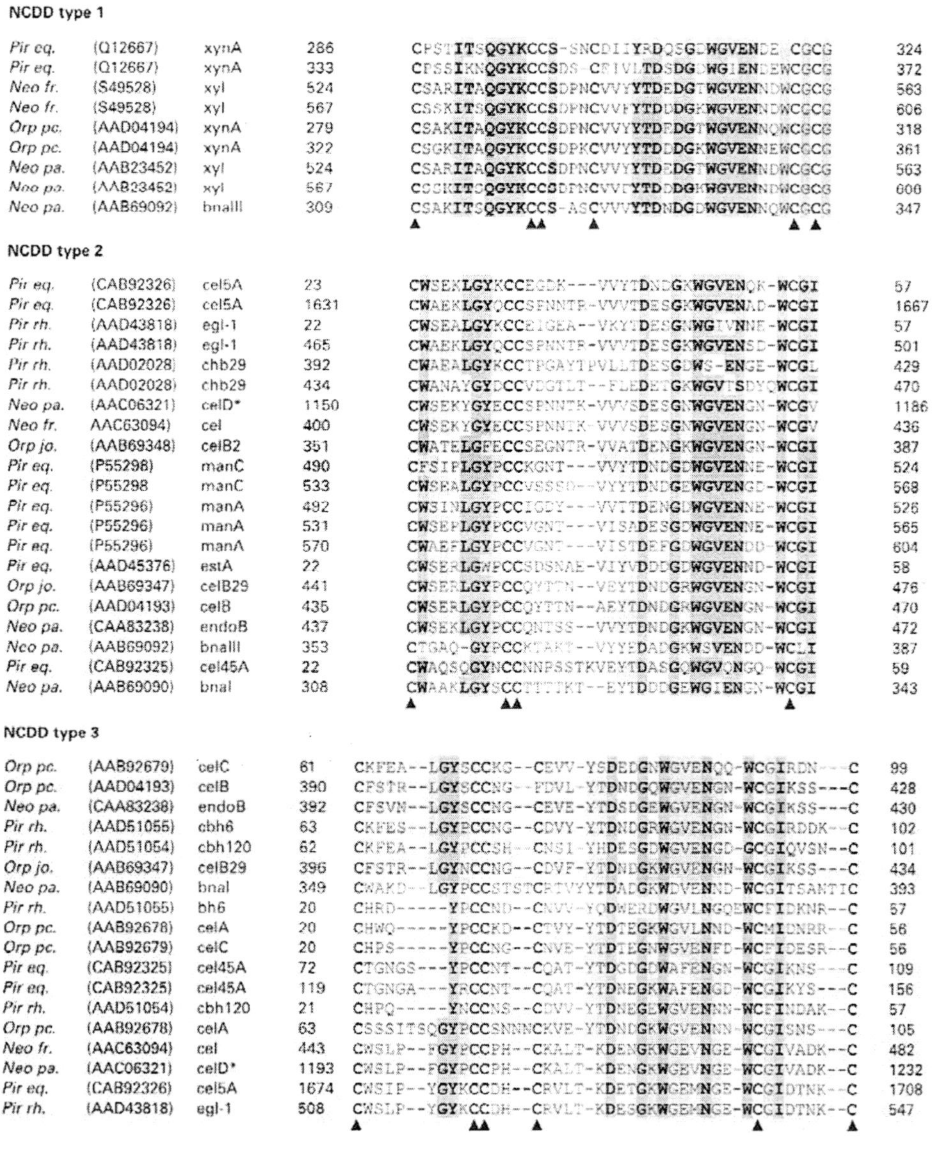

NCDD type 1

Organism	Accession	Gene	Start	Sequence	End
Pir eq.	(Q12667)	xynA	286	CPSTITSQGYKCCS-SNCDIIYRDQSGDWGVENDE CGCG	324
Pir eq.	(Q12667)	xynA	333	CPSSIKNQGYKCCSDS CPIVLTDSDGDWGIENDEWCGCG	372
Neo fr.	(S49528)	xyl	524	CSARITAQGYKCCSDPNCVVYYTDEDGTWGVENNDWCGCG	563
Neo fr.	(S49528)	xyl	567	CSSKITSQGYKCCSDPNCVVFYTDDDGKWGVENNDWCGCG	606
Orp pc.	(AAD04194)	xynA	279	CSAKITAQGYKCCSDPNCVVYYTDEDGTWGVENNQWCGCG	318
Orp pc.	(AAD04194)	xynA	322	CSGKITAQGYKCCSDPKCVVYYTDDDGKWGVENNEWCGCG	361
Neo pa.	(AAB23452)	xyl	524	CSARITAQGYKCCSDPNCVVYYTDEDGTWGVENNDWCGCG	563
Neo pa.	(AAB23452)	xyl	567	CSSKITSQGYKCCSDPNCVVFYTDDDGKWGVENNDWCGCG	606
Neo pa.	(AAB69092)	bnaIII	309	CSAKITSQGYKCCS-ASCVVVYTDNDGDWGVENNQWCGCG	347
				▲ ▲▲ ▲ ▲ ▲	

NCDD type 2

Organism	Accession	Gene	Start	Sequence	End
Pir eq.	(CAB92326)	cel5A	23	CWSEKLGYKCCEGDK---VVYTDNDGKWGVENQK-WCGI	57
Pir eq.	(CAB92326)	cel5A	1631	CWAEKLGYQCCSPNNTR-VVVTDESGKWGVENAD-WCGI	1667
Pir rh.	(AAD43818)	egl-1	22	CWSEALGYKCCEIGEA--VKYTDESGNWGIVNNE-WCGI	57
Pir rh.	(AAD43818)	egl-1	465	CWAEKLGYQCCSPNNTR-VVVTDESGKWGVENSD-WCGI	501
Pir rh.	(AAD02028)	chb29	392	CWAEALGYKCCTPGAYTPVLLTDESGDWS-ENGE-WCGL	429
Pir rh.	(AAD02028)	chb29	434	CWANAYGYDCCVDGTLT--FLEDETGKWGVTSDYQWCGI	470
Neo pa.	(AAC06321)	celD*	1150	CWSEKYGYECCSPNNTK-VVVSDESGNWGVENGN-WCGV	1186
Neo fr.	AAC63094)	cel	400	CWSEKYGYECCSPNNTK-VVVSDESGNWGVENGN-WCGV	436
Orp jo.	(AAB69348)	celB2	351	CWATELGFECCSEGNTR-VVATDENGKWGVENGN-WCGI	387
Pir eq.	(P55298)	manC	490	CFSIPLGYPCCKGNT---VVYTDNDGDWGVENNE-WCGI	524
Pir eq.	(P55298	manC	533	CWSEALGYPCCVSSSD--VYYTDNDGEWGVENGD-WCGI	568
Pir eq.	(P55296)	manA	492	CWSINLGYPCCIGDY---VVTTDENGDWGVENNE-WCGI	526
Pir eq.	(P55296)	manA	531	CWSEPLGYPCCVGNT---VISADESGDWGVENNE-WCGI	565
Pir eq.	(P55296)	manA	570	CWAEFLGYPCCVGNT---VISTDEFGDWGVENDD-WCGI	604
Pir eq.	(AAD45376)	estA	22	CWSERLGWPCCSDSNAE-VIYVDDDGDWGVENND-WCGI	58
Orp jo.	(AAB69347)	celB29	441	CWSERLGYPCCQYTTN--VEYTDNDGRWGVENGN-WCGI	476
Orp pc.	(AAD04193)	celB	435	CWSERLGYPCCQYTTN--AEYTDNDGRWGVENGN-WCGI	470
Neo pa.	(CAA83238)	endoB	437	CWSEKLGYPCCQNTSS--VVYTDNDGKWGVENGN-WCGI	472
Neo pa.	(AAB69092)	bnaIII	353	CTGAQ-GYPCCKTAKT--VYYEDADGKWSVENDD-WCLI	387
Pir eq.	(CAB92325)	cel45A	22	CWAQSQGYNCCNNPSSTKVEYTDASGQWGVQNGQ-WCGI	59
Neo pa.	(AAB69090)	bnaI	308	CWAAKLGYSCCTTTIKT--EYTDDDGEWGIENGN-WCGI	343
				▲ ▲▲ ▲	

NCDD type 3

Organism	Accession	Gene	Start	Sequence	End
Orp pc.	(AAB92679)	celC	61	CKFEA--LGYSCCKG--CEVV-YSDEDGNWGVENQQ-WCGIRDN---C	99
Orp pc.	(AAD04193)	celB	390	CFSTR--LGYSCCNG--FDVL-YTDNDGQWGVENGN-WCGIKSS---C	428
Neo pa.	(CAA83238)	endoB	392	CFSVN--LGYSCCNG--CEVE-YTDSDGEWGVENGN-WCGIKSS---C	430
Pir rh.	(AAD51055)	cbh6	63	CKFES--LGYPCCNG--CDVY-YTDNDGRWGVENGN-WCGIRDDK--C	102
Pir rh.	(AAD51054)	cbh120	62	CKFEA--LGYPCCSH--CNSI-YHDESGDWGVENGD-GCGIQVSN--C	101
Orp jo.	(AAB69347)	celB29	396	CFSTR--LGYNCCNG--CDVF-YTDNDGKWGVENGN-WCGIKSS---C	434
Neo pa.	(AAB69090)	bnaI	349	CWAKD--LGYPCCSTSTCRTVYYTDADGKWDVENND-WCGITSANTIC	393
Pir rh.	(AAD51055)	bh6	20	CHRD-----YPCCND--CNVV-YQDWERDWGVLNGQEWCFIDKNR--C	57
Orp pc.	(AAB92678)	celA	20	CHWQ-----YPCCKD--CTVY-YTDTEGKWGVLNND-WCMIDNRR--C	56
Orp pc.	(AAB92679)	celC	20	CHPS-----YPCCNG--CNVE-YTDTEGNWGVENFD-WCFIDESR--C	56
Pir eq.	(CAB92325)	cel45A	72	CTGNGS---YPCCNT--CQAT-YTDGDGDWAFENGN-WCGIKNS---C	109
Pir eq.	(CAB92325)	cel45A	119	CTGNGA---YRCCNT--CQAT-YTDNEGKWAFENGD-WCGIKYS---C	156
Pir rh.	(AAD51054)	cbh120	21	CHPQ-----YNCCNS--CDVV-YTDNEGEWGVENNN-WCFINDAK--C	57
Orp pc.	(AAB92678)	celA	63	CSSSITSQGYPCCSNNNCKVE-YTDNDGKWGVENNN-WCGISNS---C	105
Neo fr.	(AAC63094)	cel	443	CWSLP--FGYPCCPH--CKALT-KDENGKWGEVNGE-WCGIVADK--C	482
Neo pa.	(AAC06321)	celD*	1193	CWSLP--FGYPCCPH--CKALT-KDENGKWGEVNGE-WCGIVADK--C	1232
Pir eq.	(CAB92326)	cel5A	1674	CWSIP--YGYKCCDH--CRVLT-KDETGKWGEMNGE-WCGIDTNK--C	1708
Pir rh.	(AAD43818)	egl-1	508	CWSLP--YGYKCCDH--CRVLT-KDESGKWGEMNGE-WCGIDTNK--C	547
				▲ ▲▲ ▲ ▲ ▲	

Figure 8. Translated NCDD-PCR clones isolated from *Piromyces* sp. E2 (roman) and *Orpinomyces* sp. PC-2 (italic). The alignment with fixed cysteine residues reveals a division into sequence groups A to C. The first and last amino acids are encoded by the degenerate primers and are underlined. The presumed linker region is boxed. Cysteine residues are in boldface, and conserved residues are in shaded boxes. Spaces are indicated by dashes. The sequences that served as a positive control are in boldface (from Steenbakkers et al. [107]).

A scaffolding polypeptide equivalent to scaffoldins of bacterial cellulosomes has yet to be purified and sequenced from an anaerobic fungus. Fanutti et al. [53], and Fillingham et al [54], as discussed earlier, have presented evidence for such polypeptides of molecular masses 116 and 97 kDa cellulosomes from *P. equi,* and *N. patriciarum*, respestively. That the NCDD of Cel45A from *P.equi* binds to polypeptides from *P. equi, N. patriciarum,* and *Orpinmyces* sp. PC-2 is a confirmation of the presence of a scaffoldin polypeptide in the cellulosome from each species [98]. The possible presence of multiple scaffoldins in the cellulosome from *Orpinomyces* sp. PC-2 has been reported by Steenbakkers et al. [107]. They made two constructs of CelC from the *Orpinomyces* species, one containing the original NCDD (Type 3) repeat CelC(+), the other without the repeat, CelC(-). Purified cellulosomes from the *Orpinomyces* sp. subjected to SPS-PAGE separated into at least 20 different polypeptides (Fig. 9, lane A). Four of these bound CelC(+) (Fig. 9, lane B), whereas none bound CelC(-) (Fig. 9, lane C). The NCDD-binding polypeptides had molecular masses of 64, 66, 95 and 130 kDa. The proteins with the lower masses could be proteolytic products of the 130kDa protein. This appeared unlikely since the same results were obtained after storage of the cellulosomal preparation at 4°C for 4 months. Thus, it must be considered that cellulosomes from anaerobic fungi may have multiple scaffoldins. It should be noted that some bacterial cellulosome have more than one scaffoldin [100,121].

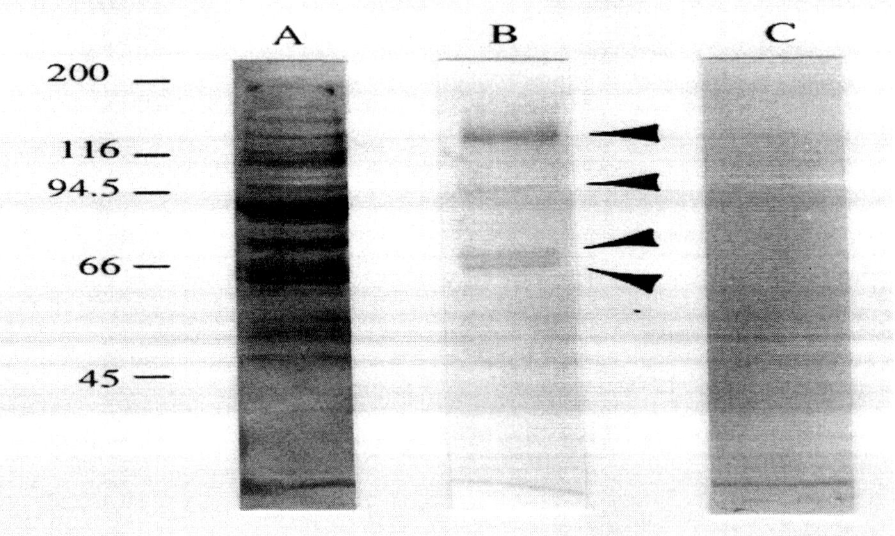

Figure 9. Analysis of CelC NCDD binding to the *Orpinomyces* sp. PC-2 (hemi) cellulase complex. The purified complex sample was boiled in SDS buffer, and the denatured polypeptides were separated by SDS-PAGE using 10% (wt/vol) acrylamide. Gels were stained with Coomassie brilliant blue (A) or labeled by using biotinylated full-length CelC (B) and CelC minus the NCDD repeat (C). The values on the left are molecular sizes in kDa. The arrowheads indicate the scaffoldin polypeptides (from Steenbakkers et al. [107]).

Genomic Organization of Cellulolytic Enzymes in Anaerobic Fungi

Thus far almost all information on the (hemi-)cellulolytic genes of anaerobic fungi is obtained from studies on cDNA clones. The information from gDNA sequences is very limited. Several authors have indicated that multiple copies exist for these genes [4, 30, 53, 61, 62, 77, 90]. In other (hemi-)cellulolytic organisms (e.g. aerobic fungi and clostridia) these enzymes do not occur as products from multicopy genes. A gDNA library of *Piromyces* sp.

E2 was screened with different probes specific for (hemi-)cellulolytic enzymes [60]. This screening resulted in three gDNA clones with genes encoding glycoside hydrolase enzymes of families 1 (β-glucosidase), 6 (exoglucosidase) and 26 (mannanase). Each clone contained two or more genes of the same family which were highly homologous. The results obtained show that the (hemi-)cellulolytic genes indicated are present in multiple copies and clustered on the genome in a head-to-tail organization. As has been suggested for other multiple-copy genes, this organization may have evolved as a result of a requirement for synthesis of large amounts of the gene products.

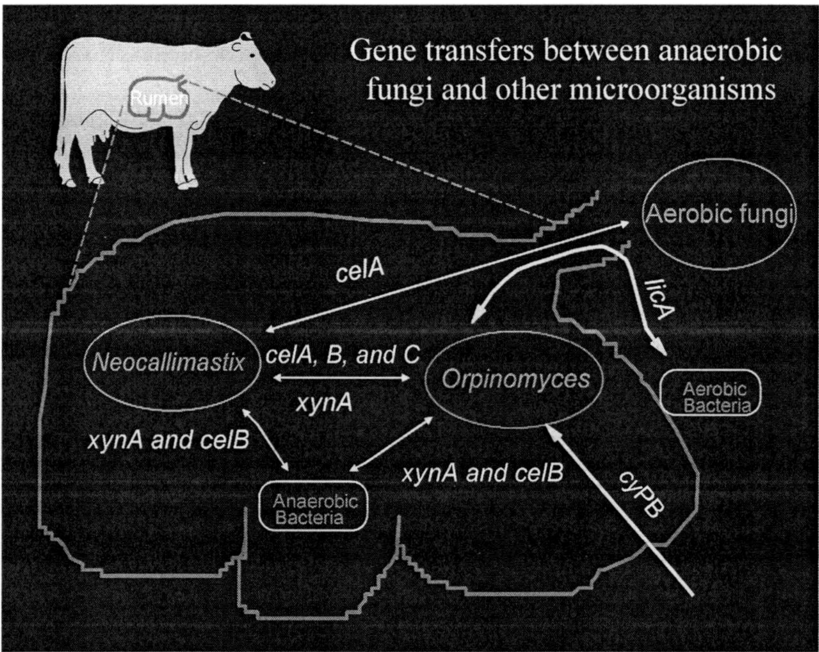

Figure 10. Demonstration of potential transfers of genes between anaerobic and aerobic fungi, aerobic and anaerobic bacteria based on sequence homology and presence or absence of introns. *celA*, *B*, and *C* are cellulases; *xynA* is xylanase; *licC* is lichenase; *cyPB* is cyclophilin.

CONCLUSIONS

Cellulotic/hemicellulolytic anaerobic fungi in the intestinal tracts of herbivorous animals play important roles in the breakdown of lignocellulosic feed and hydrolysis of plant carbohydrates. To that extent they produce several different glycoside hydrolases including endoglucanases, cellobiohydrolases, mannanases, cinnamoyl esterases, acetyl- xylan esterases, lichenases, and β-glucosidases. All of these enzymes are secreted into the environment, some as free enzymes but most of them are bound to cellulosomal complexes of from about one million to several million Da containing from 10 to more than 20 polypeptides. The cellulosomes from the anaerobic fungi appear similar in architecture to cellulosomes from *C. thermocellum* and other anaerobic bacteria. Thus, enzymes and other proteins of the cellulosomes of anaerobic fungi have dockerin modules (NCDDs) binding them to cohesins of scaffolding polypeptides. It is recognized, however, that there is no sequence homology between dockerins from bacteria and fungi. Consequently, it is likely that the scaffolding polypeptides from fungi differ from those of anaerobic bacteria. An isolation and sequencing of a scaffolding protein from an anaerobic fungus has yet to be done.

Anaerobic fungi live in close contact with bacteria and other microorganisms in the intestine including the rumen and caecum of herbivorous animals where they digest ingested plant food. A large number of glycoside hydrolases and other proteins from anaerobic fungi have been sequenced. Comparison of these sequences has led to the observation that some enzymes are homologous to enzymes from bacterial sources whereas sequences of other enzymes are similar to those of aerobic fungi. It can be concluded that enzymes of anaerobic fungi have different origins as illustrated in Fig. 10. Horizontal gene transfers seem common between microorganisms living in close contact in ecosystems such as the rumen and caecum [57,85]. Other observations indicate that gene duplication and insertion of conserved modules into genes leading to the production of multi-module proteins are common in anaerobic fungi.

ACKNOWLEDGEMENTS

Support from US Department of Energy, and Consortium for Plant Biotechnology Research is gratefully acknowledged. Thanks go also to all coworkers involved in the work with anaerobic fungi.

REFERENCES

Akhmanova, A., Voncken, F.G.J., Harhangi, H., Hosea, K.M.,Vogels, G.D., and Hackstein, J.H.P. 1998. Cytosolic enzymes with a mitochondrial ancestry from anaerobic chytrid *Piromyces* sp. E2. *Molec. Microbiol. 30*:1017-1027.

Akhmanova, A., Voncken, F.G.J.,Hosea, K.M., Harhangi, H., Keltjens, J.T., Op den Camp H., Vogels, G.D., and Hackstein, J.H.P. 1999. A hydrogenosome with pyruvate formate-lyase: anaerobic chytrid fungi use an alternative route for pyruvate catabolism. *Molecular Microbiol. 32*:1103-1114.

Akin, D.E., Lyon, C.E., Windham, W.R., and Rigsby, L.L. 1989 Physical degradation of lignified stem tissues by ruminal fungi. *Appl. Environ. Microbiol. 55*:611-616.

Ali, B.R.S., Zhou, L., Graves, F.M., Freedman, R.B., Black, G.W., Gilbert, H.J., and Hazlewood, G.P. 1995. Cellulases and hemicellulases of the anaerobic fungus *Piromyces* constitute a multiprotein cellulose-binding complex and are encoded by multigene families. *FEMS Microbiol. Lett. 125*:15-22.

Aylward, J.H., Gobius, K.S., Xue, G.-P., Simpson, G.D., and Dalrymple, B.P. 1999. The *Neocallimastix patriciarum* cellulase, CelD, contains three almost identical domains with high specific activities on Avicel. *Enz. Microbial Technol. 24*:609-614.

Barichievich, E.M., and Calza, R.E. 1990. Supernatant protein and cellulase activities of the anaerobic ruminal fungus *Neocallimastix frontalis* EB188. *Appl. Environ. Microbiol. 56*:43-48.

Barr, D.J.S., Yanke, L.J., Bae, H.D., McAllister ,T.A., and Cheng, K.-J. 1995. Contributions on the morphology and taxonomy of some rumen fungi from Canada. *Mycotaxon* LIV: 203-214.

Bauchop, T., and Mountford, D.O. 1981. Cellulose fermentation by a rumen anaerobic fungus in both the absence and the presence of rumen bacteria. *Appl. Environ. Microbiol. 42*:1103-1110.

Bayer, E.A., Morag, E., Lamed, R. 1994. The cellulosome - a treasure-trove for biotechnology. *Trends Biotech. (TIBTECH) 12*:379-386.

Bayer, E.A., Shoham, Y., and Lamed, R. 2000. Cellulose-decomposing bacteria and their enzyme systems. In M.Dworkin, S.Falkow, E. Rosenberg, K.-H.Schleifer and E. Stackebrandt (eds.), *The Prokaryotes: An Evolving Electronic Resource for the Microbiological Community, 3rd edition*.Springer-Verlag, New York.

Béguin, P., and Lemaire, M. 1996. The cellulosome: an exocellular, multiprotein complex specialized in cellulose degradation. *Crit. Rev. Biochem. Mol. Biol. 31*:201-236.

Bélaich, J.-P., Tardif, C., Bélaich, A., and Gaudin, C. 1997. The cellulolytic system of *Clostridium cellulolyticum*. *J. Biotechnol. 57*:3-14.

Biely, P., Puls, J., and Schneider, H. 1985. Acetylxylan esterases in fungal cellulolytic systems. *FEBS Lett. 186*:80-84.

Billon-Grand, G., Fiol, J.B., Breton, A., Bruyere, A., and Oulhaj, Z. 1991. DNA of some anaerobic rumen fungi: G + C content determination. *FEMS Microbiology Letters 82*:267-270.

Black, G.W., Hazlewood, G.P., Xue, G.-P., Orpin, C.G., and Gilbert, H.J. 1994. Xylanase B from *Neocallimastix patriciarum* contains a non-catalytic 455-residue linker sequence comprised of 57 repeats of an octapeptide. *Biochem. J. 299*:381-387.

Blum, D.L., Kataeva, I.A., Li, X.-L., and Ljungdahl, L.G. 2000. Feruloyl esterase activity of the *Clostridium thermocellum* cellulosome can be attributed to previously unknown domains of XynY and XynZ. *J. Bacteriol. 182*:1346-1351.

Blum, D.L., Li, X.-L., Chen, H., and Ljungdahl, L.G. 1999. Characterization of an acetyl xylan esterase from the anaerobic fungus *Orpinomyces* sp. strain PC-2. *Appl. Environ. Microbiol. 65*:3990-3995.

Blum, D.L., Li, X.-L., Chen, H., and Ljungdahl, L.G. 1999. Characterization of a feruloyl esterase from the anaerobic fungus *Orpinomyces* sp. strain PC-2. 99th Ann. *Meeting Am. Soc. Microbiol. Abstr. K-153*, p.430.

Borneman, S., and Akin, D.E. 1994. The nature of anaerobic fungi and their polysaccharide degrading enzymes. *Mycoscience 35*:199-211.

Borneman, W.S., Akin, D.E., and Ljungdahl, L.G. 1989. Fermentation products and plant cell wall-degrading enzymes produced by monocentric and polycentric anaerobic ruminal fungi. *Appl. Environ. Microbiol. 55*:1066-1073.

Borncman, W.S., Hartley, R.D., Morrison, W.H., Akin, D.E., and Ljungdahl, L.G. 1990. Feruloyl and p-coumaroyl esterase from anaerobic fungi in relation to plant cell wall degradation. *Appl. Microbiol. Biotechnol. 33*:345-351.

Borneman, W.S., Ljungdahl, L.G., Hartley, R.D., and Akin, D.E. 1991. Isolation and characterization of *p*-coumaroyl esterase from the anaerobic fungus *Neocallimastix* strain MC-2. *Appl. Environ. Microbiol. 57*:2337-2344.

Borneman, W.S., Ljungdahl, L.G., Hartley, R.D., and Akin, D.E. 1992. Purification and partial characterization of two feruloyl esterases from the anaerobic fungus *Neocallimastix* strain MC-2. *Appl. Environ. Microbiol. 58*:3762-3766.

Borneman, W.S., Ljungdahl, L.G., Hartley, R.D., and Akin, D.E. 1993 Feruloyl and p-coumaroyl esterases from the anaerobic fungus *Neocallimastix* MC-2: properties and functions in plant cell wall degradation. In *Hemicellulose and Hemicellulases* (M.P. Coughlan, G. Hazlewood, eds.) pp.85-102. Portland Press Cambride U.K.

Brownlee, A.G. 1989. Remarkably AT-rich genomic DNA from the anaerobic fungus *Neocallimastix. Nucl. Acid Res. 17,*1327-1335.

Buliga, G.S., Brant, D.A., and Fincher, G.B. 1986. The sequence statistics and solution conformation of a barley (1 → 3,1 →4)-β-D-glucan. *Carbohydr. Res. 157*:139-156.

Cantwell, B.A., Sharp, P.M., Gormley, E., and McConnell, D.L. 1988. Molecular cloning of *Bacillus* β-glucanases. In *Biochemistry and Genetics of Cellulose Degradation.* (J.-P. Aubert, P.Béguin, J. Millet, eds.) pp. 181-201. Academic Press, London.

Caldwell, D.R., and Bryant, M.P. 1996. Medium without rumen fluid for nonselective enumeration and isolation of rumen bacteria. *Appl. Microbiol. 14*:794-801.

Carvalho, A.L., Dias, F.M.V., Prates, J.A.M., Nagy, T., Gilbert, H.J., Davies, G.J., Ferreira, L.M.A., Romão, M.J., and Fontes, G.A. 2003. Cellulosome assembly revealed by the crystal structure of the cohesin-dockerin complex. *Proc.Nat.Acad.Sci. U.S. 100*:13809-13814.

Chen, H., Li, X.-L., Blum, D.L., Ljungdahl, L.G. 1998. Two genes of the anaerobic fungus *Orpinomyces* sp. strain PC-2 encoding cellulases with endoglucanase activities may have arisen by gene duplication. *FEMS Microbiol. Lett 159*:63-68.

Chen, H., Li, X.-L., Blum, D.L., Ximenes, E.A., and Ljungdahl, L.G. 2003. CelF of *Orpinomyces* PC-2 has an intron and encodes a cellulase (CelF) containing a carbohydrate-binding module. *Appl. Biochem. Biotech. 105*-108:775-785.

Chen, H., Li, X., and Ljungdahl, L.G. 1994. Isolation and properties of an extracellular β-glucosidase from the rumen fungus *Orpinomyces* sp. strain PC-2. *Appl. Environ. Microbiol. 60*:64-70.

Chen, H., Li, X.L., and Ljungdahl, L.G. 1995. A cyclophilin from the polycentric anaerobic rumen fungus *Orpinomyces* sp. strain PC-2 is highly homologous to vertebrate cyclophilin B. *Proc. Natl. Acad. Sci. USA 92*:2587-2591.

Chen, H., Li, X.L., and Ljungdahl, L.G. 1997. Sequencing of a 1,3-1,4-β-D-glucanase (lichenase) from the anaerobic fungus *Orpinomyces* strain PC-2: Properties of the enzyme expressed in Escherichia coli and evidence that the gene has a bacterial origin. *J. Bacteriol. 179*:6028-6034.

Chen, H.Z., Li, X.-L., and Ljungdahl, L.G. 1999. Purification and characterization of an endoxylanase from the anaerobic polycentric fungus *Orpinomyces* sp. strain PC-2. *SAAS Bulletin Biochem & Biotech. 12*:7-14.

Chesson, A., and Forsberg, C.W. 1988. Polysaccharide degradation by rumen microorganisms. In *The Rumen Microbial Ecosystem* (P. N. Hobson, ed.) pp. 251-284. Elsevier Applied Science, New York.

Continho, P.M., and Henrissat, B. 1999a. Carbohydrate-active enzymes server at http://afmb.cnrs-mrs.fr/i<<cazy/CAZY/index.html.

Continho, P.M., and Henrissat, B. 1999b. Carbohydrate-active enzymes: an integrated database approch. *In Recent Advances in Carbohydrate Bioengineering* (H.J.Gilbert, G. Davies, B. Henrissat, and B. Svensson eds.) pp. 3-12. The royal Society of Chemistry, Cambridge.

Coughlan, M.P., Hon-nami, K., Hon-nami, H., Ljungdahl, L.G., Paulin, J.J., and Rigsby W.E. 1985.The cellulolytic enzyme complex of *Clostridium thermocellum* is very large. *Biochem. Biophys.Res. Commun. 130*:904-909.

Cybinski, D.H., Layton, I., Lowry, J.B., and Dalrymple B.P. 1999. An acetylxylan esterase and a xylanase expressed from genes cloned from the ruminal fungus *Neocallimastix patriciarum* act synergistically to degrade acetylated xylans. *Appl. Microbiol. Biotechnol. 52*:221-225.

Dalrymple, B.P., Cybinski, D.H., Layton, I., McSweeney, C.S., Xue G.-P., Swadling, Y.J., Lowry, J.B. 1997. Three *Neocallmastix patriciarum* esterases associated with the degradation of complex polysaccharides are members of a new family of hydrolases. *Microbiology 143*:2605-2614.

Davies E.T., Li, X.-L., and Ljungdahl L.G. 2002. Optimization of recombinant protein production in high cell density fed-batch *Escherichia coli* cultures. Abstract PC43, Society for Industrial Microbiology Annual Meeting, Philadelphia, PA.

Denman, S., Xue, G.-P., and Patel, B. 1996. Characterization of a *Neocallimastix patriciarum* cellulase cDNA (celA) homologous to Trichoderma reesei cellobiohydrolase II. *Appl. Environ. Microbiol. 62:*1889-1896.

De Segura, B.G., and Fèvre, M. 1993. Purification and characterization of two 1,4-β-xylan endohydrolases from the rumen fungus *Neocallimastix frontalis*. *Appl. Environ. Microbiol. 59*:3654-3660.

Dijkerman, R., Op den Camp, H.J.M., Van der Drift, C., and Vogels, G.D. 1997. The role of the cellulolytic high molecular mass (HMM) complex of the anaerobic fungus *Piromyces* sp. strain E2 in the hydrolysis of microcrystalline cellulose. *Arch. Microbiol. 167*:137-142.

Dijkerman, R., Vervuren, M.B.F., Op den Camp, H.J.M., and van der Drift, C. 1996. Adsorption charcteristics of cellulolytic enzymes from the anaerobic fungus *Piromyces* sp. strain E2 on microcrystalline cellulose. *Appl. Environ. Microbiol. 62*:20-25.

Doi, R.H., Kosugi, A., Murashima, K., Tamura, Y., and Han, S.O. 2003. Cellulosomes from mesophilic bacteria. *J.Bacteriol.185*:5907-5914.

Doi, R.H., and Tamaru, Y. 2001. The *Clostridium cellulovorans* cellulosome: An enzyme complex with plant cell wall degrading activity. *Japan Chem. J. Forum and J. Wiley & Sons Chem. Rec. 1*:24-32.

Durand, R., Rascle, C., and Fèvre, M. 1996. Molecular characterization of xyn3, a member of the endoxylanase multigene family of the rumen anaerobic fungus *Neocallimastix frontalis*. *Curr. Genetics 30*:531-540.

Durand, R., Rascle, C., and Fèvre, M. 1999. Expression of a catalytic domain of a *Neocallimastix frontalis* endoxylanase gene (xyn3) in *Kluyveromyces lactis* and *Penicillium roqueforti*. *Appl. Microbiol. Biotechnol. 52*:208-214.

Durand, R., Rascle, C., Fischer, M., and Fèvre, M. 1997. Transient expression of the β-glucuronidase gene after biolistic transformation of the anaerobic fungus *Neocallimastix frontalis*. *Curr. Genet. 31*:158-161.

Eberhardt, R.Y., Gilbert, H.J., and Hazlewood, G.P. 2000. Primary sequence and enzymatic properties of two modular endoglucanases, Cel5A and Cel45A, from the anaerobic fungus *Piromyces equi*. *Microbiology 146*:1999-2008.

Fanutti, C., Ponyi, T., Black, G.W., Hazlewood, G.P., and Gilbert, H.J. 1995. The conserved noncatalytic 40-residue sequence in cellulases and hemicellulases from anaerobic fungi functions as a protein docking domain. *J. Biol. Chem. 270*:29314-29322.

Fillingham, I.J., Kroon, P.A., Williamson, G., Gilbert, H.J., and Hazlewood G.P. 1999. A modular cinnamoyl ester hydrolase from the anaerobic fungus *Piromyces equi* acts synergistically with xylanase and is part of a multiprotein cellulose-binding cellulase-hemicellulase complex. *Biochem. J. 343*:215-224.

Fujino, Y., Ogata, K., Nagamine, T., and Ushida, K. 1998. Cloning, sequencing, and expression of an endoglucanase gene from the rumen anaerobic fungus *Neocallimastix frontalis* MCH3. *Biosci. Biotechnol. Biochem 62*:1795-1798.

Gettins, P.G.W. 2002. Serpin structure, mechanism and function. *Chem. Rev.102*:4751-4804.

Gilbert, H.J., Hazlewood, G.P., Laurie, J.I., Orpin, C.G., and Xue, G.P. 1992. Homologous catalytic domains in a rumen fungal xylanase: evidence for gene duplication and prokaryotic origin. *Mol Microbiol. 6*:2065-2072.

Gomez de Segura, B.G., and Fèvre, M. 1993. Purification and characterization of two 1,4-β-xylan endohydrolases from the rumen fungus *Neocallimastix frontalis*. *Appl. Environ. Microbiol. 59*:3654-3660.

Hackstein, J.H.P., Akhmanova, A., Boxma, B., Harhangi, H., and Voncken, F.G.J. 1999. Hydrogenosomes: Eukaryotic adaptations to anaerobic environments. *Trends in Microbiol. 7*:441-446.

Harangi, H.R., Akhmanova, A., Steenbakkers, P.J.M., Jetten, M.S.M., van der Drift, C., and Op den Camp, H.J.M. 2003. Genomic DNA analysis of genes encoding (hemi-)cellulolytic enzymes of the anaerobic fungus *Piromyces* sp. E2. *Gene 314*:73-80.

Harangi, H.R., Freelove, A.C.J., van Dinther, M., Steenbakkers, P.J.M. Akhmanove, A., van der Drift, C., Gilbert, H.J., and Op den Camp, H.J.M. 2002. Cel6A, a major exoglucanase from the cellulosome of the anaerobic fungi *Piromyces* sp. E2 and Piromyces equi. *Biochem. Biophys. Acta 1628*:30-39.

Harangi, H.R., Steenbakkers, P.J.M., Akhmanova, A., Jetten, M.S.M., van der Drift, C., and Op den Camp, H.J.M. 2002. A highly expressed family 1 β-glucosidase with transglycosylation capacity from the anaerobic fungus *Piromyces* sp. E2. *Biochim. Biophys. Acta 1574*:293-303.

Hazlewood, G.P., and Gilbert, H.J. 1998. Structure-function relationships in the cellulose-hemicellulose system of anaerobic fungi. In *Carbohydrases from Trichoderma reesei and Other Microorganisms*. (M. Claeyssens, W. Nerincks, K. Piens, eds.) pp. 147-155. Cambridge: The Royal Society of Chemistry.

Hebraud, M., and Fèvre, M. 1990. Purification and characterization of an aspecific glycoside hydrolase from the anaerobic ruminal fungus *Neocallimastix frontalis*. *Appl. Environ. Microbiol. 56*:3164-3169.

Ho, Y.W., and Barr, D.J.S. 1995. Classification of anaerobic gut fungi from herbivores with emphasis on rumen fungi from Malaysia. *Mycologia 87*:655-677.

Ikegawa, T., Mayama, S., Nakayashiki, H., and Kato, H. 1996. Accumulation of diferulic acid during the hypersensitive response of oat leaves to *Puccinia coronata* f. sp. *avenae* and its role in the resistance of oat tissue to cell wall degrading enzymes. *Physiol. Mol. Plant Pathol. 48*:245-255.

Irving, J.A., Steenbakkers, P.J.M., Lesk, A.M., Op den Kamp, H.J.M., Pike, R.N., and Whisstock, J.C. 2002. Serpins in prokaryotes. *Mol. Biol. Evol. 19*: 1881-1890.

Kataeva, I.A., Uversky, V.N., and Ljungdahl, L.G. 2003. Calcium and domain interactions contribute to the thermostability of domains of the multimodular cellobiohydrolase, CbhA, a subunit of the *Clostridium thermocellum* cellulosome. *Biochem. J. 69*:3119-3128.

Kataeva, I.A.,Uversky, V.N., Brewer, J.M., Schubot, F., Rose, J.P., Wang, B.-C., and Ljungdahl L.G. 2004. Interactions between immunoglobulin-like and catalytic modules in *Clostridium thermocellum* cellulosomal cellobiohydrolase CbhA. *Protein Engineering, Design and Selection. 17*:759-769.

Lamed, R., Setter, E., Kenig, R., and Bayer, E.A. 1983. The cellulosome- a discrete cell surface organelle of *Clostridium thermocellum* which exhibits separate antigenic, cellulose-binding and various cellulolytic activities. *Biotechnol. Bioeng. Symp. 13*:163-181.

Lee, J.M.T., Hu, Y., Zhu, H., Cheng, K.J., Krell, P.J., and Forsberg, C.W. 1993. Cloning of a xylanase gene from the ruminal fungus *Neocallimastix patriciarum* 27 and its expression in Escherichia coli. *Can. J. Microbiol. 39*:134-139.

Li, J., and Heath, I. B. 1992. The phylogenetic relationships of the anaerobic chytridiomycetous gut fungi (*Neocallimasticaceae*) and the *Chytridiomycota.* I. Cladistic analysis of rRNA sequences. *Can. J. Bot. 70*:1738-1746.

Li, J., Heath, I. B., and Packer, L. 1993. The phylogenetic relationships of the anaerobic chytridiomycetous gut fungi (*Neocallimasticaceae*) and the *Chytridiomycota.* II. Cladistic analysis of structural data and description of *Neocallimasticales* ord. nov. *Can. J. Bot. 71*:393-407.

Li, X., and Calza, R.E. 1991. Kinetic study of a cellobiase purified from *Neocallimastix frontalis* EB188. *Biochim. Biophys. Acta 1080*:148-154.

Li, X., and Calza, R.E. 1991. Purification and characterization of an extracellular β-glucosidase from the rumen fungus *Neocallimastix frontalis* EB188. *Enzyme Microb. Technol. 13*:622-628.

Li, X.-L., Chen, H.Z., He, Y., Blum, D.L., and Ljungdahl, L.G. 1997. High molecular weight cellulase/hemicellulase complexes of anaerobic fungi. 97th Ann. Meeting Am. *Soc. Microbiol. Abstr*. O-31, p.424.

Li, X.-L., Chen, H., and Ljungdahl, L.G. 1997. Two cellulases, CelA and CelC. From the polycentric anaerobic fungus *Orpinomyces* strain PC-2 contain N-terminal docking domains for a cellulase-hemicellulase complex. *Appl. Environ. Microbiol. 63*:4721-4728.

Li, X.-L., Chen, H., and Ljungdahl, L.G. 1997 Monocentric and polycentric anaerobic fungi produce structurally related cellulases and xylanases. *Appl. Environ. Microbiol. 63*:628-635.

Li, X.-L., Ljungdahl, L.G., Ximenes, E.A., Chen, H., Felix, C.R., Cotta, M.A., and Dien, B.S. 2004. Properties of recombinant β-glucosidase from polycentric fungus *Orpinomyces* PC-2 and its application for cellulose hydrolysis. *Appl. Biochem. Biotech. 113*:233-250.

Li, X.-L., Ximenes, E.A., Chen, H., Cotta, M.A., and Ljungdahl, L.G. 2004. Cloning and sequencing of two highly homologous cellulase genes, celH and celI from the anaerobic fungus *Orpinomyces* strain PC-2. In *Biotechnology of Lignocellulose Degradation and Biomass Utilization*. (K.Ohmiya, K. Sakka, S. Karita, T. Kimura, M. Sakka, Y. Onishi, eds.) pp. 638-641. Uni Publishers Co.,Ltd.Tokyo, Japan.

Liu, J.-H., Selinger, L.B., Hu, Y.-J., Moloney, M.M., Cheng, K.-J., and Beauchemin K.A. 1997. An endoglucanase from the anaerobic fungus *Orpinomyces joyonii*: chracterization of the gene and its product. *Can. J. Microbiol. 43*:477-485.

Liu, J.H., Selinger, L.B., Tsai, C.F., and Cheng, K.J. 1999. Characterization of a *Neocallimastix patriciarum* xylanase gene and its product. *Can. J. Microbiol. 45*:970-974.

Ljungdahl, L.G., Coughlan, M.P., Mayer, F., Mori, Y., and Hon-nami, K. 1988. Macrocellulase complexes and yellow affinity substance from *Clostridium thermocellum*. *Methods Enzymol. 160*:483-500.

Ljungdahl, L.G., and Eriksson, K.-E. 1985. Ecology of microbial cellulose degradation. *Adv. Microbial. Ecology 8*:237-299.

Ljungdahl, L.G., Li, X.-L., and Chen, H. 1998. Evidence in anaerobic fungi of transfer of genes between them and from aerobic fungi, bacteria and animal hosts. In *Thermophiles: The keys to molecular evolution and the origin of life?* (J. Wiegel and M.W.W. Adams, eds.) pp. 187-197. Taylor and Frances, Inc. Philadelphia, PA, USA.

Lowe, S.E., Theodorou, M.K., and Trinci, A.P.J. 1987. Growth and fermentation of an anaerobic rumen fungus on various carbon sources and effect of temperature on development. *Appl. Environ. Microbiol. 53*:1210-1215.

Marvin-Sikkema, F.D., Driessen, A.J.M., Gottschal, J.C., and Prins R.A. 1994. Metabolic energy generation in hydrogenosomes of the anaerobic fungus *Neocallimastix*: Evidence for a functional relationship with mitochondria. *Mycol. Res. 98*:205-212.

Marvin-Sikkema, F.D., Gomes, T.M.P., Grivet, J.-P., Gottschal, J.C., and Prins R.A. 1993. Characterization of hydrogenosomes and their role in glucose metabolism of *Neocallimastix* sp. L2. *Arch. Microbiol. 160*:388-396.

Mayer, F., Coughlan, M.P., Mori, Y., Ljungdahl, L.G. 1987. Macromolecular organization of the cellulolytic enzyme complex of *Clostridium thermocellum* as revealed by electron microscopy. *Appl. Environ. Microbiol. 53*:2785-2792.

Millward-Sadler, S.J., Hall, J., Black, G.W., Hazlewood, G.P., and Gilbert H.J. 1996. Evidence that the *Piromyces* gene family encoding endo-1,4-mannanases arose through gene duplication. *FEMS Microbiol. Lett. 141*:183-188.

Mountfort, D.O., Orpin, C.G. 1994. Editors. Anaerobic Fungi: Biology, Ecology, and Function. Marcel Dekker, Inc New York.

Orpin, C.G. 1975. Studies on the rumen flagellate *Neocallimastix frontalis*. *J. Gen. Microbiol. 91*:249-262.

Orpin, C.G. 1977. The rumen flagellate *Piromonas communis*: Its life-history and invasion of plant material in the rumen. *J. Gen. Microbiol. 99*:107-117.

Orpin, C.G. 1981. Isolation of cellulolytic phycomycete fungi from the caecum of a horse. *J. Gen. Microbiol. 123*:187-296.

Orpin, C.G., Letcher, A.J. 1979. Utilization of cellulose, starch, xylan, and other hemicelluloses for growth by the rumen phycomycete *Neocallimastix frontalis*. *Curr. Microbiol. 3*:121-124.

Prates, J.A.M., Tarboureich, N., Charnock, S.J., Fontes, C.M.G.A., Ferreira, L.M.A., and Davies, G.J. 2001. The structure of the feruloyl esterase module of xylanase 10B from *Clostridium thermocellum* provides insights into substrate recognition. *Structure 9*:1183-1190.

Qui, X., Selinger, B., Yanke, L.-J., and Cheng, K.-J. 2000. Isolation and analysis of two cellulase cDNAs from *Orpinomyces joyonii*. *Gene 245*:119-126.

Raghothama, S., Eberhardt, R.Y., Simpson, P., Wigelsworth, D., Hazlewood, G.P., Nagy, T., Gilbert, H.J., and Williamson M.P., 2001. Characterization of a cellulosome dockerin domain from the anaerobic fungus *Piromyces equi*. *Nature Struct. Biol. 8*:775-778.

Ralph, J., Grabber, J.H., and Hatfield, R.D. 1995. Lignin-ferulate cross-links in grasses: active incorporation of ferulate polysaccharides esters into ryegrass lignins. *Carbohydrate Research 275*:167-178.

Rincón, M.T., Martin, J.C., Aurilia, V., McCrae, S.I., Rucklidge, G.J., Reid, M.D., Bayer, E.A., Lamed, R., and Flint, H.J. 2004. ScaC, an adaptor protein carrying a novel cohesin that expands the dockerin-binding repertoire of the *Ruminococcus flavefaciens* 17 cellulosome. *J.Bacteriol. 186:*2576-2585.

Sakon, J., Irwin, D., Wilson, D.B., and Karplus, P.A. 1997. Structure and mechanism of endo/exocellulase E4 from *Thermomonospora fusca*. *Nat. Struct. Biol. 4:*810-818.

Schimming, S., Schwarz, H., and Staudenbauer, W.L. 1992. Structure of *Clostridium thermocellum* gene licB and the encoded β-1,3-1,4-glucanase: a catalytic region homologous to *Bacillus lichenases* joined to the reiterated domain of clostridial cellulases. *Eur. J. Biochem 204*:13-19.

Schubot, F.D., Kataeva, I.A., Blum, D.L., Shah, A.K., Ljungdahl, L.G., Rose, J.P., Wang, B.-C. 2001. Structural basis for the substrate specificity of the feruloyl esterase domain of the cellulosomal xylanase Z from *Clostridium thermocellum*. *Biochemistry 40*:12524-12532.

Selinger, L.B., Forsberg, C.W., Cheng, K.-J. 1996. The rumen: A unique source of enzymes for enhancing livestock production. *Anaerobe 2:*263-284.

Shimon, L.J.W., Bayer, E.A., Morag, E., Lamed, R., Yaron, S., Shoman, Y., Frolow, F. 1997. A cohesin domain from *Clostridium thermocellum*: the crystal structure provides new insights into cellulosome assembly. *Structure 5*:381-390.

Steenbakkers, P.J.M. 2003. The cellulosome of the anaerobic fungus *Piromyces* sp. strain E2. PhD thesis. Department. of Microbiology and Evolutionary Biology, Faculty of Science, University of Nijmegen, Nijmegen, The Netherlands.

Steenbakkers, P.J.M., Li, X.-L., Ximenes, E.A., Arts, J.G., Chen, H., Ljungdahl, L.G., and Op den Camp, H.J.M. 2001. Noncatalytic docking domains of cellulosomes of anaerobic fungi. *J. Bacteriol. 183*:5325-5333.

Steenbakkers, P.J.M., Ubhayasekera, W., Goosen, H.J.A.M., van Lierop, H.J.M., van der Drift, Vogels, G.D., Mowbray, S.L., and Op den Camp, H.J.M. 2002. An intron-containing glycoside hydrolase family 9 cellulase gene encodes the dominant 90 kDa component of the cellulosome of the anaerobic fungus *Piromyces* sp. strain E2. *Biochem. J., 365*:193-204.

Teunissen, M.J. 1992. Anaerobic fungi and their cellulolytic and xylanolytic enzymes. *PhD thesis*. Department of Microbiology and Evolutionary Biology, Faculty of Science, University of Nijmegen, Nijmegen, The Netherlands.

Teunissen, M.J., Hermans, J.M.H., Huis in't Veld, J.H.J., and Vogels G.D. 1993. Purification and characterization of a complex-bound and a free β-1,4-endoxylanase from the culture fluid of the anaerobic fungus *Piromyces* sp. strain E2. *Arch.Microbiol. 159*:265-271.

Trinci, A.P.J., Davies, D.R., Gull, K., Lawrence, M.I., Nielsen, B.B., Rickers, A., and Theodorou, M.K. 1994. Anaerobic fungi in herbivorous animals. *Mycol. Res. 96*:129-152.

Voncken, F. 2001. Hydrogenosomes eukaryotic adaptations to anaerobic environments. PhD thesis. Department of Microbiology and Evolutionary Biology, Faculty of Science, University of Nijmegen, Nijmegen, The Netherlands.

Williamson, G., Kroon, P.A., and Faulds C.B. 1998. Hairy plant polysaccharides: a close shave with microbial esterases. *Microbiology 144:*2011-2023.

Wilson, D.B., Irwin, D.C. 1999. Genetics and properties of cellulases. *Adv. Biochem. Engineerig/Biotechnology 65*:1-21.

Wilson, C.A., McCrae, S.I., and Wood, T.M. 1994. Characterization of a β-D-glucosidase from the anaerobic rumen fungus *Neocallimastix frontalis* with particular reference to attack on cello-oligosaccharides. *J. Biotech. 37*:217-227.

Wilson, C.A., and Wood, T.M. 1992. The anaerobic fungus *Neocallimastix frontalis*: isolation and properties of a cellulosome-type enzyme fraction with the capacity to solubilize hydrogen-bond-ordered cellulose. *Appl. Microbiol. Biotechnol. 37*:125-129.

Wood T.M., and Wilson C.A. 1995. α-(-4-O-methyl)-D-glucuronidase activity produced by the rumen anaerobic fungus *Piromonas communis*: a study of selected properties. *Appl. Microbiol. Biotechnol. 43*:893-900.

Wubah, D.A. 2004. Anaerobic zoosporic fungi associated with animals. *In Biodiversity of Fungi: Inventory and Monitoring Methods* (G.M. Mueller, G.F.Bills, M.S. Foster, eds.). pp 501-510, Elsevier Academic Press. Burlington, MA, USA.

Wubah, D.A., Fuller, M.S., and Akin, D.E. 1991. Isolation of monocentric and polycentric fungi from the rumen and feces of cows in Georgia. *Can. J. Bot. 69*:1232-1236.

Ximenes, E.A., Chen, H., Kataeva, I.A., Cotta, M.A., Felix, C.R. and Ljungdahl, L.G., and Li X.-L. 2005. A mannanase, ManA, of the polycentric anaerobic fungus *Orpinomyces* sp. strain PC-2 has carbohydrate binding and docking modules. *Can. J. Microbiol.* *51*:559-568.

Xu, Q., Bayer, E.A., Goldman, M., Kenig, R., Shoham, Y., and Lamed, R. 2004. Architecture of the *Bacteroides cellusolvens* cellulosome: Description of a cell surface-anchoring scaffoldin and a family 48 cellulase. *J. Bacteriol. 186*:968-977.

Xue, G.-P., Gobius, K.S., and Orpin, C.G. 1992. A novel polysaccharide hydrolase cDNA (celD) from *Neocallimastix patriciarum* encoding three multi-functional catalytic domains with high endoglucanase, cellobiohydrolase and xylanase activities. *J. Gen. Microbiol. 138*:2397-2403.

Xue, G.-P., Orpin, C.G., Gobius, K.S., Aylward, J.H., Simpson, G.D. 1992. Cloning and expression of multiple cellulase cDNAs from the anaerobic rumen fungus *Neocallimastix patriciarum* in *Escherichia coli*. *J. Gen. Microbiol. 138*:1413-1420.

Yarlett, N., Orpin, C.G., Munn, E.A., Yarlett, M.C., and Greenwood, C.A. 1986. Hydrogenosomes in the rumen fungus *Neocallimastix patriciarum*. *Biochem J. 236*:729-739.

Yarlett, N., Rowlands, C., Yarlett, N.C., Evans, J.C., and Lloyd, D. 1987. Respiration of the hydrogensome-containing fungus *Neocallimastix patriciarum*. *Arch. Microbiol. 148*:25-28.

Zhou, L., Xue, G.-P., Orpin, C.G., Black, G.W., Gilbert, H.J., and Hazlewood, G.P. 1994. Intronless celB from the anaerobic fungus *Neocallimastix patriciarum* encodes a modular family A endoglucanase. *Biochem. J. 297*:359-364.

INDEX

A

B

C

D

E

F

G

H

I

N

O

P

R

S

T

U

V

W

X

Y

Z